Forest Growth and Yield Modeling

Forest Growth and Yield Modeling

Aaron R. Weiskittel, PhD

Assistant Professor of Forest Biometrics and Modeling, Irving Chair of Forest Ecosystem Management, School of Forest Resources, University of Maine, Orono, ME 04469, USA

David W. Hann, PhD

Professor of Forest Modeling, Department of Forest Engineering, Resources, and Management, College of Forestry, Oregon State University, Corvallis, OR 97331, USA

John A. Kershaw, Jr., PhD

Professor of Forest Mensuration, Faculty of Forestry and Environmental Management, University of New Brunswick, PO Box 4400, 28 Dineen Dr., Fredericton, New Brunswick E3B 5A3, Canada

Jerome K. Vanclay, PhD

Professor of Sustainable Forestry, School of Environmental Science and Management, Southern Cross University, PO Box 157, Lismore, NSW 2480, Australia

WILEY-BLACKWELL

A John Wiley & Sons, Ltd., Publication

Wiley-Blackwell is an imprint of John Wiley & Sons, formed by the merger of Wiley's global Scientific, Technical and Medical business with Blackwell Publishing.

Registered office: John Wiley & Sons, Ltd, The Atrium, Southern Gate, Chichester, West Sussex, PO19 8SQ, UK

Editorial offices: 9600 Garsington Road, Oxford, OX4 2DQ, UK
The Atrium, Southern Gate, Chichester, West Sussex, PO19 8SQ, UK
111 River Street, Hoboken, NJ 07030-5774, USA

For details of our global editorial offices, for customer services and for information about how to apply for permission to reuse the copyright material in this book please see our website at www.wiley.com/wiley-blackwell.

Library of Congress Cataloging-in-Publication Data

Forest growth and yield modeling / Aaron R. Weiskittel... [et al.].
p. cm.
Includes index.
ISBN 978-0-470-66500-8 (hardback)
1. Trees–Growth–Computer simulation. 2. Forest productivity–Computer simulation.
3. Trees–Growth–Mathematical models. 4. Forest productivity–Mathematical models. I. Weiskittel, Aaron R.
SD396.F66 2011
634'.0441–dc23

2011014943

A catalogue record for this book is available from the British Library.

This book is published in the following electronic format: ePDF 9781119998525; Wiley Online Library 9781119998518; ePub 9781119971504; Mobi 9781119971511

Set in 10.5/13 pt Times Roman by Thomson Digital, Noida, India

First Impression 2011

Contents

Preface

In 1994, Jerome Vanclay published a comprehensive and definitive text on forest growth and yield modeling. Since then, significant changes in data availability, computing power, and statistical techniques have largely changed the state of forest growth and yield modeling in a rather short time period. This new book attempts to build on the successful approach of the 1994 book and provide a broad perspective on all aspects of forest growth and yield modeling.

Most foresters, students, and even researchers treat forest growth and yield models as incomprehensible and outdated black boxes that are frustrating to use and with predictions that can be inaccurate. Yet, growth and yield predictions are still central to answering a variety of practical and research questions on a daily basis, often with little appreciation of how the models actually operate, their key assumptions, and the difficulty of the task at hand. As with the previous edition, this book attempts to make growth models more accessible to a wider audience by exploring their individual components, discussing aspects of their construction, and, most importantly, describing their limitations. Specific attention is given to individual tree growth models because they are the tool most commonly used for practical decisions. For each type of growth model, several example growth models from different regions of the world are described in detail so that the differences between modeling approaches are better illustrated and the black-box nature of specific models is lessened.

The text is intended for practitioners, researchers, and students alike. Given their relative lack of coverage in other books, two detailed chapters on measuring site productivity and competition are given, which could be used in several undergraduate and graduate-level university courses. There are also individual chapters that describe whole-stand/size-class, individual tree, process-based, and hybrid models. The key growth model components discussed in detail are increment equations, static equations, mortality, and regeneration/recruitment. Other chapters include combining models of different resolutions, modeling silvicultural treatments, and potential future directions. Finally, chapters on model evaluation, model development, and model use are given to guide future efforts. The extensive bibliography should serve as a useful guide for specific references on more advanced topics.

A team of authors with a diverse background and expertise was assembled to provide a comprehensive and international perspective. The book's original author, Dr. Jerome Vanclay, provided expertise on all aspects of growth models, particularly models developed in the southern hemisphere and for tropical forests. Dr. John Kershaw, a co-author on Wiley's fourth edition of the *Forest Mensuration* book, brought a perspective on growth and yield models used in North America, particularly Canada. Dr. David Hann has an extensive career working on forest growth and yield models and is the developer of the ORGANON growth model, which is widely used in the US Pacific Northwest. Finally, Dr. Aaron Weiskittel has spent time working with process and hybrid models and is currently developing an individual tree growth model for the northeastern North America.

Acknowledgements

The authors acknowledge the help of many friends and colleagues during the preparation of this book. Drs. Peter Marshall (University of British Columbia), David Marshall (Weyerhaeuser Company), and Quang Cao (Louisiana State University) provided useful feedback on the content and organization of the text. Greg Johnson (Weyerhaeuser Company), Jim Goudie (Canadian Forest Service), Dr. Robert Monserud (retired, US Forest Service), Dr. Rongxia Li (University of Maine), University of Washington Stand Management Cooperative (SMC), US Forest Service, and New Brunswick Department of Natural Resources provided data useful in the construction of several figures.

Dr. Phil Radtke (Virginia Polytechnic Institute and State University), Dr. Peter Gould (US Forest Service), Dr. Laura Leites (Pennsylvania State University), Dr. Jeff Gove (US Forest Service), Dr. Martin Ritchie (US Forest Service), Dr. Jeremy Wilson (University of Maine), Dr. Thomas Ledermann (Austria Federal Forest Office), Dr. Laura Kenefic (US Forest Service), Dr. Duncan Wilson (Oklahoma State University), Matthew Russell (University of Maine), Joshua Sjostrom (University of Maine), Elijah Shank (University of Maine), and Elizabeth McGarrigle (University of New Brunswick) reviewed several chapters. Special thanks to Drs. Robert Curtis (US Forest Service), Oscar García (University of Northern British Columbia), and Dave Hyink (retired, Weyerhaeuser Company) for extensive reviews of the book. Special thanks are due Clare Lendrem for an outstanding job as copy-editor.

Finally, thanks to the encouragement, patience, and understanding of friends and family.

1 Introduction

All models are an abstraction of reality that attempt to conceptualize key relationships of a system. Models can be both quantitative and conceptual in nature, but all models are integrators of multiple fields of knowledge. Consequently, models generally have several important and varied uses. Forest growth and yield models are no different. Foresters often have a general sense of a stand's developmental trajectory and what can be done to alter it. However, it generally takes years of experience to achieve this level of expertise and, even then, quantifying the predictions can be difficult. Forest growth models attempt to bridge this gap by providing model users the ability to predict the future condition of the forest. Ultimately, growth models are the quantitative generalizations on the knowledge of forest stand development and their response to silvicultural treatments.

Forest growth and yield models have a long, and rapidly expanding, history of development (Figure 1.1, 1.2). Their development and use has particularly increased in the last two decades, due in part to the greater availability of personal computers to perform both data analysis and complex simulations (Figure 1.2). This has resulted in a wide array of modeling approaches, each with their own advantages and disadvantages. In particular, models differ in the type of data used and the method of construction. This book attempts to provide an overview of the primary concepts involved in forest modeling, the various techniques used to represent the determinants of growth, and the techniques needed to both develop and use a growth model properly.

Although the concepts of forest growth and yield have long been a part of forestry, they have been defined and named in various ways, particularly in the US (Bruce, 1981). In this book, *increment* is defined as the difference between tree or stand dimensions from one time period to the next, while *growth* is the final dimension from one time period to the next. In other words, increment is determined by either solving a growth equation or by observing growth at two points of time (Bruce, 1981).

Forest Growth and Yield Modeling, First Edition. Aaron R. Weiskittel, David W. Hann, John A. Kershaw, Jr. and Jerome K. Vanclay.

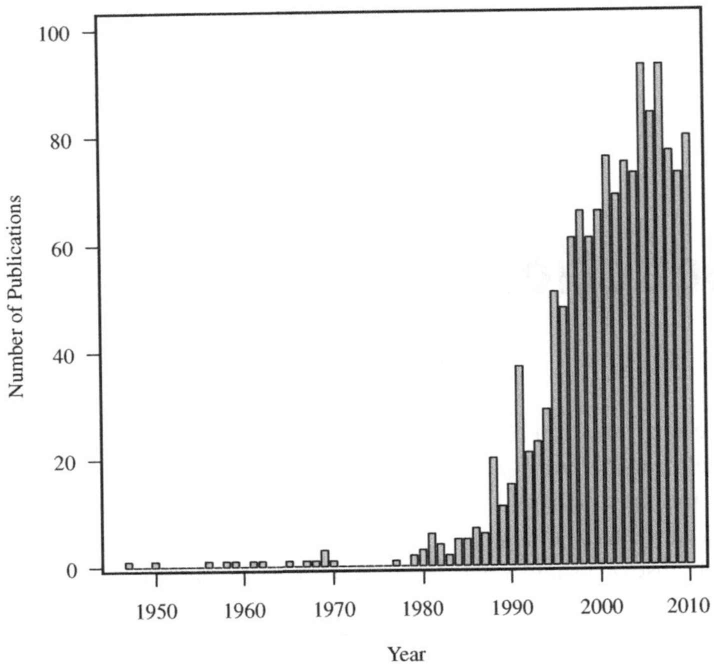

Figure 1.1 Number of publications on growth and yield, by publication year, based on a keyword search of the CAB Direct database (www.cabdirect.org, accessed December 21, 2010).

This book is focused on models that predict the development of a single forest stand (Figure 1.3). Although a distinction between empirical and mechanistic models is often made (e.g. Taylor *et al.*, 2009), this is not a useful metric of differentiation, as all models are on a spectrum of empiricism. Instead, this book groups forest stand development models into four broad categories: (1) statistical models; (2) process; (3) hybrid; and (4) gap (Table 1.1; Figure 1.4).

Statistical models rely on the collection and analysis of data that will characterize the targeted population in a manner that allows statistical variability to be estimated for parameters. The primary intent of statistical models is for prediction of forest stand development and yield over time. Process models represent key physiological processes (e.g. light interception, photosynthesis), often for understanding and exploring system behavior, which are then combined to characterize both tree and stand development. Hybrid models merge features of statistical and process models and are used both for understanding and for prediction. Gap models are designed to explore long-term ecological processes, generally for understanding interactions that control forest species succession. Models that integrate the development of multiple forest stands, such as landscape models, exist (e.g. Mladenoff, 2004), but will not be covered in this book.

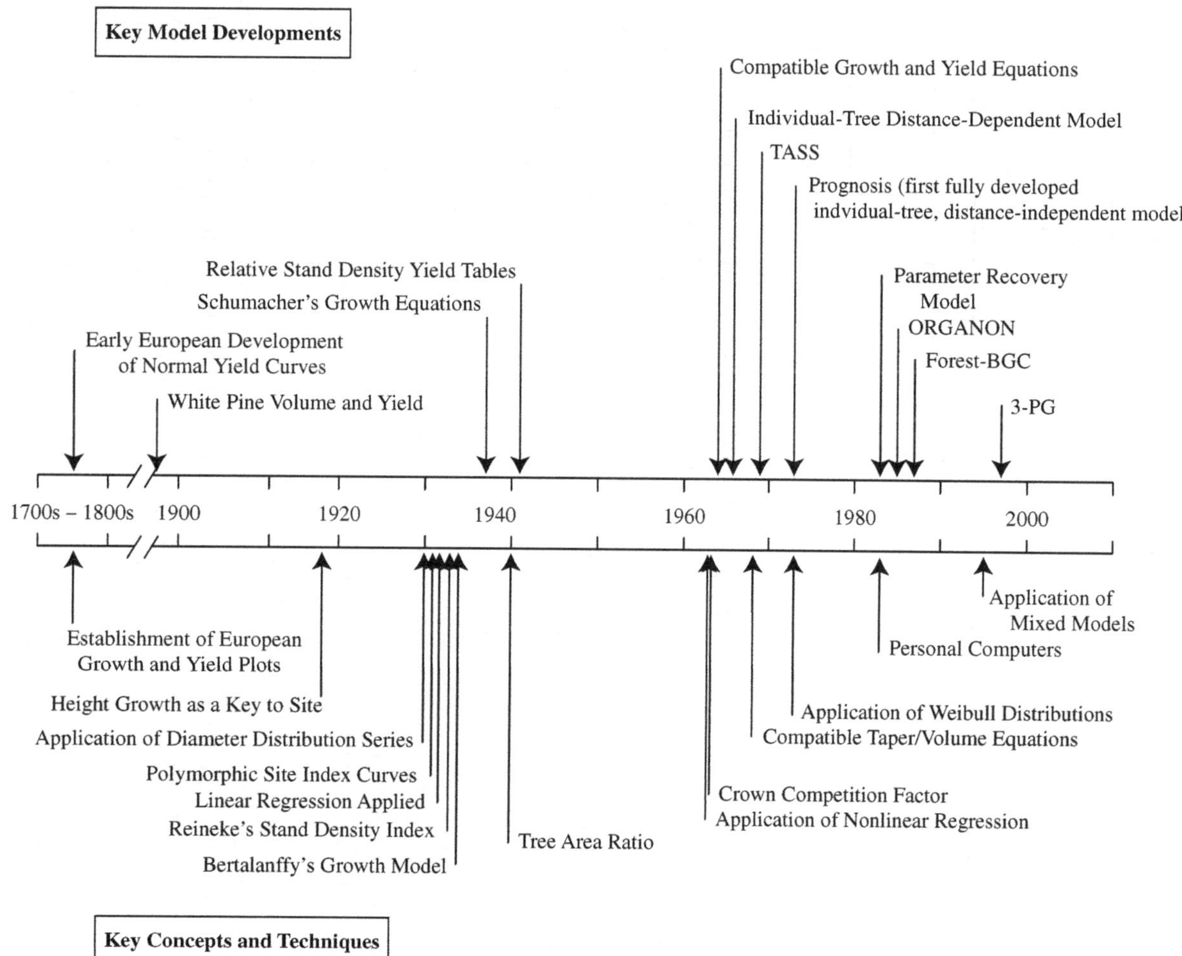

Figure 1.2 Key milestones in model development and associated concepts and techniques.

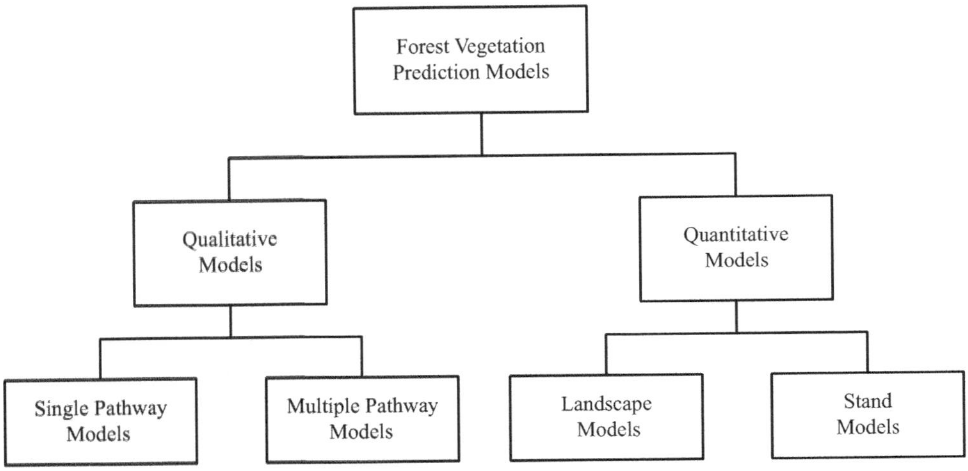

Figure 1.3 Types of forest vegetation prediction models. Adapted from Taylor *et al.* (2009).

Within any given model category, models differ in their resolution (both spatial and temporally), spatial dependence, and degree of determinism. Spatial resolution refers to the basic unit for predictions, with the simplest being a whole-stand approach (Chapter 4), and the individual-tree approach is the most detailed (Chapter 5). A size-class model is a compromise between the whole-stand and individual-tree approaches (Chapter 4). Some process models even have a spatial resolution of an individual leaf within a tree crown. In addition, a significant amount of effort has been made in combining predictions from models with different spatial resolutions (Chapter 10).

Temporal resolution is the basic time step for model predictions. Several process models have daily or even hourly time steps, while statistical models generally have 1- to 10-year temporal resolutions. Models also vary in their use of spatial information. Distance-dependent or spatially explicit models require spatial location information; often individual-tree x–y coordinates are needed. Distance-independent or spatially implicit models do not require this information.

Finally, models differ in their use of deterministic approaches, which means that a particular function will always return the same output return value for any given set of input values. In contrast, stochastic approaches incorporate some purely random element and will give different return values in successive runs with any given set of input values. Stochasticity can be an important element of forest modeling, as some relevant factors like natural disturbances that ultimately govern the growth and yield of a particular stand can be random or unpredictable. However, a model with too many stochastic elements can make interpretation a challenge.

Stochasticity is one approach for addressing the variability that is inherent in all aspects of modeling. Even models in fundamental sciences like physics and chemistry have purely random elements. However, biological systems are even more variable and models need to

Table 1.1 Categories of quantitative single stand forest development models and their definition, use, advantages, and disadvantages.

Type of model	Definition	Important uses	Advantages	Disadvantages	Key references
Statistical	Utilize empirical data and statistical techniques like regression to derive quantitative relationships	Update forest inventories; compare forest silvicultural treatments; estimate sustainable harvests	Robust; long history of development; rely on data generally available; output geared for operational decisions; can represent a wide range of conditions and sampling schemes	Require high quality empirical data; can extrapolate poorly; generally insensitive to climate	Taylor *et al.* (2009)
Process	Represent key plant physiological processes like photosynthesis, which are then scaled to the stand-level to estimate growth	Understand the underlying mechanisms influencing growth; test hypotheses about plant behavior; predict potential forest productivity	Can theoretically extrapolate to novel situations; sensitive to climate; mechanistic	Dependent on several difficult-to-measure parameters; input data not widely available; high computational demand; output often unusable for operational decisions	Mäkelä *et al.* (2000a); Landsberg (2003)
Hybrid	Combine statistical and process approaches in attempt to take advantage of the strengths of both approaches	Predict growth using climatic factors; prediction of novel forest silvicultural treatments	Robust; sensitive to climate; minimize the number of required parameters; can use traditional forest inventory data	Accuracy improvements can be minimal when compared to a purely statistical approach; climate and soils input data not widely available	Monserud (2003)
Gap	Rely heavily on ecological theory and interpretation of species dynamics relative to both competition and environmental conditions	Predict long-term forest succession; test ecological theories	Incorporate a variety of natural disturbance agents; long time scales	Prediction accuracy is often low compared to statistical models; difficult to initialize with forest inventory data; several subjective parameters	Bugmann (2001); Shugart (2002)

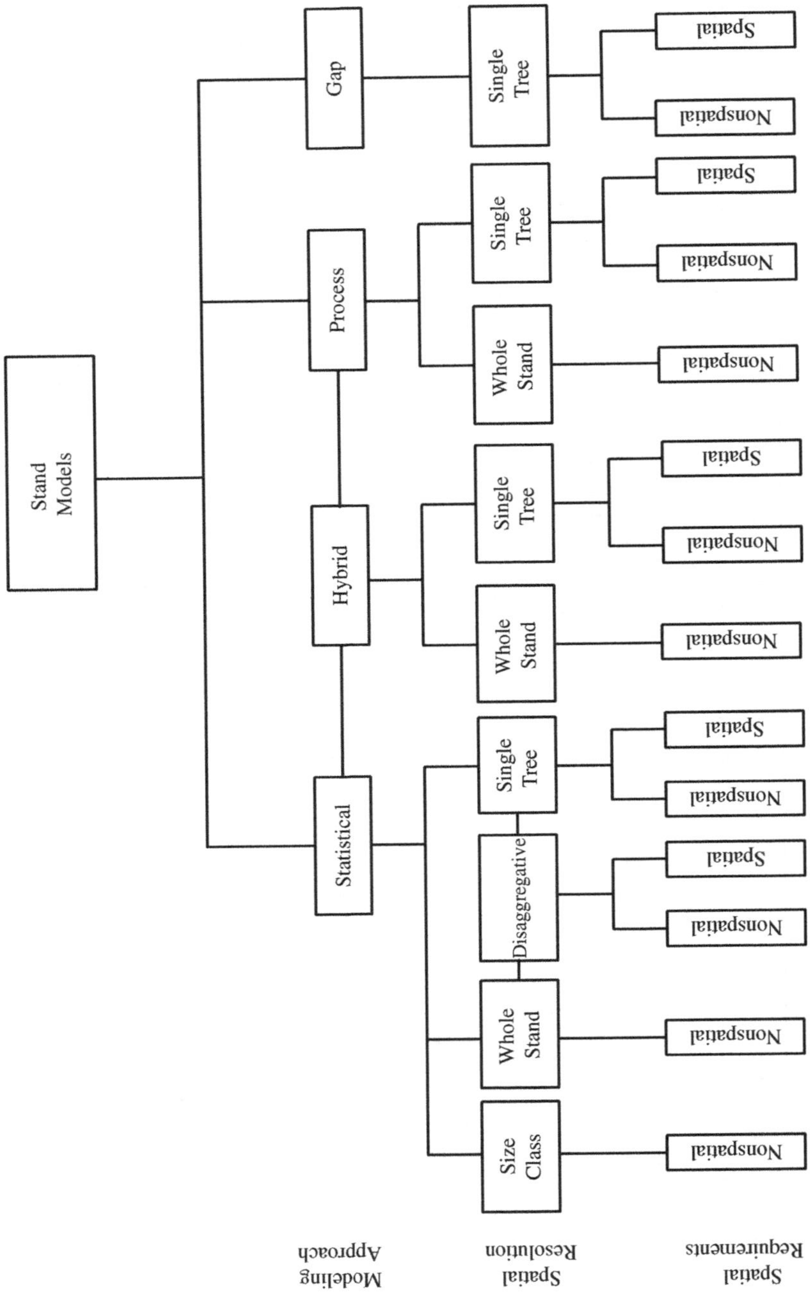

Figure 1.4 Types of forest vegetation prediction models that are focused on the stand-level.

recognize the important sources. Therefore, the models examined in this book have a framework that is based upon our current biological knowledge and are parameterized with the knowledge that the parameters are uncertain.

Forest growth models have several components. At minimum, forest stand development models must represent growth (Chapter 6) and mortality (Chapter 8). Models must also have components that relate the traditional tree measurements of diameter and height to other attributes like total volume or biomass with the use of static equations (Chapter 7). Comprehensive growth models include components to predict regeneration and ingrowth (Chapter 9) and representation of silvicultural treatments (Chapter 11). In addition, understanding the key biological determinants of growth and yield, namely competition (Chapter 2) and site potential productivity (Chapter 3), is important.

1.1 Model development and validation

As with most fields, forest modeling is both an art and a science. Ideally, the development of any model involves a comprehensive understanding of the system and an approach for detecting the crucial relationships. This often means that modelers must be multidisciplinary. In addition, model development is often an iterative and collaborative effort between modelers, fundamental scientists, and model users. The process of modeling is an assessment of current understanding of forests, information needed for management, and crucial knowledge gaps.

Consequently, research questions can often be generated by assessment of model strengths and weaknesses. This also illustrates an important modeling distinction, namely the use of models for prediction versus understanding, which will be further discussed below. Although there are important general modeling philosophies like Occam's principle of parsimony, which suggests that models should be as simple as possible, but as complex as necessary (Kimmins *et al.*, 2008), achieving this is often easier said than done.

Regardless of modeling approach, empirical data of one type or another will be required for either model construction (Chapter 14) or model evaluation (Chapter 15). Data can often vary greatly in its quality and overall usefulness for modeling. Among others, data quality is influenced by how well the data represents the population of interest, the variables collected, and the degree of measurement error, which is often an overlooked yet important determinant of predictability (e.g. Hasenauer and Monserud, 1997). The statistical tools used to construct models are continually changing and evolving. Chapter 14 provides a brief overview of the key statistical techniques in order to give a better context to statistical forest growth and yield models.

To be useful for a given purpose, a model must be representative of reality to some degree. Consequently, a variety of methods have been used to verify model predictions (Chapter 15). This has ranged from simple statistical tests to complex stochastic simulations. Each has their own merits, but, in general, models must be verified using multiple approaches to ensure full reliability. If model predictions are found to be inadequate, a larger question quickly becomes how to fix or re-calibrate the model. This can often be a complicated undertaking, but emerging approaches may simplify the process.

1.2 Important uses

Models are tools designed to be used in a variety of ways (Chapter 16). The key uses of any well-developed model are prediction and education in its broadest sense (Figure 1.5). In forestry, some key prediction roles of growth models are (1) update forest inventories; (2) assess alternative forest silvicultural systems; (3) determine the influence of disturbance agents like insects or disease; (4) estimate sustainable yield of forest products; and (5) generalize regional trends. Growth and yield information is required to make all major forest management decisions. Some of the basic decisions that require accurate growth and yield information include: (1) even-aged stand-level decisions; (2) uneven-aged stand-level decisions; (3) forest- or ownership-level decisions; and (4) regional and national decisions. The type of information needed from a forest growth and yield model

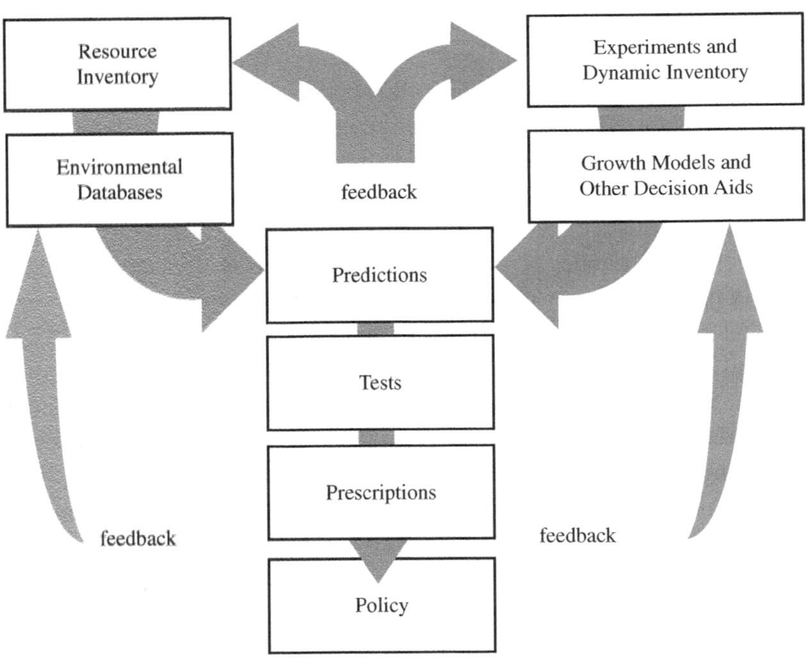

Figure 1.5 The role of growth models in decision making, forest management, and the formation of forest policy. Adapted from Nix and Gillison (1985).

Table 1.2 Uses of growth and yield models to aid in key forest management decisions.

Type of decision	Important factors to consider	Reference
Even-aged stand-level	Planting density; thinning strategy; fertilization strategy; species or species mix; rotation length	Hann and Brodie (1980)
Uneven-aged stand-level	Sustainable diameter distribution; cutting cycle length; species mix; fertilization strategy; conversion strategy	Hann and Bare (1979)
Forest or ownership level	Schedule of stand treatments; allowable harvest; wildlife habitat; aesthetics	Bettinger *et al.* (2009)
Regional or national level	Carbon sequestration potential; allowable harvest; wildlife habitat	Bettinger *et al.* (2005)

to make these different decisions depends on the spatial and temporal level at which information is needed (Table 1.2).

For example, a silviculturist would primarily use a growth and yield model to project the development of the stand under alternative treatment strategies such thinning or fertilization regimes. A forest planner would likely use a growth and yield model to stratify individual stands in a forest into homogeneous units, project the development of each stratum, and use a harvest scheduler to determine the optimal silvicultural system and allowable harvest. A policy-maker would generally use a growth and yield model to depict regional or national trends like carbon sequestration potential or sustainable harvest levels to set effective policies. In fact, growth models were used in the United States, by the Chicago Climate Exchange and the California Climate Action Registry, to set standards for carbon credit trading and greenhouse gas registries at regional and national scales.

Additional uses of models are the visualization of management alternatives and the assessment of forest stand dynamics on wildlife habitat and streamside conditions for fish habitat. Consequently, the implications of basing decisions on a growth and yield model at any level are often quite significant, which both model developers and users need to be aware of.

There are several complex issues facing the practice of forestry today, like assessing the effects of climate change, forest carbon neutrality, and long-term sustainability. Answering these open questions with empirical data is often difficult, requires long-term investment, or is impossible. Consequently, growth and yield models are widely used by scientists as research tools to test hypotheses and understand system behavior. For example, the ORGANON growth and yield model (Hann, 2011) has been widely used by scientists to answer several research questions on a broad array of topics ranging from forest management, planning, and economics, to conservation issues (Table 1.3).

Models are good research tools as they allow the construction of what-if scenarios and experimentation with different parameter settings. In addition, the development and

Table 1.3 Examples of the applied uses of the ORGANON growth and yield model.

Study	Purpose
Forest management	
Maguire *et al.* (1991)	Examine the influence of alternative management on wood quality
Welty *et al.* (2002)	Assess strategies for managing riparian zones
Wilson and Oliver (2000)	Strategies for density management to ensure stability
Sessions *et al.* (2004)	Manage the consequences of wildfire
Forest planning	
Johnson *et al.* (2007a)	Develop large-scale, long-term plans for usage of forested landscapes
Sessions *et al.* (2000)	Develop mature forest habitat
Shillinger *et al.* (2003)	Predict future timber supply
Johnson *et al.* (2007b)	Large-scale assessment of socioeconomic effects on forest structure and timber production
Economics	
Birch and Johnson (1992)	Determine the economic impact of green tree retention
Fight *et al.* (1993)	Conduct a financial analysis of pruning alternatives
Busby *et al.* (2007)	Evaluate the opportunity cost of forest certification
Latta and Montgomery (2004)	Create cost-effective older stand structures
Wildlife	
Hayes *et al.* (1997)	Evaluate response of wildlife to thinning
Calkin *et al.* (2002)	Managing for wildlife biodiversity
Lichtenstein and Montgomery (2003)	Assessing influence of timber management on wildlife biodiversity
Andrews *et al.* (2005)	Strategies for creating northern spotted owl nesting sites
Nalle *et al.* (2004)	Strategies for joint management of timber and wildlife

construction of any growth model often leads to new and interesting research questions. This is because model development largely requires making and testing key assumptions, assessing patterns, and providing full disclosure, which are all basic tenets of the scientific method. In other words, developing a model requires the processes or system being modeled to be conceptualized and understood.

Forest growth and yield models are useful tools for education, a role that ORGANON has often played (Marshall *et al.*, 1997). This is because models require hands-on interaction, synthesis of multiple concepts, and critical thinking skills to assess the

Table 1.4 Model name, type, resolution, distance dependence, stochasticity, region, primary species, and reference for example models considered in the text.

Model name	Model type	Resolution		Distance dependent	Stochastic	Region/country	Primary species[a]	Reference
		Spatial	Temporal					
3-PG	Hybrid	Whole stand	Monthly	No	No	Several	DF, LP, EG, EN, NS, RP, SP, SS, WL	Landsberg and Waring (1997)
BALANCE	Process	Individual tree	One year	Yes	No	Germany	EB, NS	Grote and Pretzsch (2002)
CABALA	Hybrid	Whole stand	Monthly	No	No	Australia	EG	Battaglia *et al.* (2004)
CenW	Process	Whole stand	Monthly	No	User's choice	Australia	ED, RP	Kirschbaum (1999)
DFSIM	Statistical	Whole stand	Five year	No	No	Pacific Northwest, United States	DF	Curtis *et al.* (1981)
FIBER	Statistical	Size class		No	No	Northeast, United States	AB, BF, RS, BS, WS, EH, NC, SM, RM, YB, PB, QA, RO, TA, WA, WP	Solomon *et al.* (1995)
Forest-BGC	Process	Whole stand	Daily	No	No	Several	—	Running and Coughlan (1988); Running and Gower (1991)

(continued)

Table 1.4 *(Continued)*

Model name	Model type	Resolution		Distance dependent	Stochastic	Region/country	Primary species[a]	Reference
		Spatial	Temporal					
JABOWA	Gap	Individual tree	One year	No	Yes	Northeast, United States	AE, AB, BC, BF, BP, BW, BT, BN, RS, BS, WS, EH, NC, SM, RM, YB, PB, QA, RO, TA, WA, WO, WP	Botkin *et al.* (1972a, b)
ORGANON	Statistical	Individual tree	Five year, One year	No	User's choice	Pacific Northwest, United States	PM, BM, BO, LO, GC, OO, DF, RA, TO, WL, GF, IC, PY, PP, PL, WH, WF	Hann (2011)
CROBAS	Hybrid	Size class	One year	No	No	Finland; Quebec	JP, SP, NS	Mäkelä (1997)
PROGNAUS	Statistical	Individual tree	Five year	No	User's choice	Austria	NS, WF, EL, TP, SP, EB	Monserud *et al.* (1997)
Scube	Statistical	Whole stand	One year	No	No	British Columbia, Canada	WS, ES	García (2011)
SILVA	Statistical	Individual tree	Five year	Yes	Yes	Germany	BP, NS, WF, EB, SP, SO	Pretzsch *et al.* (2002)
SORTIE	Gap	Individual tree	One year	Yes	Yes	Northeast, United States; Quebec and British Columbia, Canada	AB, EH, JP, LP, SM, RM, TA, WA, WP, WH, YB	Pacala *et al.* (1993; 1996); Coates *et al.* (2003)
TASS	Statistical	Individual tree	One year	Yes	Yes	Pacific Northwest, Canada	DF, WS	Mitchell (1969; 1975)

[a]See Appendix 1 for species codes.

appropriateness of output. In addition, combining model prediction with visualization tools (Chapter 16) allows visual demonstration of key concepts like stand structure and stratification, which can be difficult to achieve with just words or in the field.

1.3 Overview of the book

Forest growth modeling is an evolving and comprehensive field that can be difficult to describe fully. Previous books on forest growth modeling have either become outdated (e.g. Vanclay, 1994), focused primarily on one geographic region (e.g. Hasenauer, 2006), or are specific to a particular modeling approach (e.g. Landsberg and Sands, 2011). This book attempts to provide a comprehensive overview of forest models from multiple perspectives in order to be useful to model developers, scientists, students, and model users alike.

The book is divided into 17 individual chapters that give an overview of the key concepts determining growth and yield (Chapters 2, 3), the different types of modeling approaches (Chapters 4, 5, 12, 13), and the various dimensions of developing, validating, and using a growth model (Chapter 14, 15, 16). Example models are described in detail for each modeling approach to illustrate key differences and provide information on some of the more widely used models (Table 1.4).

In particular, the components of statistical, distance-independent, individual-tree models are discussed in detail (Chapters 6, 7). Attention is given to this type of modeling approach because it has been widely adopted and extensively used for operational management planning. For example, statistical, distance-independent, individual-tree models are currently available and used throughout the United States (Crookston and Dixon, 2005), western Canada (e.g. Temesgen and LeMay, 1999), central Canada (e.g. Bokalo *et al.*, in review), eastern Canada (e.g. Woods and Penner, 2007), central Europe (e.g. Monserud *et al.*, 1997), and northern Europe (e.g. Hynynen *et al.*, 2002). The approach has been preferred because it can be used in a wide range of stand structures, particularly in uneven-aged (Peng, 2000) and mixed species stands (Porté and Bartelink, 2002). Throughout the book, specific attention is given to the ORGANON growth and yield model of the United States Pacific Northwest (Hann, 2011), as it has a long history of continuous development, is applicable to a large number of conifer and hardwood species in a wide array of stand conditions, and has been rigorously tested.

It is our hope that the book can help promote a more comprehensive understanding of forest models, and guide future modeling efforts.

2

Indices of competition

2.1 Introduction

The degree that a site is occupied by trees, and the resulting level of competition between those trees, is a primary factor that drives the growth and yield of a stand (Curtis, 1970). Trees compete for a variety of resources including light, water, nutrients, and physical space. Competition among plants is summarized as the action on, and reaction to, their living environment (Ford and Sorrensen, 1992). As trees grow, they modify their surrounding environment and alter the ability of their neighbors to acquire resources. Consequently, competition is highly dynamic, both spatially and temporally. Tomé and Burkhart (1989) identified three primary components of competition: (1) the general environment of the tree; (2) micro-environmental and genetic influences; and (3) the influence of local neighbors.

There are many different types of competition. One of the most important distinctions is the mode of competition, which can be *aboveground* and *belowground*. Aboveground competition is primarily for light, while belowground competition is focused on soil water and nutrient acquisition. Two other types of competition generally distinguished in forestry are the species of competitor trees and the symmetry of competitive interactions. *Intra-specific* is competition between individuals of the same species, while *inter-specific* is competition between individuals of different species. Forest growth and yield models have generally just focused on intra-specific competition (Liu and Burkhart, 1994), but understanding and predicting influences of inter-specific competition is becoming increasingly important due to an emphasis on mixed-species management.

One- and *two-sided* competition refer to the evenness of competition in a stand (Weiner, 1990). In one-sided competition, larger trees are not affected by their smaller neighbors; while resources are shared (equally or proportionally to size) by all trees in two-sided

Forest Growth and Yield Modeling, First Edition. Aaron R. Weiskittel, David W. Hann, John A. Kershaw, Jr. and Jerome K. Vanclay.

competition (Soares and Tomé, 2003). It is commonly assumed that one-sided competition is driven by the availability of aboveground resources, and two-sided is more reflective of belowground competition (Casper and Jackson, 1997), but this is an oversimplification of the various complex processes involved, and there are likely to be several mechanisms driving competition symmetry (Schwinning and Weiner, 1998). Further, it is not possible to directly measure severity of competition at any particular point in time (Curtis, 1970). However, many growth and yield models consider both one- and two-sided indices of competition with the implicit assumption that both aboveground and belowground competition is being represented by them. In reality, modelers are attempting to quantify the level of competition experienced by a tree, and its social position within the stand, rather than describe its symmetry.

Although competition acts on individual trees, it is often described at the stand-level in forestry. Common indices of competition in forestry are ones that describe stand *density* or *stocking*. Stand density is a quantitative measure of the degree of crowding and resulting level of competition existing within the stand (Zeide, 2005). Stocking is a relative measure (usually expressed as a proportion or a percentage) of the adequacy of the stand's density for meeting the target management density (Curtis, 1970). Stockability is the inherent capability of a site to support a maximum stand density. Various indices are used to describe density, stocking, and stockability, which are incorporated to varying degrees in growth models.

Competition influences forest growth and dynamics in multiple ways. At the tree-level, competition reduces growth, particularly diameter growth, and increases both crown recession and the probability of mortality. Competition can also cause stagnation, stratification, and instability. Consequently, an index of competition must account for: (1) the *number* of individuals; (2) the *size* of the individuals; and (3) the *distribution* of individuals. From a growth and yield modeling perspective, Munro (1974) identified two major classes of competition indices: distance-independent and distance-dependent indices. These are discussed in this chapter from both a one- and a two-sided competition perspective. Comprehensive assessment of competition indices as they relate to growth and yield models is also to be found in Biging and Dobbertin (1992; 1995) and Pretzsch (2009).

2.2 Two-sided competition

2.2.1 Distance-independent

Common distance-independent indices of two-sided competition are number of trees per unit of area and total basal area. Although number of trees per unit of area is the true measure of "density," it is not a sufficient index of competition (Zeide, 2005). Basal area (BA) is a more effective measure since it incorporates number of trees in a stand and their diameters, but is not a true measure of competition unless it is combined with some measure of stand development.

In addition, basal area treats all species as equal contributors to competition, and there are multiple pathways to the same value (Figure 2.1A). For example, a basal area of $50\,m^2\,ha^{-1}$ can be from over 700 trees ha^{-1} that are 30 cm in diameter at breast height (DBH) or nearly 6400 trees ha^{-1} that are 10 cm in DBH. In addition, Curtis (1970) and Zeide (2005) suggested that basal area is only an effective index of competition when it is compared to basal area from a normal yield table that varies with species, site, and age, making it a relative density measure.

Curtis (1970) proposed that including a measure of site quality and tree size along with basal area in a growth equation effectively mimics an index of relative density. This may help explain why basal area is one of the most commonly used indices of two-sided

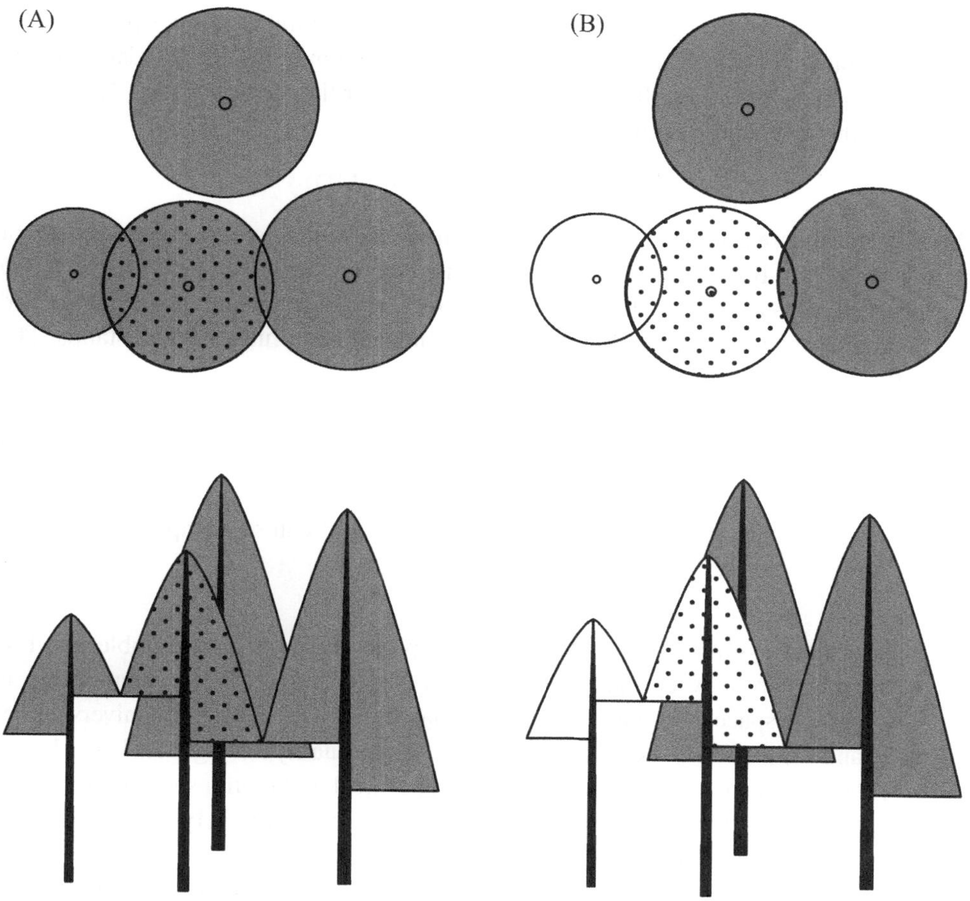

Figure 2.1 Illustration of one- (B) and two-sided (A) as well as distance-independent (top) and distance-dependent (bottom) measures of competition. The individual with the small dots is the subject tree, and all trees colored gray are considered competitors.

competition in distance-independent growth models. Like basal area, the quadratic mean diameter (QMD) is not a true measure of competition, but widely used to describe stand structure. There are several practical and historical reasons why QMD has been preferred over the simple arithmetic mean (Curtis and Marshall, 2000).

By combining the basic stand variables of number of trees, basal area, and QMD, various alternative measures of stand density have been constructed to describe competition primarily in even-aged stands. Although there is quite an assortment of these different stand density measures, they are essentially the same thing in various forms and, from a practical standpoint, differ only by scale factors. Consequently, this chapter provides an overview of a few selected measures, particularly Reineke's (1933) stand density index (SDI) and relative density (Drew and Flewelling, 1979; Curtis, 1982).

Reineke's (1933) SDI combines both the frequency of trees per unit of area and their average size as expressed by QMD. The SDI expresses the number of trees per unit area which that stand would have at a standard average DBH. Commonly, the standard average DBH is 10 inches for the English system and 25.4 cm for the metric system. The SDI is defined by the following relationships:

$$\log(\text{TPH}) = a + b \cdot \log(\text{QMD}) \qquad [2.1]$$

where TPH is the trees per hectare, QMD is quadratic mean diameter (i.e. the diameter of the tree of average basal area), and a and b are parameters. Reineke (1933) found that b was relatively constant at -1.605 for several species, and independent of site quality and age. Parameter a varies by species and can be estimated from fully stocked stands. The SDI is then calculated as:

$$\text{SDI} = \text{TPH} \cdot \left(\frac{\text{QMD}}{25.4}\right)^{1.605} \qquad [2.2]$$

Using the species-specific parameter a, the maximum SDI can be computed as:

$$\text{SDI}_{\max} = 10^{(a + b \cdot \log(25.4))} \qquad [2.3]$$

The most critical assumptions of this relationship are that: (1) a predictable species-specific relationship between maximum size and density exists; (2) the relationship is linear in log–log space; (3) the slope of the maximum size–density is universal (i.e. – 1.605); and (4) the relationship is independent of site quality and age (Jack and Long, 1996). Although this relationship relies on several critical assumptions that may or may not be met, it is widely used and successfully incorporated in a variety of stand density management diagrams (Jack and Long, 1996).

From SDI and $\text{SDI}_{\max}$, a relative density measure is easily computed as the ratio of SDI to $\text{SDI}_{\max}$ (Long, 1985). Rather than using SDI, Drew and Flewelling (1979) developed a relative density index based on mean tree volume and the $-\frac{3}{2}$ rule of self-thinning (see Chapter 8). Curtis (1982) developed a relative density index based on basal area and quadratic mean diameter. Although these indices of relative competition are different, they are conceptually similar (Figure 2.2). For example, relative densities of 25, 35, and 60% of SDI_{max} are associated with the onset of competition, lower limit of full site occupancy, and lower limit of self-thinning, respectively (Long, 1985). Based on their approach of estimating relative density, Drew and Flewelling (1979) suggest that relative densities of 0.15, 0.40, and 0.55 correspond to the onset of competition, lower limit of full site occupancy, and the zone of imminent competition mortality, respectively. Curtis (2010) suggests that trees less than 4 cm DBH should be excluded from the computation of any relative density measure.

SDI is an effective index of competition in pure, even-aged stands, but may require modification in stands with an irregular DBH distribution. In addition, Reineke's (1933) SDI lacks additivity, or, in other words, an individual's contribution to the SDI cannot be determined independently of all of the other trees in the stand (Zeide, 1983). One alternative to Reineke's (1933) SDI is the summation method (Stage, 1968; Long and Daniel, 1990; Shaw, 2000). Curtis (2010) reviewed the different methods for estimating this value and presented this general equation:

$$SDI_{sum} = \frac{0.02512 \cdot \sum_{k=1}^{n}(DBH_k^{1.6})}{Area} \qquad [2.4]$$

where SDI_{sum} is the additive SDI, which is always less than SDI. However, the ratio of SDI_{sum} to SDI is relatively constant for most even-aged stands, and rapidly decreases as DBH distributions become more irregular (Ducey and Larson, 2003). Although the ratio is used as an indicator of stand complexity (e.g. Long and Shaw, 2005), it is highly sensitive to sampling methods used to obtain the forest inventory data, particularly the minimum DBH measured (Ducey, 2009a).

SDI and SDI_{max} are difficult to determine for mixed species stands. Woodall *et al.* (2005) suggested that SDI_{max} be determined in mixed species stands as a function of the mean specific gravity of the wood of trees on the plot. Similarly, Ducey and Knapp (2010) modified SDI_{sum} to account for differences in species composition:

$$SDI_{sum} = \sum_{k=1}^{n}\left((\beta_0 + \beta_1 \cdot SG_{spp,k})\left(\frac{DBH_k}{25.4}\right)^{1.605} \cdot EXPF_k\right) \qquad [2.5]$$

where $SG_{spp,k}$ is the species specific gravity and $EXPF_k$ is the expansion factor for the kth tree, and the β_is are estimated parameters. This measure is both additive and anamorphic, but its relationship to growth has yet to be determined. However, Pretzsch and Biber (2010) used a modified approach similar to equation [2.5] in their analysis of mixed species stand growth in central Europe. In addition, SDI and SDI_{max} are widely used in many variants of Forest Vegetation Simulator (FVS) to estimate mortality (Crookston and Dixon, 2005), while relative density is used in the MELA model to predict individual-tree growth and survival (Hynynen *et al.*, 2002).

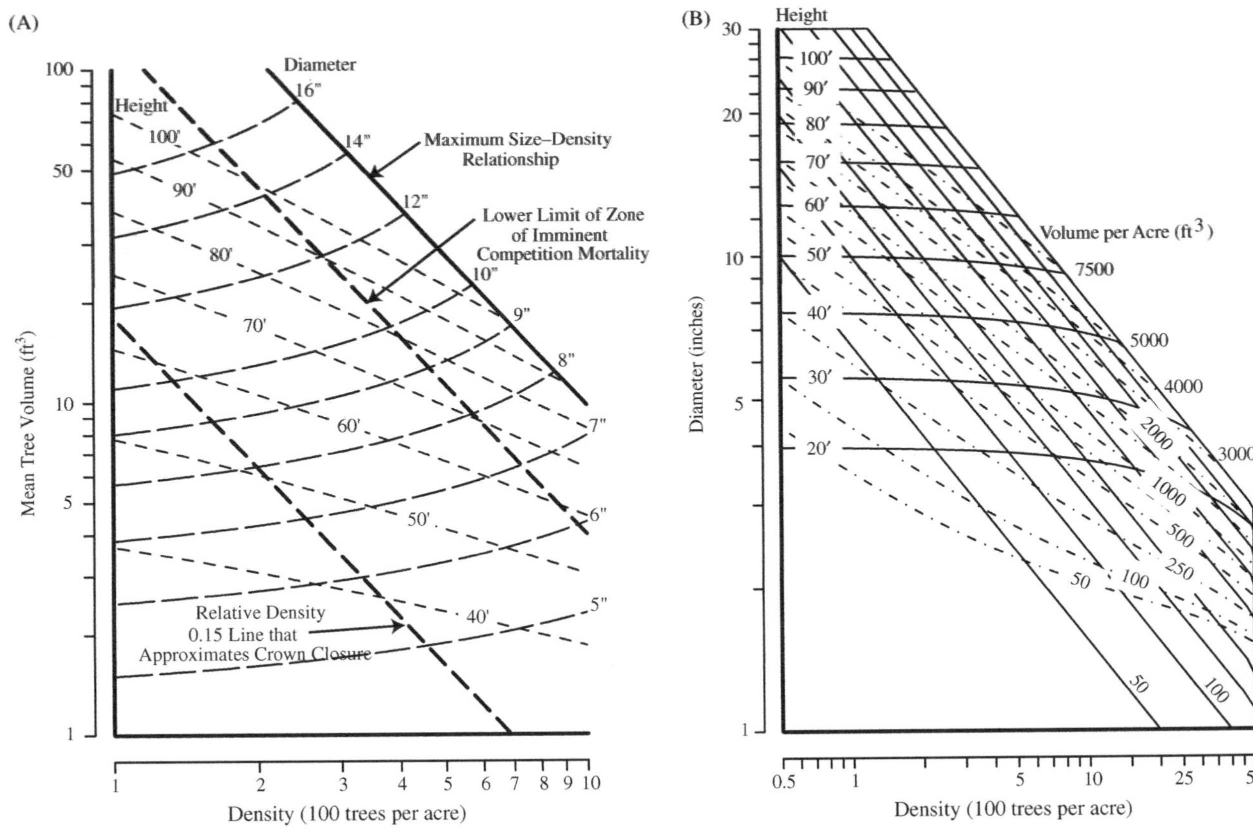

Figure 2.2 Stand density management diagrams for coastal Douglas-fir. Panel (A) uses mean volume as the tree size metric, while Panel (B) utilizes quadratic mean diameter and the exponent −1.6 rather than $\frac{-3}{2}$. Panels (A) and (B) were adapted from Drew and Flewelling (1979), and Long *et al.* (1988), respectively. Note: both graphs use English units.

Crown competition factor (CCF) was suggested by Krajicek *et al.* (1961) as an effective density measure because it accounts for differences in species composition and diameter distributions. CCF estimates the area available to the average tree in a stand relative to the maximum area it could use if it were open-growth (Krajicek *et al.*, 1961). CCF is considered independent of site quality and stand age and is applicable in both even- and uneven-aged stands (Krajicek *et al.*, 1961). However, CCF requires an equation for predicting maximum crown width for each species in the stand (see Chapter 7). Although not a measure of crown closure (Curtis, 1970), complete crown closure of unthinned stands is generally assumed to occur from a CCF of 100 to the maximum for the species (Krajicek *et al.*, 1961), which can be over 500 for some species like Douglas-fir and western hemlock (e.g. Hann *et al.*, 2003).

A related measure is the tree-area ratio (TAR) of Chisman and Schumacher (1940), which depicts the ground area occupied by a tree as a parabolic relationship to its DBH. Summing this relationship for each tree in the stand can serve as a measure of growing space utilization:

$$\mathrm{TAR} = \beta_0 \cdot N + \beta_1 \cdot \sum_{k=1}^{N} \mathrm{DBH}_K + \beta_2 \cdot \sum_{k=1}^{N} \mathrm{DBH}_K^2 \qquad [2.6]$$

where N is the number of trees per unit area. In the case where data are collected from sample plots, the tree-area ratio is calculated using:

$$\mathrm{TAR} = \beta_0 \cdot \sum_{k=1}^{n} \mathrm{EXPF}_k + \beta_1 \cdot \sum_{k=1}^{n} (\mathrm{EXPF}_k \cdot \mathrm{DBH}_k) + \beta_2 \cdot \sum_{k=1}^{n} (\mathrm{EXPF}_k \cdot \mathrm{DBH}_k^2) \qquad [2.6b]$$

The parameters (β_0, β_1, β_2) are often determined using regression and data from "fully" stocked stands. Consequently, TAR often expresses the proportion of full stocking achieved by that stand. However, if the parameters are determined from stands with a range of densities, TAR reflects the proportion of stocking compared to the average stocking of the data. Curtis (1970) illustrated the mathematical equivalence of CCF and TAR in pure species stands, and Stage and Ledermann (2008) showed that plots of TAR and CCF parallel those of SDI. Curtis (1970) found relatively low correlations between TAR and both age and site index.

These previous indices of two-sided, distance-independent competition are only based on tree diameter and are insensitive to changes in tree height. Instead, relative spacing indicates the degree of competition as an expression of the number of trees and dominant height (Wilson, 1946):

$$\mathbf{RS} = \frac{(A \cdot N)^{\frac{1}{2}}}{\mathrm{HT}_{\mathrm{dom}}} \qquad [2.7]$$

where RS is relative spacing, A is unit area (43 560 ft^2 or 10 000 m^2), N is number of trees per unit area, and HT_{dom} is average height of dominant trees. For even-aged stands, RS decreases as stands grow older and the number of trees decreases due to mortality. Like TAR and CCF, RS is also closely related to SDI (Avery and Burkhart, 2002). RS is used in several static equations (e.g. Ducey, 2009b) and in stand growth and mortality equations (e.g. Qin and Cao, 2006), as well as in developing thinning schedules (Wilson, 1979).

Measures such as stand volume and bole area incorporate a variety of competition-related attributes including number of trees, size, and tree form. Lexen (1943) found stand bole surface area to be an effective predictor of growth. Likewise, Inoue (2009) found a strong relationship between mean stem surface area and the maximum size–density line. However, stand volume and bole surface area are not independent of stand age or site quality, which makes their interpretation difficult.

2.2.2 Distance-dependent

Compared to distance-independent indices of two-sided competition, relatively fewer distance-dependent indices of two-sided competition exist. One widely used index is the Clark and Evans (1954) average distance to the nearest neighbor. This index describes the spatial heterogeneity of trees within a stand by estimating how much they deviate from a random pattern, on average. It is estimated as:

$$\mathbf{R} = \mathbf{2} \cdot \frac{\sum_{k=1}^{N} w_k^2}{N} \cdot \sqrt{N} \qquad [2.8]$$

where w_k is the distance from the kth tree to its nearest neighbor. A truly randomly distributed population has a Clark and Evans (1954) index of 1, while an index value of 0 indicates maximum aggregation of individuals.

A related idea is the notion of point density rather than stand density. Most of the previously discussed indices of two-sided competition focus on average stand conditions, while more specific indices might be needed to describe the local competition environment in a stand, and its variability. Spurr (1962) introduced this concept by modifying variable radius sampling to account for the size and spatial distribution of trees. Although Spurr's (1962) point density index is distance dependent, several approaches are used to account for point density in a distance-independent manner (see Stage and Wykoff, 1998).

2.3 One-sided competition

One-sided competition attempts to describe an individual tree's social status within a stand, and to quantify its surrounding local competitive environment. Consequently, indices of one-sided competition are usually unique for each individual in a stand. Numerous indices of both distance-independent and distance-dependent one-sided competition exist. The focus of this section is on ones that are currently widely used in operational growth and yield models.

2.3.1 Distance-independent

2.3.1.1 Absolute

Commonly used distance-independent indices of one-sided competition used in growth models are basal area in larger trees (BAL) and crown competition factor in larger trees (CCFL). Both indices are estimated by summing the basal area or CCF of trees with DBH values larger than the subject tree (Figure 2.1B). These indices, particularly BAL, are widely used in individual-tree static, increment, and mortality equations (Chapters 6–8). Limitations of BAL include that it treats each species as an equal competitor based solely on its DBH and does not account for differences in stand structure. Consequently, some authors contend that a modified BAL works better than one that is unmodified. For example, Schröder and von Gadow (1999) suggested taking Stage's (1973) percentile in the distribution of tree basal area and dividing it by RS to form an index of competition, which they found was superior to unmodified BAL.

Other absolute indices of one-sided competition are crown-based indices such as crown closure at the top of the tree (CCH; Hann and Ritchie, 1988) or a certain percentage (p) of subject tree height (CC_p; Wensel *et al.*, 1987). Both CCH and CC_p require a crown profile equation for each species. To calculate CCH of a particular tree, crowns widths at the height of a subject tree are calculated for all trees in the sample. The crown widths are then converted to an area, summed, and divided by ground area to express the value as a percentage. Wensel *et al.* (1987) found that the most effective CC_p was at 66% of the subject tree height, which was further supported by Biging and Dobbertin (1995). However, Hann and Ritchie (1988) found CCH to be more effective than any CC_p. Two potential reasons for these differences are that the analyses by Wensel *et al.* (1987) as well as Biging and Dobbertin (1995) did not include indices of both one- and two-sided competition in their equations, and the crown profile equations they used were developed using trees that were measured when laying on the ground, which can result in narrower crown width predictions.

Regardless, these height-ranked, crown-based variables are generally more effective than DBH-ranked, basal area based variables (Bravo *et al.*, 2001) and even some distance-dependent competition indices (Biging and Dobbertin, 1995). However, height-ranked competition variables such as CCH do require that the height of all sample trees be measured without error, which is rarely done in practice and has likely limited their application.

2.3.1.2 Relative

Relative indices of one-sided competition were widely used in the past. These were calculated by dividing a subject tree's DBH or height by the average or maximum value for the stand. Some of the more common include the ratio of DBH to mean stand DBH (Glover and Hool, 1979), tree basal area to mean basal area (Daniels *et al.*, 1986), and tree height to mean dominant height (Daniels *et al.*, 1986). Stage (1973) proposed "percentile in the distribution" of tree basal area per unit cross-sectional area at DBH (i.e. 100 × (BAL/BA)) as an indicator of relative dominance.

Although relative measures of tree social position are often well correlated with growth (Daniels *et al.*, 1986), they have important limitations as indices of competition. First, relative measures only depict a tree's competitive position in the current stand, while ignoring the amount of competitive pressure within the stand. For example, a relative tree size of 0.5 can have very different meanings in a young, fully stocked stand compared to an older, understocked one. Second, relative tree measures do not perform logically when a stand is thinned. Wykoff (1990) noted that the percentile in the distribution of tree basal area used by Stage (1973) was artificially reduced for most residual trees when a thinning from below was conducted and, consequently, an absolute measure like BAL was preferred.

2.3.2 Distance-dependent

Distance-dependent indices of one-sided competition require the spatial locations of all trees and have a long history of development. Theoretically, distance-dependent indices provide a better description of competition because they assess a tree's local environment and the proximity of competitors. While numerous distance-dependent indices of one-sided competition exist, they can be grouped into five general classes: (1) size–distance (SD) relationships; (2) competitive influence zones (CIZ); (3) growing space (GS) indices; (4) open-sky view (OSV); and (5) light-interception (LI) methods (Figure 2.3). Each class is described below.

2.3.2.1 Size–Distance Relationships

Size–distance indices estimate competition from the size and distance to each potential competitor (Figure 2.3C, D). They are based on the idea that competition is greater as neighboring trees increase in size and proximity. The index may be computed in the horizontal (e.g. based on stem or crown diameters of neighboring trees) or the vertical plane (e.g. based on heights of neighboring trees). In either case, the reference height may be varied to obtain more meaningful estimates (e.g. breast height, crown base of subject tree, top of subject tree, etc.). Size–distance indices were found useful in even-aged pure and mixed-species forests in Finland (Pukkala, 1989; Vettenranta, 1999), as well as an uneven-aged stand (Mailly *et al.*, 2003).

These are the most commonly used distance-dependent indices of competition, due to the relative ease of estimation, and are computed in various ways (see Tomé and Burkhart, 1989). Some of the most important differences in computing size–distance indices

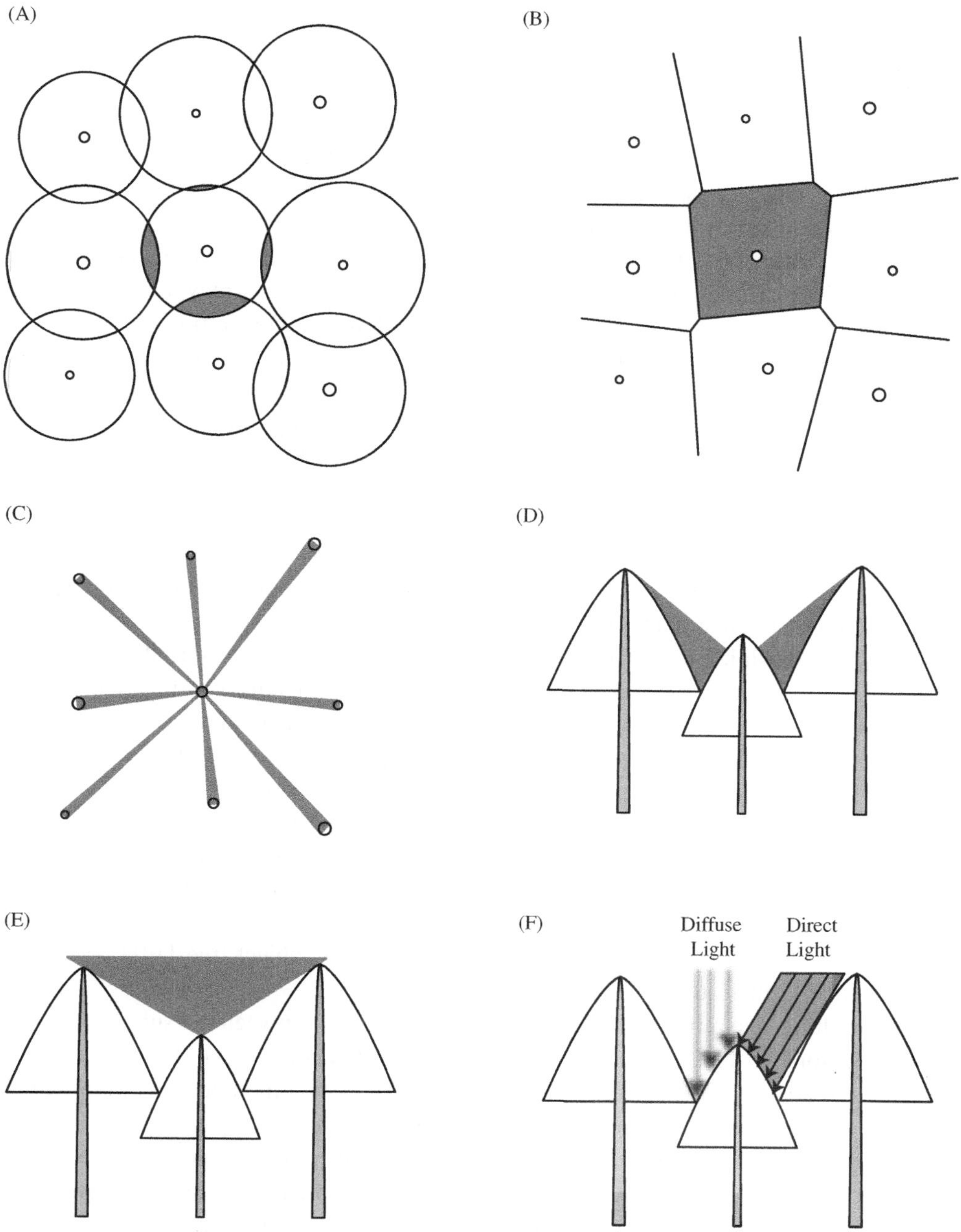

Figure 2.3 Distance-dependent measures of competition including: (A) competitive influence zone; (B) area potentially available; (C, D) size–distance; (E) open-sky view; and (F) light interception.

include: (1) the size attribute used in the index; (2) how competitor trees are selected; (3) the manner in which the size attribute of the subject tree is compared to that of competitor trees; and (4) the weighting of competitors.

Various size variables are used including DBH (Lemmon and Schumacher, 1962a), total tree height (Corral Rivas *et al.*, 2005), bole volume (Weiner, 1984), and crown cross-sectional area (Biging and Dobbertin, 1992). DBH is frequently used as the size variable because it is commonly available and other measures are not vastly superior (e.g. Corral Rivas *et al.*, 2005). DBH also is generally found to be superior to tree basal area (Daniels, 1976; Tomé and Burkhart, 1989). However, crown-based indices may work better for some species when this information is available (Biging and Dobbertin, 1992).

To limit edge-effects and computation time, it is necessary to specify a search-radius within which competition is assessed. Opie (1968) identified competitors using various fixed- and variable-radius plots, and found the predictive ability of the index was altered by the method utilized, but circular plots of a constant radius tended to perform best. Likewise, Hegyi (1974) defined all trees within a fixed distance of the subject tree as competitors. The optimal fixed distance has varied from 2–3 m (e.g. Hegyi, 1974; Weiner, 1984) to over 15 m (e.g. Opie, 1968), with larger sizes generally being preferred (Alemdag, 1978). Daniels (1976) found that selecting competitors with a variable-radius plot was more effective than a fixed-radius one, with a smaller basal area factor (BAF) being preferred; while Lin (1974) used just the four nearest competitors. Rather than use a horizontal search-radius, Ford and Diggle (1981) selected competitors as those individuals taller than a line at a 45° angle from the top of the subject plant, and based their index on the angle subtended by competitors. This vertical approach has generally worked better than the horizontal one (Biging and Dobbertin, 1992; Corral Rivas *et al.*, 2005), and it is the approach used in the SILVA growth model (Pretzsch *et al.*, 2002).

The ratio of competitor tree size divided by subject tree size is the most common method used to compare size attributes once the competitors are identified (e.g. Hegyi 1974; Martin and Ek 1984; Biging and Dobbertin, 1992). However, Tomé and Burkhart (1989) found that the difference between competitor and subject tree DBH was more effective than using ratios.

The resulting size ratios are often weighted by distances between the subject tree and its competitors, so that closer ones have a greater influence. Hegyi (1974) scaled his ratio by the inverse of distance, while Martin and Ek (1984) modified this ratio with a negative exponential function of distance divided by the sum of competitor and subject tree DBH. Weighting is generally assumed to provide superior estimates of competition, but Biging and Dobbertin (1992) found that some unweighted measures of crown-based size–distance were superior compared to ones weighted by distance. The final size–distance index of competition for the subject tree is calculated by summing the weighted (or unweighted) size ratios (or size differences) across all identified competitors on the plot.

2.3.2.2 Competitive Influence Zones

Competitive influence zones are based on two assumptions: (1) the area over which a tree competes for resources is proportional to its size; and (2) each tree has a potential area of influence over which it obtains or competes for site resources (Bella, 1971). Competition experienced by each tree is assessed by the degree of overlap between influence zones. This index is expressed as either an absolute area in terms of m^2, or scaled as a relative area. Like size–distance ratios, competitive influence zones can be weighted by tree size or other factors like shade tolerance (e.g. Ek and Monserud, 1974).

Most competitive influence zone indices assume that the zone is circular about the trees. The radius of the zone is a function of tree size and is often based on the maximum crown width of a similar-sized open-grown tree (Corral Rivas *et al.*, 2005). A competitor is any surrounding tree whose influence zone overlaps with the subject tree's influence zone (Figure 2.3A). Thus, competitive influence zones avoid many of the complications associated with selecting competitor trees that are experienced in size–distance indices. Competitive influence zone indices can be overly influenced by competitors smaller than the subject tree. In an attempt to avoid this, Bella (1971) estimated the competitive pressure based on the ratio between the competitor and subject tree's DBH, while Ek and Monserud (1974) used the product of the subject tree's crown radius and its height as an estimate of crown volume to adjust their competitive influence zone index.

2.3.2.3 Growing Space Indices

Growing space indices are similar to competitive influence zone indices, but express competition as the area not occupied by competitors rather than the fraction of area occupied by competitors. One of the most commonly used growing space indices is area potentially available (APA; Brown, 1965). APA of each tree is usually calculated by dividing the total (simulated) plot area among the trees according to their size and position (Figure 2.3B). In other words, APA creates polygons around each tree by the intersections of the perpendicular bisectors of the distances between a subject tree and its competitors. Thus, APA avoids the assumption that competitive pressures are symmetrically distributed around a tree as is often presumed in competitive influence zone indices.

Like competitive influence zone indices, one limitation of growing spacing indices like APA is that resources are assumed to be homogeneously distributed around a tree, and smaller sized trees can have a similar influence to larger ones. Moore *et al.* (1973) modified APA so that it was weighted by the relative size of competitors, which was found to improve predictions in a mixed-species, uneven-aged hardwood stand. Other weighting schemes for APA were suggested by Alemdag (1978) and Pelz (1978). However, weighting still fails to address the potential of greater competitive pressures due to effects of larger trees being masked by nearer smaller neighbors. Weighting may also cause area in the stand to not be allocated to a tree.

Other efforts to improve APA have focused on accounting for differences in shape and position of an individual within a polygon, expanding it from two- into three-dimensions, and using curved rather than straight lines to define boundaries. Mead (1966) identified two concepts to describe the shape of a polygon and tree position within it: (1) *eccircularity* (the extent to which a polygon is elliptical); and (2) *abcentricity* (degree that a tree is displaced from the center of the area). However, Daniels *et al.* (1986) found that using either estimates of eccircularity or abcentricity did not improve predictions of individual-tree growth in a loblolly pine plantation, but suggested that they might be more effective in stands with a more irregular spatial pattern.

Competition is a three-dimensional process, and APA generally only addresses the horizontal competitive pressures, which are assumed to be related to the vertical competitive pressures. Pelz (1978) developed a three-dimensional APA by multiplying the two-dimensional measure by height, which was found to work better in a young hardwood stand. To also address vertical competitive pressures, Wimberly and Bare (1996) estimated a "layered" APA by filtering potential competitor trees by crown class. For example, APA for suppressed individuals was calculated using all neighbors, while APA for codominants was estimated using only codominant and dominant neighbors.

Generally, the edges of APA polygons are often treated as straight. In low-density stands, this can lead to unrealistic estimates of APA. Nance *et al.* (1988) used curved boundaries that were constrained by subject tree maximum crown width, and truncated when this contacted the polygons of other trees, which led to significant improvement compared to a traditional straight-line APA estimate. Additional refinements to APA may be needed to reflect differences in competitive pressures asserted by different species in mixed-species stands.

Another growing space index is exemplified in the work of Mitchell (1975), which is based on a three-dimensional depiction of individual tree crowns. By representing competition at this level of detail, it avoids many of the limitations associated with APA. Essentially, in these crown-based indices, individual branch growth is extended until interference from a tree's neighbors, and branches begin to die as they lose foliage, which influences a tree's growth and competitive vigor. Thus, the competitive pressures around a tree vary vertically as well as radially, the growing space is constrained by biological limits, and weighting isn't necessary. A limitation of the approach is that they require very detailed crown data for parameterization, and can be computationally demanding, even with today's computing power.

2.3.2.4 Open-Sky View Indices

Competition for light is often assumed to be the primary determinant of tree growth. Open-sky view approaches determine the proportion of unobscured sky above each tree

(Figure 2.3E). These indices require that each competitor tree be depicted three-dimensionally. Due to differences in light quality, open-sky view indices sometimes weight parts of the sky differently, such as higher weights for the overhead sky and less for the horizons (Chan *et al.*, 1986). Commonly, the reference point for these indices is the top of a tree, but the centroid of the crown or areas of recent crown development of the subject tree are also used. This allows open-sky indices to effectively represent different components of competition. For example, estimating open sky from the top of the tree would primarily depict competition from larger trees, while determining open sky from the base of the tree would include both larger and smaller neighbors.

Three primary methods are used to generate open-sky indices: (1) ray tracing (Brunner, 1998); (2) use of hemispherical photographs taken from the tree location (e.g. MacDonald *et al.*, 1990); and (3) simulating the fraction of open sky seen from the tree location and its neighbors (e.g. Van Pelt and North, 1996). All these methods are computationally demanding and are integrated into growth models in varying ways.

Ray tracing involves simulating radiation from various points in the sky and projecting its passage through a canopy. Open-sky view is estimated by counting the fraction of paths free from obstruction. In hemispherical photographs, an extreme angle of view (e.g. 180°) is provided with a fisheye lens. Various software tools exist to analyze hemispherical photographs, such as HemiView (Rich *et al.*, 1999) and GLA (Frazer *et al.*, 1999), which provide estimates of open sky. The limitation of hemispherical photographs is defining thresholds that separate sky from trees. This can be done automatically or manually; both methods have important advantages and disadvantages (Jarčuška, 2008). Given the logistical difficulty of obtaining hemispherical photographs for larger trees, the use of this technique is primarily limited to saplings or lower levels in the canopy.

In the third approach, rather than use hemispherical photographs taken in the field, computer programs are used to simulate virtual hemispherical photographs. This approach is used in the SORTIE gap model (Coates *et al.*, 2003), and exemplified in the work of Van Pelt and North (1996). This avoids the difficulties in separating sky from tree, the expense of obtaining hemispherical photographs, and is used regardless of tree size. However, trees are often represented as simplified geometric shapes, which limits their accuracy due to the dynamic nature of tree crowns. For example, the virtual hemispherical photographs of Silbernagel and Moeur (2001) could only explain 50% of the variation in canopy openness estimates derived from actual hemispherical photographs.

Significant progress has been made in representing trees more realistically (see Godin, 2000). Although this approach is very similar to ray tracing, it is often a more simplified calculation since the subject tree's view is just divided into sky or obstruction rather than multiple rays being simulated. In addition, application of this technique to dominant overstory trees is problematic because the competitive impact of other trees is not sufficiently represented, as illustrated in Hanus (2003). The light-interception techniques described below provide an approach that can overcome this limitation.

2.3.2.5 Light-Interception Indices

Similar to open-sky view indices, light-interception indices of competition three-dimensionally depict forest canopies and focus on competition for radiation as the primary determinant of growth. However, rather than estimating the amount of open sky available to an individual tree, these indices calculate the amount of light absorbed by a tree over a specified time period (Figure 2.3F). This avoids having to weight parts of the sky differently or selecting a certain reference point on a subject tree, as the approach is often more holistic. However, this also generally increases the calculation complexity. Four methods are used to estimate light interception: (1) determine the fraction of crown exposed to direct sunlight; (2) projecting obstructions onto the crown surface; (3) ray tracing; and (4) solving radiation transport equations for a stand.

Hatch *et al.* (1975) presented an index of exposed crown surface area that depended on height to crown base (HCB), and relative size of the subject tree and adjacent trees. The measure wasn't found to be more effective than a competitive influence zone index based on DBH only (Hatch *et al.*, 1975). However, the index of Hatch *et al.* (1975) treated tree crowns as cones and used a fixed solar altitude. This index could easily be improved with today's technology and better representation of tree crowns. For example, Duursma and Mäkelä (2007) used crown surface area, ratio of leaf area to crown surface area, a light extinction coefficient, and a parameter describing mean solar angle to estimate annual light interception for a single tree. Estimating exposed crown area from aerial photographs, Wyckoff and Clark (2005) found the measure to be nearly as effective a predictor of tree growth as tree basal area. Rather than estimating light interception, Kuuluvainen and Pukkala (1987) projected obstructions onto the crown of a subject tree to estimate its light exposure, which was done multiple times per day and throughout the year. Granberg (1988) further refined their approach.

Multiple spatially dependent approaches to ray tracing are presented with varying degrees of complexity (Brunner, 1998; Stadt and Lieffers, 2000; Groot, 2004). For example, some models require information on leaf area density, leaf inclination angles, and the degree of foliage clustering (Groot, 2004). The ray-tracing method has three primary steps: (1) representation of the crown surface in three dimensions; (2) determining the intersection of light rays with tree crowns; and (3) modifying the transmission of light through crowns (Groot, 2004).

Generally, crowns are treated as symmetrical geometric solids (e.g. Stadt and Lieffers, 2000); however, more flexible approaches are available (Brunner, 1998; Groot, 2004). Estimation of light rays and tree crown intersections can be determined using a gridded approach (e.g. Brunner, 1998) or numerical methods (Groot, 2004), with the latter being significantly more efficient. Once light intersects a crown, it is absorbed, transmitted, or reflected. Generally, only transmission is simulated because of the difficulty in separating the components of radiation transfer (Groot, 2004). Transmission is often estimated based on the path length of a ray through the canopy, which can be further modified to account for differences in foliage density and the elevation angle of the ray (Groot, 2004). Ray-tracing simulations are often conducted using multiple time periods during a day and over the course of a year. The intensity of competition is calculated as the difference between amount of light absorbed by a subject tree and the maximum light available overhead. Although the computational demand of the ray-tracing approach is still high even for today's computers, it is shown to correlate strongly with observed growth (Brunner and Nigh, 2000).

An alternative to ray tracing is to use radiation transport theory to estimate intensity and quality of incoming light at any point in a stand (e.g. Hanus 2003). Ross (1981) gives an extensive summary of the theory. The approach works by dividing a stand into multiple small cells (termed voxels), and differential equations of radiation transport are solved for each cell. Like ray tracing, this approach is quite complex and, again, computationally demanding, which has limited its application in growth models. In addition, recent research has emphasized the importance of diffuse radiation (e.g. Rocha *et al.*, 2004), which is very difficult to predict because of its complex interaction with atmospheric conditions.

2.4 Limitations

Competition is a continuous, complex, and dynamic process, which is indirectly assessed by the various indices described in this chapter. Most of these indices focus on aboveground competition as the primary determinant of growth (particularly distance-dependent indices like open-sky view and light interception), when belowground competition is likely to be equally or more important (Coomes and Grubb, 2000). Furthermore, there is often a poor correlation between aboveground and belowground growth on an individual tree (e.g. Eis, 1986), which suggests that different factors influence their behavior. Thus, a single index of competition cannot holistically represent all components of competition. Competitive pressures also change with site conditions, in that below-ground competition becomes more important as soil resources decrease.

Competitive interactions between two species are also highly complex. Perry *et al.* (2008) classified interactions between two species into five classes: (1) predation and parasitism (one gains and one loses); (2) competition (both lose); (3) amensalism (one loses, the other is unaffected); (4) mutualism (both gain); and (5) neutralism (both unaffected). Most indices described in this chapter only focus on competition, while the other aspects of species interactions are not addressed. For example, red alder can be a strong competitor for light, which slows Douglas-fir growth (D'Amato and Puettmann, 2004), but red alder can significantly enhance soil nitrogen availability (Bormann *et al.*, 1994), which ultimately benefits both species. There are also several other important limitations of competition indices that are necessary to recognize in growth models.

2.4.1 Low predictive power

Compared to tree size, competition indices often explain a limited amount of variation in growth because the two are highly interrelated. Perry (1985) identified four reasons for the low predictive power of competition indices: (1) tree size is an integrator of competition over the entire life of a tree; (2) response to competition is highly variable because of differences in genetics and environmental conditions; (3) interactions between individuals can be co-operative rather than competitive; and (4) indices based on two-sided competition may not reflect competition properly when light is limiting. In addition, trees are quite plastic with respect to their morphology and physiology, which can reduce the degree of size asymmetry in competition (Schwinning and Weiner, 1998). Thus, indices of competition often fail to express the complexity of competition (Soares and Tomé, 1999).

Multiple solutions are suggested for improving the representation of competition in growth models. For example, rather than rely on absolute growth rate, relative growth rate is found to express effects of competition better (Larocque and Marshall, 1993). However, MacFarlane and Kobe (2006) did not find relative measures of growth more effective than absolute ones and suggested they be compared on a study-by-study basis. The most common approach is to include indices of both one- and two-sided competition in growth equations (Hann and Hanus, 2002a; Larocque, 2002; Hann *et al.*, 2003). This ensures that the availability of both aboveground and belowground resources is represented, albeit indirectly. Canham *et al.* (2004) separated effects of shading and crowding for two conifer species in British Columbia and found that growth declined more steeply as a function of crowding than shading, but effects of crowding were sensitive to the species of the competitor trees. However, the species in Canham *et al.* (2004) were both relatively tolerant of shade, and so the results may not hold for other species, since shading might be more important for intolerant species. In central Europe, Pretzsch and Biber (2010) found that competition moves from predominately one-sided to mostly two-sided as the site quality moves from fertile to poor. To ensure that this is properly represented, they suggested that an interaction between the competition index and measure of site quality must be included in any growth model.

In addition to species composition and site quality, the symmetry of competition likely varies as stands develop over time. In young stands, competition is primarily two-sided, but the importance of one-sided competition increases after stands close and differentiation begins. For example, Filipescu and Comeau (2007) found that the levels of competition and its relationship to growth differed between young and older stands, which had to be addressed in their growth model. Due to the complexity of these interactions, future efforts to improve competition indices will likely require the further development of process-based modeling approaches (Larocque, 2002).

2.4.2 Distance-independent vs. distance-dependent

Multiple studies have attempted to find the best index of competition for predicting growth (e.g. Biging and Dobbertin, 1992; 1995; Corral Rivas *et al.*, 2005; Filipescu and Comeau,

2007). No clear trends have emerged from these analyses, even when just distance-dependent indices are compared (Alemdag, 1978; Weiner, 1984; Daniels *et al.*, 1986). For example, Daniels *et al.* (1986) suggested that APA was the most effective distance-dependent index of competition for predicting tree growth, while others have found size–distance indices to be superior (Tomé and Burkhart, 1989; Biging and Dobbertin, 1992; Filipescu and Comeau, 2007). However, most analyses that focused on distance-dependent competition indices were conducted in even-aged, single-species stands, and have not considered open-sky view or light-interception indices, which may be more effective. For example, Hanus (2003) found that light interception was a more effective predictor of western hemlock and Douglas-fir height growth across a range of size classes and stand structures, compared to various other distance-dependent indices of competition.

Although the spatial arrangement of trees can significantly influence their growth (e.g. Pukkala and Kolström, 1991), most empirical analyses find distance-dependent indices of competition not to be superior to distance-independent ones (Martin and Ek, 1984; Cole and Lorimer, 1994; Biging and Dobbertin, 1995; Wimberly and Bare, 1996). These findings can likely be attributed to various limitations of distance-dependent indices of competition.

First, most distance-dependent indices of competition struggle to define the area of competitive effect to capture variation in the local competitive environment (Ledermann and Stage, 2001). Some studies found that the correlation between a distance-dependent index of competition and individual-tree growth was higher as the distance used to define competitors was increased (Opie, 1968; Alemdag, 1978). As the search distance for competitors increases, the differentiation between local and stand-level competition blurs.

Second, distance-dependent indices of competition generally focus on one-sided competition, which essentially ignores belowground competition (Larocque, 2002). Effectively representing belowground competition is difficult because tree roots can extend far beyond the crown of a tree (Curtis, 1964a; Eis, 1974; 1987). A tree's effective root system can be further extended through root grafting (Eis, 1972) and the presence of large mycorrhizal mats in the stand (Griffiths *et al.*, 1996). Therefore, competition for moisture and nutrients is often broader than competition for light, and may extend way beyond the immediate surrounding trees. Furthermore, competition for moisture and nutrients is often considered to be size symmetric, while competition for light is often considered to be size asymmetric (Schwinning and Weiner, 1998).

Third, several distance-dependent indices of competition focus primarily on tree diameter when crown-based indices are generally more effective, even when spatial information is not available (Biging and Dobbertin, 1995; Bravo *et al.*, 2001). Also, most distance-dependent indices fail to account for the species of competitor trees. This might partly explain why crown-based indices are often more effective, because they implicitly represent species differences (Bravo *et al.*, 2001). All distance-dependent indices of competition require the use of edge correction techniques, which can influence their effectiveness (e.g. Martin *et al.*, 1977). Given the general lack and high expense of obtaining stem map data required by distance-dependent competition indices, this indicates that distance-independent indices will suffice for most operational growth and yield models.

2.4.3 Influence of sampling design

Indices of both one- and two-sided competition are sensitive to the sample design used to estimate them. This in large part explains why using a sampling scheme that is different from the one used to develop a model can cause imprecision in predictions (Hann and Zumrawi, 1991). Generally, larger plots are more effective for both distance-dependent (Opie, 1968; Alemdag, 1978; Schreuder and Williams, 1995) and distance-independent indices (Ledermann and Eckmüllner, 2004; Lappi, 2005). In addition, a plot has to be sufficiently large enough to address both the current and future distributions of tree size (Ledermann and Eckmüllner, 2004). Determining a plot size that accomplishes this is difficult. Lappi (2005) suggested that the plot size that produced the highest correlation was similar to the size of the tree's influence zone, but an even larger size was needed for unbiased predictions. Stage and Ledermann (2008) suggested that the optimum plot size for estimating competition indices was generally smaller than those used for determining stand volume.

In comparing fixed- and variable-radius sampling, Stage and Ledermann (2008) indicated that variable-radius sampling was more effective and negated any benefits of distance-dependent indices of competition in comparison to distance-independent ones. Most growth models, however, are developed using fixed-radius plots that are relatively small (0.05–0.1 ha), which might explain why most comparative studies find that distance-dependent indices of competition fail to significantly improve growth predictions (Stage and Ledermann, 2008). Furthermore, growth models are often developed using a single plot in a stand, which can underestimate the true variance of the competition indices, particularly in natural stands (García, 2006). Consequently, multiple, randomly placed plots are necessary to assess two-sided competition effectively. These can also be used to quantify what Stage and Ledermann (2008) term "semi-distance-independent" indices of competition, which were found to be just as effective as several distance-dependent indices.

Stage and Wykoff (1998) proposed an approach for correcting competition indices for differing plot sizes. This was accomplished by modeling the sampling properties of the competition estimates depending on plot size and assuming a random (Poisson) spatial pattern for tree locations. Model error variances were then combined with these estimates and used to dynamically modify model coefficients during simulation. However, the assumption of a Poisson spatial pattern is not valid for many stand structures and phases of development (Lappi, 2005). In addition, the approach does not

Table 2.1 Example growth and yield models, the competition variables used, and the type of model form in which the competition is used.

Growth model	Competition variable[a]		Model form
	One sided	Two sided	
Whole stand			
DFSIM (Curtis *et al.*, 1981)	—	BA, RD	Realized
Scube (García, 2011)	—	BA, TPH, CC	Realized
Size class			
FIBER (Solomon *et al.*, 1986)	—	BA_I, BA_R	Realized
Vanclay (1989a)	—	BA	Realized
Distance-dependent individual tree			
SILVA (Pretzsch *et al.*, 2002)	KKL, KMA, NDIST	—	Potential × modifier
TASS (Mitchell, 1975)	FV/FV_{max}	TPH	Realized
Distance-independent individual tree			
ORGANON (Hann, 2011)	CCFL, BAL, CCH	BA	Realized, potential × modifier
PROGNAUS (Monserud *et al.*, 1997)	BAL	BA	Realized
CACTOS (Wensel *et al.*, 1987)	CC_{66}	BA	Potential × modifier

[a] Where BA is total basal area; BA_I is total initial basal area before harvest; BA_R is total basal area after removal; BAL is basal area in larger trees; CCFL is crown competition factor in larger trees; CC is canopy crown closure; CCF is crown competition factor; CCH is crown closure at the top of the subject tree; CC_{66} is crown closure at 66% of subject tree's height; FV is tree foliar volume; FV_{max} is the tree foliar volume in the absence of competition; KKL is the crown competition index of Pretzsch *et al.* (2002); KMA is the ratio between the sum of crown surface area of coniferous competitors to that of all competitors; NDIST is the horizontal distance between the competition center of gravity to the stem center of a subject tree; RD is Curtis (1982) relative density; and TPH is trees per hectare.

allow differentiation in DBH caused by variation in tree response to competition and, consequently, results in unrealistic predictions of diameter distributions over time (Ledermann and Eckmüllner, 2004).

2.5 Summary

Stand growth and dynamics is a byproduct of individual-tree competition and differences in species reaction to it. Competition is generally for available resources such as light and moisture, but can also include physical abrasions caused by swaying tree crowns

(Goudie *et al.*, 2009) or chemical interference tactics employed by species such as black walnut and *Eucalyptus* (Jose *et al.*, 2006). Consequently, competition is described in multiple ways, including aboveground or belowground, one- or two-sided, and inter- and intra-specific. Growth and yield models represent competition by using various distance-independent and distance-dependent indices (Table 2.1).

Despite their long history of development and the mathematical complexity of some indices, almost all are indirect estimates of competition and fail to fully represent its complexity. Some reasons for this are that most indices generally represent only one aspect of competition (one- vs. two-sided), belowground processes are still poorly understood, tree size is a good integrator of past competition, and their effectiveness is sensitive to the sampling design used to characterize competition. To effectively represent competitive processes, growth models need to account for tree size, include both one- and two-sided indices of competition, and assess possible interactions between competition indices and tree size and site quality.

For most practical purposes, knowledge of tree spatial location is not necessary and representation of within-stand variability is of greater importance. However, some remote sensing technologies may make tree spatial information more widely available in the future, and allow for the further development of sophisticated approaches like light-interception indices of competition. Overall, future refinements in competition indices will require a better understanding of tree physiological response to competition, and more mechanistic approaches. Redefining our notion of competition may also be necessary (Ledermann and Eckmüllner, 2004). For example, Wykoff (1997) used an alternative definition of competition by including the subject tree itself rather than just focusing on its perceived competitors. Regardless of the difficulty in representing competitive processes properly, growth and yield models cannot ignore its importance and should ensure that biological expectations of competition are met.

3
Forest site evaluation

3.1 Introduction

Forest site quality is defined as the physical and biological factors that characterize a site's ability to support tree growth (Skovsgaard and Vanclay, 2008). Forest site quality is influenced by a complex array of factors and is defined both qualitatively and quantitatively. Two primary measures used in forest site evaluation are *phytocentric* and *geocentric*. Phytocentric measures are tree-based metrics that attempt to characterize a site based on attributes of the stand or components of individual trees. Geocentric measures rely on physical site properties such as climate and soils to assess forest sites. Depending on scale and how close the indicator is related to the stand volume production, both indirect and direct measures of forest site evaluation exist (Skovsgaard and Vanclay, 2008) (Table 3.1). For example, soil texture and nutrient availability are direct geocentric measures of forest site, while total volume is a direct phytocentric measure (Skovsgaard and Vanclay, 2008).

It is also important to distinguish between potential and realized measures of forest site evaluation. Phytocentric measures are often realized measures of forest site evaluation because they rely on tree-based attributes, which are influenced by a variety of factors that are not easily assessed, such as past natural disturbances. Potential productivity can be expressed as the site's maximum ability to support total biomass of a particular species or any species. Both assume that potential productivity is independent of current stand structure. Often, site productivity in forestry refers to the ability of a stand to produce aboveground wood volume.

Forest site evaluation is a necessary component of any growth and yield model because only regional averages would be depicted if site differences were not modeled. In addition, estimates of site productivity must be consistent, as any error may propagate through

Forest Growth and Yield Modeling, First Edition. Aaron R. Weiskittel, David W. Hann, John A. Kershaw, Jr. and Jerome K. Vanclay.

Table 3.1 Methods used for assessing forest site productivity. Adopted from Skovsgaard and Vanclay (2008).

	Direct	Indirect
Phytocentric	Volume measurements	Site index; habitat type
Geocentric	Soil texture; soil moisture; available radiation	Aspect; elevation; latitude; longitude

growth, mortality, and recruitment functions to affect all modeling results. Although multiple measures of forest site quality exist, some attributes of a good measure are that it should be: (1) reproducible and consistent over long periods of time; (2) indicative of site and not strongly influenced by stand structure, composition, or management practices; (3) correlated with site productivity potential; and (4) at least as good as any other measure.

3.2 Phytocentric measures of site quality

3.2.1 Site index

One of the most common phytocentric measures of forest site quality is site index, which is the expected height at a certain reference age. Site index has a long history of development and use, as its utility was noted towards the end of the nineteenth century (Batho and García, 2006). Among the reasons for its widespread adoption are the relative ease with which site can be estimated from field observations, its proven efficacy in predicting volume growth and yield, and the strength of the relationship between tree height and age in even-aged forests. Some key assumptions in the use of site index as a measure of site quality are that: (1) tree height is independent of stand conditions; (2) dominant trees that have not experienced any suppression or other damage are easily identified; and (3) tree height is an effective integrator of the key biological determinants of growth.

Although these assumptions are generally met, there are some significant exceptions, which can limit the utility of site index as a productivity measure. For example, several studies found tree height to be significantly influenced by stand density (Lynch, 1958; Cieszewski and Bella, 1993; MacFarlane *et al.*, 2000; Flewelling *et al.*, 2001). The severity of this influence is related to species tolerance and site conditions; however, adjustments to ensure unbiased estimates can be made (e.g. Flewelling *et al.*, 2001). To ensure that estimated site index is representative of the site, trees that are undamaged must be selected, which are generally considered the dominant trees in a stand. Although some undetectable damage will be present in any site tree, this influence is assumed to be small. Nigh and Love (1999) investigated this assumption and unexpectedly found that the majority (50–75%) of their site trees had past damage. However, this damage did not significantly influence their site index estimates (Nigh and Love, 1999), which suggests that it is a relatively robust measure.

Physiologically, height growth will generally be favored over diameter growth for any stand-grown tree. In addition, height growth of certain species is often unaffected by stand species composition if the tree is in a dominant position. For species that recover from previous suppression, this may explain why site index is an effective measure of site quality in both even- and uneven-aged stands (Hann, 1998; Hann and Hanus, 2002a). However, adjustments are sometimes needed to account for prior growth suppression (Seymour and Fajvan, 2001).

Despite its wide utility, site index has some limitations as a productivity measure. The most important include that it: (1) is difficult to apply in mixed species or uneven-aged stands; (2) can change over time for a given stand; (3) is highly sensitive to measurement errors; (4) is rather imprecise in very young or older stands; (5) cannot be applied to areas that have no trees; (6) cannot be used to compare productivity potential between species; and (7) requires a well-constructed dominant height equation.

For use in growth models, site index estimates must be available for each species in a stand. This means that either enough site trees for each species in a mixed stand must be measured or a conversion equation developed and applied. Due to the expense of collecting tree age and height data, conversion equations are often used to estimate site index for each species in a mixed stand. In developing conversion equations, geometric regression, which is described well by LeDuc (1987), is often used to convert site index estimates from one species to another (Nigh, 1995a; Wang, 1998a). Rather than just fit conversion equations where multiple estimates of site index across species in a mixed stand were available, Wang (1998a) first stratified the data based on ecological zone, and found the approach to be much more effective. Although site index conversion equations are developed for a variety of species (e.g. Nigh, 2002; Johansson, 2006), caution is advised in their application because species can have synergistic as well as competitive interactions, which could alter their growth dynamics in pure and mixed species stands and lead to biased site index estimates (Nigh, 1995a). On the other hand, Wang (1998a) only found significant differences in site index between two species in 4 of the 29 ecoregions studied. When multiple species are present in a stand, as commonly encountered in hardwood stands, a composite site index based on the generally available species is an effective approach (Curtis and Post, 1964).

In uneven-aged stands, individuals can spend a significant portion of their life with reduced height increment due to overstory competition, which often leads to an underestimation of site index. The severity of this problem depends on a species' ability to respond to release, its shade tolerance, and amount of time since release. For example, Hann (1998) discovered that the height of Douglas-fir could recover from early suppression, the effect of which Hann and Hanus (2002b) tested and found no significant difference in Douglas-fir site index estimates from even- and uneven-aged stands. A variety of methods have been proposed when a recovery does not occur, including adjusting tree age into an effective age (Seymour and Fajvan, 2001; Hamel *et al.*, 2004), developing equations for different successional stages (Nigh, 1998), and incorporating a measure of stand structure in a height–age equation (Ouzennous *et al.*, 2008). In contrast to these other methods, Stage (1963) simply estimated site index from the annual height increment on a tree at a certain height (e.g. 16.8 m) combined with the number of rings in a defined radius of the pith at breast height, to circumvent the effects of early suppression.

An ideal measure of forest site productivity is consistent over time. For a given stand, site index often varies significantly due to changes in genetics, climate, and influence of management practices. Monserud and Rehfeldt (1990) found that genetic variation explained over 40% of the original variation in site index, which suggests that site index can change between generations. Environmental conditions have a significant influence on site index estimates, as these affect the genetics of the stand in the long term and the realized growth in the short term (Monserud and Rehfeldt, 1990). Bontemps *et al.* (2009) illustrated this, as they found that an accelerated dominant height increment for European beech in France was evident for much of the twentieth century and has reached over 60% of the normal rate.

Certain forest silvicultural treatments like soil bedding can permanently increase site productivity, but most forest silvicultural treatments such as vegetation control or fertilization just accelerate stand development without altering the final volume that may be attained (Fox, 2000). In addition, measures of stand height have a reduced ability as an indicator of site productivity for thinned stands (Skovsgaard and Vanclay, 2008). Regardless of the long-term influence of these practices, site index often is quite sensitive to forest management practices. Consequently, estimates of current site index are a complex byproduct of genetics, climate, past management practices and their interactions, which make it difficult to assess the real long-term potential of any given location with the use of site index.

Determining site index requires the measurement of the height and age of individual trees, both of which are subject to measurement error. Not only does measurement error bias the resulting site index estimates (see Chapter 14), it can also result in biased site index equations themselves (Goelz and Burk, 1996). The number of height–age measurements required in a stand depends on the specifications of the site index equation being used. One type of site equation is applied to stand-level estimates of average dominant height and age. Varying definitions of average dominant height exist, which affects how many trees to measure and often results in significantly different estimates of site index (Sharma *et al.*, 2002). Furthermore, plot size can have a substantial impact upon measures of average dominant height, such as top height, which can bias estimated site index (García, 1998; Magnussen, 1999).

The other type of site index equations is applied to individual-tree measurements of dominant height and age. In such a case, the equation is used to estimate site index for several trees in a stand and then averaged. The number of recommended site trees measured in a stand varies by site index equation and generally ranges from 2 to 10. Ultimately, the number depends on desired precision (Monserud, 1984).

Site index estimates are based on a reference age, which is typically set at some age less than the anticipated rotation age (Goelz and Burk, 1996). The accuracy of site index estimates decreases significantly as stands diverge from the site index reference age. For example, Goelz and Burk (1992) found that top height at index age is often underestimated for high sites and overestimated for low sites. Site index estimation is particularly problematic in young planted stands because they often have not fully expressed site potential for growth due to planting shock and other site establishment problems. An examination of the basic site index data used in Hann *et al.* (2006) shows that site index of Douglas-fir plantations did not stabilize until the average breast-height age exceeded 30 years (Figure 3.1).

Phytocentric measures of site productivity rely on trees to assess site quality. Consequently, trees must be present to determine site quality. Site index estimates are only

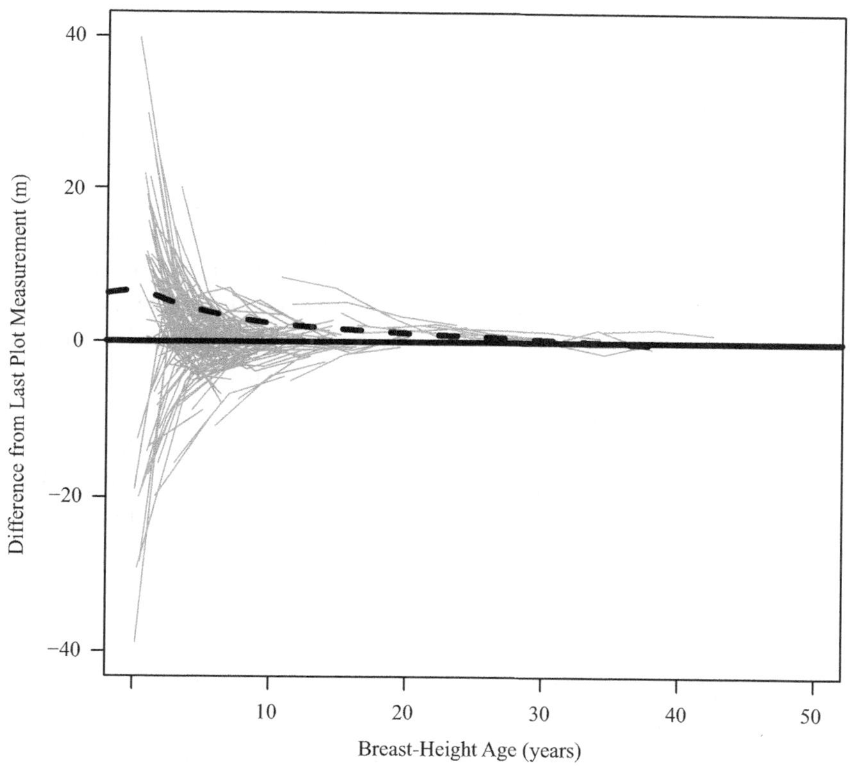

Figure 3.1 Changes in estimated coastal plantation Douglas-fir site index (base age 50 years) on periodically remeasured permanent research plots in Oregon and Washington (data from the University of Washington Stand Management Cooperative). The dashed line is the lowess regression line.

applicable to the species present and cannot be used to compare productivity estimates. This limitation is well illustrated in the establishment of fast-growing plantations in South America, where either no trees were previously present or the native forest was relatively unproductive. Both of these situations would make it difficult to estimate the productivity potential for a fast-growing and exotic species like *Eucalyptus*.

Finally, estimating site index requires a species-specific height–age equation. Numerous equations have been developed for a variety of species and regions (e.g. Carmean *et al.*, 1989), and they differ in many ways. The most important differences include: (1) the type of data; (2) the model form; and (3) the parameter estimation method used in their development.

The guide curve method was one of the first techniques used to develop site index equations. In this method, single pairs of height–age measurements are collected on a large number of independently sampled trees or stands using temporary plots. Each pair is then plotted on graph paper with age on the x-axis and dominant height on the y-axis, and an average guide curve or equation is determined, along with the associated site index at a predetermined base age. This guide curve/equation is then scaled up and down to determine the height–age curves for selected values of site index. This method always assumes that all site indices are equally represented across the full range of ages, and it usually assumes that the shape of the height–age curve is independent of site index (i.e. anamorphic). These assumptions are often violated and this has virtually eliminated the use of this technique (Monserud, 1985).

The stem analysis technique uses felled trees to determine past height and age relations on each individual tree (Curtis, 1964b). Although this method is still used today (e.g. Subedi and Sharma, 2010), it suffers from two major problems. First, it is common practice to eliminate any trees that exhibit past height damage, even though that height damage cannot be seen from the exterior of the tree. This practice leads to equations that are biased. The second major problem is that all trees must be older than the base age and are assumed to have been dominant throughout their life. Raulier *et al.* (2003) found that a site index developed from stem analysis data significantly over predicted dominant height growth of permanent sample plots. This was likely due to tree mortality and changes in tree social position, since approximately one fifth of the dominant trees were replaced every 10 years (Raulier *et al.*, 2003), which is similar to the findings of Hu and García (2010). Consequently, remeasured plots offer several advantages in that they capture patterns of stand dynamics, represent the development of dominant trees that experience height damage, and will produce a curve that is more rapidly asymptotic compared to an individual tree (Raulier *et al.*, 2003). However, this method is subject to potentially higher measurement error, significantly greater investment in time and funding, and often restricted to even-aged stands. García (2005a) developed techniques to make full use of both stem analysis and remeasured plot data when both are available.

Polymorphic site index model forms have the property that the shape of the height–age curve depends on site index (Type I) (Figure 3.2). As a result, differences between any two

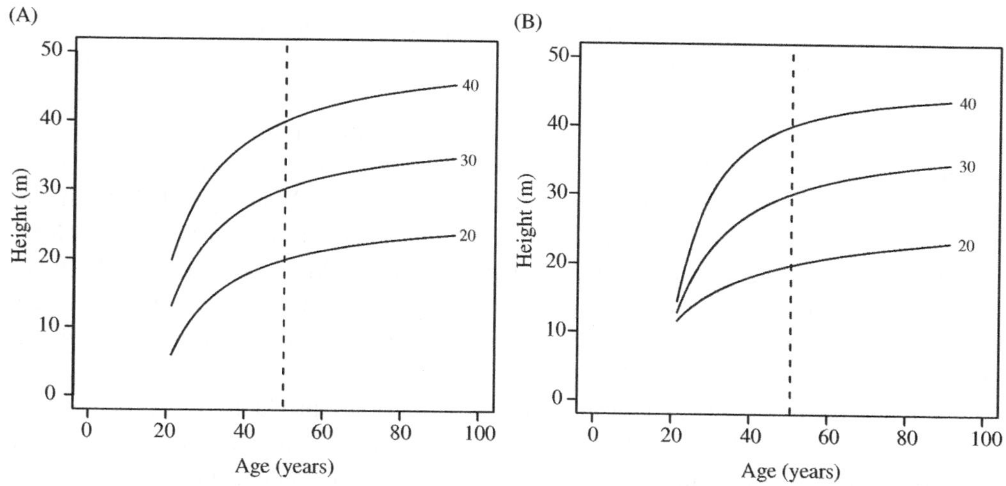

Figure 3.2 Example site index curves (base age 50 years) for white pine in southern Appalachians: (A) anamorphic curve (Doolittle and Vimmerstedt, 1960); (B) polymorphic curve (Beck, 1971).

curves are not proportional to the ratio of their site indices as is the case with anamorphic, or proportional, model forms. Polymorphic equations are often preferred because they are more flexible and can often behave like an anamorphic model form if appropriate. Another type of polymorphism (Type II) is when there may be different curve shapes within the same site index level (Krumland and Eng, 2005). This is often dealt with by either developing site index systems for each case separately or including indicator variables (e.g. Monserud, 1984). Often detecting Type II polymorphism is more difficult than it is for Type I.

Base age is the reference age, in years, for determining site index, and it can be expressed as either total age, breast-height age, or plantation age. The base age is often set near average rotation length for a species, resulting in base ages ranging from 15 to 100 years. As mentioned before, base age can have significant implications for the accuracy of predictions since predictions generally become more biased the further a stand is from base age. Base-age invariant models were developed to avoid this problem (Bailey and Clutter, 1974), and their usage is increasing (Krumland and Eng, 2005). Base-age invariance not only allows the use of various types of data for both parameterization and application, but also provides predictions of total height at ages earlier or later than the current age (Krumland and Eng, 2005). They, therefore, are considered by many to be superior to base-age specific equations.

The parameters of site index equations are estimated in a variety ways. Currently, the generalized algebraic difference approach (GADA; Cieszewski and Bailey, 2000) is one of the most common means used for estimating them (see Chapter 14). This method involves: (1) recognizing one or more parameters in the base equation as being site specific;

(2) solving the equation for these site-specific parameters and exchanging all of the general height and age terms in the solution with their initial values; and (3) replacing the site-specific parameters with the solution from Step 2 (Krumland and Eng, 2005). Assorted data structures have been used to estimate parameters for a GADA site index equation (Wang *et al.*, 2007a), but the choice of which is used can have a significant influence on the derived values (Cieszewski and Strub, 2007). A variety of GADA model forms also exist (Cieszewski *et al.*, 2006; 2007; Cieszewski and Strub, 2008). Other methods for estimating site index equation parameters are ordinary least squares (Furnival *et al.*, 1990), nonlinear regression (Weiskittel *et al.*, 2009a), nonlinear mixed effects (Wang *et al.*, 2007b; 2008a), and stochastic differentials with maximum likelihood (García, 2005a; Hu and García, 2010).

Site index is the most widely used measure of forest site productivity, particularly in statistical growth models. Hybrid and process models have attempted to avoid the use of site index because of its numerous disadvantages. However, as described below, a more effective measure of forest site productivity remains elusive.

3.2.2 Plant indicators

The presence of certain overstory and/or understory plant species to indicate productivity is based on the proposition that distinct plant communities have evolved based upon the environment they have experienced and, therefore, similar environmental conditions can be identified by recognizing the distinct plant communities that exist on them. Like site index, the utilization of this approach for evaluating forest productivity has a long history (see Cajander, 1949) and was well summarized by Daubenmire (1976). The use of plant indicators is purported to have several advantages: (1) vegetation reflects the sum of all environmental factors important to plants; (2) several vegetation characteristics have potential significance as ecological indicators beyond just productivity; and (3) the system is helpful in distinguishing the most basic ecological units of the landscape (Daubenmire, 1976). The approach is based on several premises including: (1) the species with the highest competitive powers are the best indicators; (2) understory vegetation stabilizes more quickly than other strata; and (3) key indicator plants are the same for all sites that have similar growing environments within a region (Pregitzer *et al.*, 2001).

The use of plant vegetation as a measure of site productivity is categorized into two main approaches based on: (1) potential climax vegetation (also known as subdivision, European or Braun-Blanquet method; e.g. Pfister and Arno, 1980) and (2) indicator plants (also known as ordination, the Anglo-American or Clements method; e.g. Cajander, 1949). For example, one system with broad usage in the US Rocky Mountain area is *habitat typing*, and is based on the overstory/understory species expected at climax (Pfister and Arno, 1980). While there are correlations between habitat type and site index (e.g. Monserud, 1984), there is often a substantial range in site indices for a given habitat type (Figure 3.3). Therefore, other topographic, physiographic, geologic, and/or landform

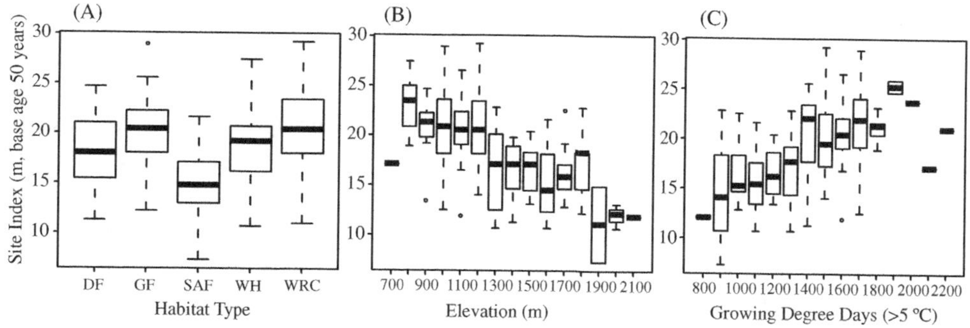

Figure 3.3 Boxplot of observed Monserud (1984) inland Douglas-fir site index (base age 50 years) in northern Rocky Mountains based on habitat type (A), elevation (B), and climate (C). The habitat types are Douglas-fir (DF), grand fir (GF), subalpine fir (SAF), western hemlock (WH), and western red cedar (WRC).

information is used to further refine this estimator of productivity. The use of understory indicator plants currently is the predominant method for assessing site productivity in coastal British Columbia (Klinka *et al.*, 1989).

Carmean (1975) lists the following reasons why plant indicators are not more widely used as productivity variables in the United States: (1) topographic, geologic, and soil features often explain the same site characteristics as plant indicators; (2) overstory species composition affects the vigor and composition of the understory, even on similar soils; (3) overstory trees are affected by characteristics of soil horizons deeper than those affecting the understory; (4) the abundance, vigor, and composition of the understory are all affected by the density of the overstory; and (5) many understory species die back during winter, making their use as indicators difficult or impossible during that time. In addition, the presence of both overstory and understory plants is significantly influenced by both forest silvicultural treatments and natural disturbances (e.g. Nolet *et al.*, 1995). Finally, habitat type classification remains a rather imprecise science, as nonmathematical classifications are unavoidably subjective and the outcome of mathematical approaches depends heavily on the algorithm chosen.

3.2.3 Other phytocentric measures

A variety of other phytocentric measures of forest site productivity are used to varying degrees. For example, maximum mean annual increment (MAI) is generally considered a suitable measure of site productivity (Skovsgaard and Vanclay, 2008), but it is difficult to determine without the use of a simulation model. The use of process models to estimate maximum MAI has been demonstrated (e.g. Battaglia and Sands, 1997). However, like site index, the use of maximum MAI is still strongly influenced by past management, and varies significantly from species to species. Regardless, maximum MAI is used as a measure of forest productivity in both the United States (Hanson *et al.*, 2002) and Scotland (Tyler *et al.*, 1996).

Similar to site index, height at a given index diameter rather than age was proposed as a measure of site productivity for mixed species and uneven-aged stands (Vanclay and Henry, 1988; Huang and Titus, 1993). However, the method is quite sensitive to the selection of an index diameter (Goelz and Burk, 1992), and Wang (1998b) found it to be an inadequate measure of site quality because it was not related to site index or ecological measures of site quality. This is likely because diameter growth is much more sensitive to stand density than is height growth.

The growth intercept method uses a measure of periodic height increment near breast height for dominant trees to indicate productivity (e.g. Powers and Oliver, 1978). The periodic height increment is determined by measuring the distance between a specified number of whorls (or, the length of a specified number of internodes). Therefore, the method is only applicable to species with distinct whorls. The number of internodes used varies between three and five depending upon the species. The use of growth intercept offers two primary advantages: (1) it can be used to assess site productivity of younger stands; and (2) it is more easily and accurately measured than total height and age. The primary disadvantages of growth intercept are: (1) it is only representative of early increment, which might not represent development of the tree over its lifetime; and (2) it can be more sensitive to climate fluctuations and management practices over its short measurement period. Regardless, several equations have been developed to predict site index from growth intercept (Nigh, 1995b; Mailly and Gaudreault, 2005; Guo and Wang, 2006).

Yield-based measures of site productivity were proposed as a practical alternative to site index (Schmoldt *et al.*, 1985), but have seen relatively little application in practice. The yield measures of Schmoldt *et al.* (1985) were determined by fitting a yield equation to a series of repeat measures of total volume for a given site. Like yield-based measures, Skovsgaard and Vanclay (2008) assert that stand volume growth per unit of height growth is an effective measure of site productivity. This value is determined from stem number, quadratic mean diameter, and stand basal area (Skovsgaard and Vanclay, 2008). Although this measure was tested for a variety of species (Sterba, 1987; Hasenauer *et al.*, 1994), its general utility remains unknown (Skovsgaard and Vanclay, 2008). However, its application to stands thinned to widely varying densities or established at a range of initial number of trees would likely be limited.

Growth of individual trees also is an effective measure of site productivity. Growth indices rank site productivity using individual-tree growth data collected from remeasured permanent plots. Vanclay (1989b) initially developed a growth index for mixed-species tropical forests using an individual-tree-based diameter increment function (Figure 3.4). The use of this type of measure was found to be effective in other forests as well (Trasobares and Pukkala, 2004). The approach involves fitting a growth equation to individual trees and estimating a site-specific parameter either through the use of indicator variables or mixed effects analysis. Differences between sites are then ranked and relativized.

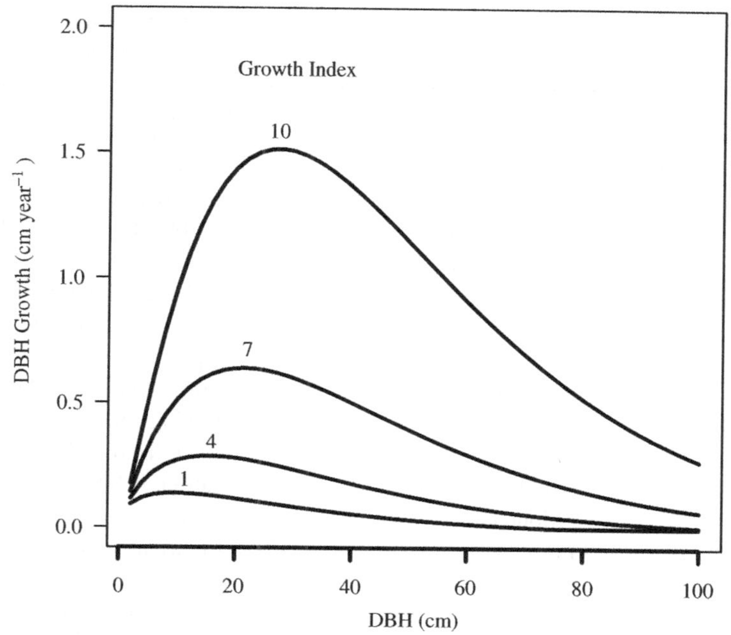

Figure 3.4 Growth index for *Cardwellia sublimis* trees in stands with 30 $m^2 ha^{-1}$ basal area and 10 $m^2 ha^{-1}$ basal area in larger trees (based on Vanclay, 1989b).

All phytocentric measures of forest site productivity assume that trees are effective representatives of site productivity. However, site and tree genotype interactions are complex and suggest that a variety of factors can influence a tree's productivity. Despite these drawbacks, these measures are the most commonly used in forestry.

3.3 Geocentric measures of site productivity

Geocentric measures of forest site productivity rely on different earth-based measures of site conditions to assess productivity. The most common geocentric measures are classified into three main groups: physiographic, climatic, and soils-based attributes. Although geocentric measures attempt to avoid the limitations of phytocentric measures, they are often species specific since species can respond differently to environmental factors. Furthermore, they are often specific to a particular region. For example, the key influential geocentric factors for a well-studied species like Douglas-fir vary from study to study, making it difficult to draw any broad generalizations (Table 3.2). Numerous studies have attempted to relate observed site index to an array of geocentric measures, with varying degrees of success (Carmean, 1975). This section reviews some of the commonly used geocentric measures.

Table 3.2 Summary of various studies that have examined the relationship between site productivity, physiographic features, climate, and soil attributes in Douglas-fir.

Study	Location	Influential factors
McArdle *et al.* (1961)	Pacific Northwest (Oregon and Washington)	Slope, aspect, elevation, and annual precipitation
Schlots *et al.* (1956)	Pacific Northwest (Oregon and Washington)	Soil depth and degree of horizonation
Steinbrenner (1981)	Pacific Northwest (Oregon and Washington)	Annual precipitation, effective soil depth, elevation, and position on slope
Klinka and Carter (1990)	Pacific Northwest (coastal British Columbia)	Evapotranspiration (May and June), growing season water deficit, and mineralizable nitrogen
Monserud *et al.* (1990)	Northwest (Idaho and Montana)	Elevation, habitat type, precipitation, and longitude
Curt *et al.* (2001)	France	Soil nutrient status, elevation, water seepage index, soil water storage capacity, and aspect
Dunbar *et al.* (2002)	Ireland	Soil parent material, aspect, and wind exposure
Fontes *et al.* (2003)	Portugal	Elevation, aspect, moisture deficit, soil base saturation, clay content, and total nitrogen content
Nigh *et al.* (2004)	Interior British Columbia	Mean annual temperature, annual heat index, and the difference between the mean temperature in the warmest and coldest months

3.3.1 Physiographic measures

The most common physiographic measures of forest site productivity are slope, aspect, elevation, latitude, and longitude. Research has often found that productivity frequently differs between northern and southern slopes (e.g. Coble *et al.*, 2001), while steeper slopes can be more productive than flat ground (e.g. McArdle *et al.*, 1949). Slope, aspect, and elevation are the primary physiographic measures of site productivity commonly used in the Forest Vegetation Simulator (FVS; Crookston and Dixon, 2005). Since the influence of slope and aspect can be interactive, various transformations of them are proposed. Stage (1976) used trigonometric functions to capture interactions between slope and aspect. This transformation was later refined to include elevation and was shown to be more effective than the original transformation (Stage and Salas, 2007). Likewise, various transformations of latitude, longitude, and elevation have been proposed, with one of the most widely used being the Hopkins (1938) index. Wang *et al.* (2004) found latitude, longitude, and elevation much more effective predictors of lodgepole pine site index in Alberta than soil moisture and nutrient variables.

Topographic and landform measures also are used to assess site productivity. For example, McNab (2010) found that a landform index, which is the algebraic sum of inclinations to the horizon sampled in equal increments of azimuth from a location in the stand, was the most effective variable for predicting productivity of two species in the Appalachian Mountains of the United States. Other topographic variables found to be useful predictors are distance to a ridge, vertical angle to the top of a sun-blocking ridge, concavity of slope, and slope position. These variables are relatively easy to derive using digital elevation models (DEMs). They are combined into various indices such as the integrated moisture index (IMI; Iverson *et al.*, 1997) and topographic wetness index (TWI; Beven and Kirkby, 1979), which have been used to varying degrees in the ecological literature.

3.3.2 Climatic measures

Commonly, physiographic variables are just surrogates for local climate. However, climate information is not widely used to assess forest site productivity because of the relative lack of weather stations in forested situations and the inability to estimate climate conditions for any given site. Tools to estimate climate at a given site are becoming increasingly available and used. For example, the Rehfeldt (2006) climate model and DAYMET (Thornton *et al.*, 1997) provide users with the ability to estimate high-resolution monthly or daily climate data, respectively. Regardless of climate data availability, climate was found to be an effective predictor of site productivity in several studies. Generally, variables that express both temperature and moisture availability are more effective than using one or the other by itself. For example, Weiskittel *et al.* (in press) found that interaction between the ratio of summer to total precipitation and growing degree days $>5\,^{\circ}\mathrm{C}$ was the most effective of the 35 climatic variables examined for predicting site index across the Western United States. Other studies have also found growing degree days to be an effective measure of productivity (Farr and Harris, 1979; Ung *et al.*, 2001; Hamel *et al.*, 2004; Monserud *et al.*, 2006).

Rather than relate climate to site index, Pokharel and Froese (2009) reported that mean annual temperature was just as effective as site index in an individual-tree basal area growth equation. On the other hand, Crookston *et al.* (2007) found that substituting climate variables for the site descriptors used in an individual-tree diameter-increment equation resulted in strange model behavior, which was attributed to the limited distribution of data points in climate space. Ung *et al.* (2001) developed a biophysical site index based primarily on climatic variables, which was just as effective as traditional site index for growth and yield predictions in the boreal forest of eastern Canada. Ung *et al.*

(2001) also found that climatic variables were effective for predicting shade-intolerant species productivity, but the climatic variables needed to be combined with a measure of successional stage to correctly predict the productivity of shade-tolerant species.

A similar approach was recently utilized by Benavides *et al.* (2009) in a mountainous region of northern Spain. This work builds on Paterson's (1956) climate–vegetation potential (CVP) index, which was designed to predict maximum growth potential in terms of volume production over large areas (Johnston *et al.*, 1967). This index was based on evapotranspiration, annual temperature range, mean annual precipitation, length of growing season, and mean monthly temperature of the warmest month.

Disadvantages of using climatic variables as measures of forest site productivity are that: (1) they are often derived from latitude, longitude, and elevation, which can result in imprecise local estimates; (2) short-term weather events can have more of an influence on estimated productivity than long-term climate "normals"; (3) climate varies strongly from year to year and decade to decade; (4) variables are often highly correlated, which can make it difficult to find the most influential variable; and (5) climate is influential at larger geographic scales, while other factors may control productivity at the local scale.

3.3.3 Soil measures

Recognizing that soil properties are one of the basic factors affecting productivity, there are a large number of studies that have related soil properties to a measure of stand productivity. Most of these studies have predicted site index as a function of soil attributes (see Carmean, 1975). Some of the most important soil factors that influence productivity are parent material, depth, texture, and nutrient availability. Often, a very limited relationship between soil attributes and site index is found (e.g. Monserud *et al.*, 1990). This is attributed to the high variability of most soil properties (especially nutrient availability) and the complex interactive relationships between them.

Often multiple soil variables are needed to effectively assess site productivity. For example, Grigal (2009) developed a soil-based productivity measure that was driven by water availability (depth to water, drainage), nutrients (base saturation, organic matter), and site (bulk density, stone volume), and Ritchie and Hamann (2008) found that water-holding capacity of the soil (depth, texture, type of bedrock) was effective at characterizing the productivity of Douglas-fir saplings. Likewise, Briggs (1994) used effective soil rooting depth and drainage class to delineate forest productivity classes for sites in northern Maine, which were then assigned an average site index value (Williams *et al.*, 1991). Turner *et al.* (1990) found that a soil classification system provided better estimates of *Pinus radiata* productivity than did other more widely used classifications.

The primary disadvantage of using soils as a basis for productivity is that regional maps of soil attributes are often unavailable, incomplete, or at resolutions that are not useful for forest management planning. Techniques are used to map soil attributes from DEMs

Table 3.3 Example growth and yield models, the site productivity variables used, and the type of model form in which the site productivity measure is used.

Growth model	Measure of site productivity		Model form
	Phytocentric	Geocentric	
Whole stand			
DFSIM (Curtis *et al.*, 1981)	Site index	—	Realized
Scube (García, 2011)	Site index	—	Realized
Size class			
FIBER (Solomon *et al.*, 1986)	Site index	Habitat type; elevation	Realized
Vanclay (1989a)	Growth index	Soil parent material	Realized
Distance-dependent individual tree			
SILVA (Pretzsch *et al.*, 2002)	Site index	Temperature difference between the coldest and warmest month of the year; number of days with a mean temperature >10 °C; mean growing season temperature; relative soil water retention; growing season precipitation; soil nutrient supply; NO_x concentration; atmospheric CO_2 concentration	Potential × modifier
TASS (Mitchell, 1975)	Site index	—	Realized
Distance-independent individual tree			
ORGANON (Hann, 2011)	Site index	—	Realized, potential × modifier
PROGNAUS (Monserud *et al.*, 1997)	—	Elevation; slope; aspect; depth of soil and humus layers; relief; soil moisture class; soil and vegetation type	Realized
CACTOS (Krumland and Wensel, 1981)	Site index	—	Potential × modifier

(e.g. Frazier *et al.*, 2009), but their accuracy largely depends on the DEM resolution, and a very high resolution is often needed to achieve desirable results (Vaze *et al.*, 2010).

3.4 Summary

Measures of forest site productivity are necessary for development of effective growth and yield models. A wide variety of approaches to assessing forest site productivity are used, but each has their own advantages and disadvantages (Table 3.3). Site index is the most common means for quantifying site productivity. However, unbiased estimates of site index require accurate measurement of tree age and height, a height–age growth equation, and the assumption that the trees were free to grow over the course of their life. Other important measures of site productivity used in growth and yield models are habitat type, and physiographic and climatic variables. They can be just as effective as site index (e.g. Pokharel and Froese, 2009), but generally require multiple parameters to ensure proper behavior.

Due to the difficulty of identifying the most influential factors controlling forest productivity, the use of both process-based models (e.g. Ollinger *et al.*, 1998) and remote sensing tools (e.g. Smith *et al.*, 2002) are increasingly used. Although these techniques perform well, process-based models are found to be no more effective at predicting site index than climate variables (Weiskittel *et al.*, in press), and, like site index, remote sensing tools only assess past realized productivity. The increasing availability of high-resolution climate data and DEMs promises further development of soil water availability maps (e.g. Schwärzel *et al.*, 2009) and topographic indices (e.g. McNab, 2010) as measures of forest productivity. Likely, one measure of forest site productivity will not be predominant, given the vast range of species and regional differences in productivity.

4

Whole-stand and size-class models

4.1 Introduction

Forest management decisions are often based on stand-level attributes such as total basal area and number of trees. Consequently, whole-stand models are some of the oldest and most widely used growth and yield models. *Whole-stand models* are those growth and yield models in which the basic units of modeling are stand and site parameters such as age, basal area, site index, stocking, and stand volume. They require relatively little information to simulate stand growth, but provide rather general information about future stand conditions. In contrast, *size-class models* commonly employ a class of trees as the basic unit for modeling and are a compromise between whole-stand and individual-tree approaches. Although size is the most often used variable to form classes in size-class models, other attributes like species and age are also used. Most size-class models originate from the classical method of stand table projection. This chapter examines several such whole-stand and size-class modeling approaches ranging from simple yield tables to sophisticated matrix methods.

4.2 Whole-stand models

Whole-stand models are easy to develop and apply and are often highly accurate in single species, even-aged stands. However, they have a limited ability to represent complex stand structures, mixed species, and forest silvicultural treatments accurately, due to the fact that they describe with a few simple averages or totals. The whole-stand models discussed in this chapter include yield tables, compatible equations, and state-space models.

Forest Growth and Yield Modeling, First Edition. Aaron R. Weiskittel, David W. Hann, John A. Kershaw, Jr. and Jerome K. Vanclay.

4.2.1 Yield tables and equations

Yield tables are one of the oldest approaches to yield estimation (Vuokila, 1965; Curtis, 1972). They are essentially summaries of expected yields tabulated by stand age and site index. Yield tables may be depicted in tabular form, graphically, or even as an equation (Figure 4.1). Modern yield tables often include not only yield, but also stand height, mean diameter, number of stems, stand basal area, and current and mean annual volume increments. Three classes of yield tables are generally distinguished: *normal*, *empirical*, and *variable density* yield tables. In reality, these do not form discrete classes, but rather reflect a continuum from normal yield tables, through variable density yield tables and other approaches, to single-tree growth models (e.g. Leary, 1991). Normal yield tables provide estimates of expected yields tabulated by stand age and site index for ideal, fully stocked, or "normal" forest stands. These were usually based on data from temporary plots and analyzed using graphical techniques. Vuokila (1965) and Curtis (1972) discussed the various approaches used in Europe and North America for the construction of normal yield tables.

Yields were generally tabulated by age and site index, but could also be presented as alignment charts (e.g. Reineke, 1927). Eichorn's (1904) rule, which states that a given mean height of a stand is matched by the same volume in all site classes, was used in the construction of many European normal yield tables (Assmann, 1970). Normal yield tables may provide reliable estimates of potential yields for even-aged stands similar in character to those used in developing the table, but may be less satisfactory for natural stands where

Yield Table

Age	Site Index (m)				
	12	15	18	21	24
10	18	24	31	37	44
20	36	48	60	73	87
30	54	71	90	109	128
40	71	94	118	143	169
50	88	116	146	177	209
60	104	138	173	210	248
70	120	159	199	242	286
80	136	179	225	273	323
90	151	200	250	304	359
100	166	219	275	333	394

Yield Curves

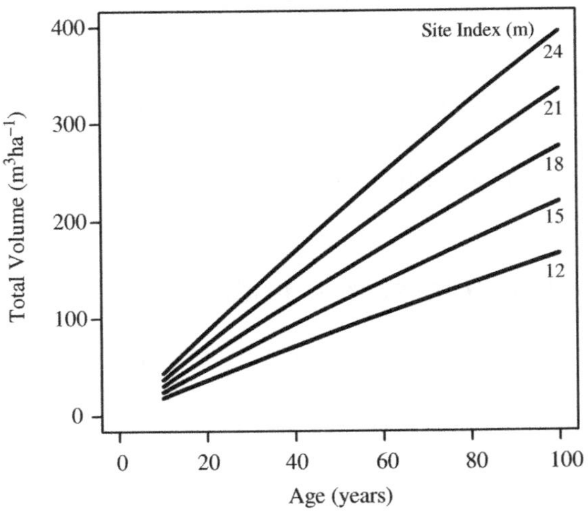

Yield Equation: $V = 32.6 \cdot \text{Site}^{1.27} \cdot (1 - e^{-0.00243 \cdot \text{Age}})$

Figure 4.1 Yield table, curves, and equation for upland oak species in the central United States (based on Schnur, 1937).

age may vary considerably within stands. Empirical yield tables are like normal yield tables, but are based on plots that are of average rather than normal stocking.

More sophisticated calculation and analytical techniques enabled additional variables to be included in yield calculations. Stand density was an obvious choice for a third variable in yield tables since it enables data from partially stocked plots to be used, and means that the yield table can be applied to any stand of the same species and site index. Several variable-density yield tables have been published (MacKinney and Chaiken, 1939; Myers, 1977; Longchang *et al.*, 1991). These tables give height, stems per hectare, mean diameter, stand basal area, mean stem volume, volume per hectare, and mean annual and cumulative volume production at five-year intervals for many species–site–silvicultural system combinations. This is the most practical, realistic, and useful approach for yield tables.

Although primarily used for single species, even-aged stands, the yield table approach has also been applied to mixed stands, especially for selection forests in central Europe. There are several ways to build compact tables for natural forests. The basal area of the dominant species may be expressed as a percentage of total stand basal area in mixed species forests (e.g. MacKinney *et al.*, 1937), and a "main stand" may be identified in uneven-aged forests (e.g. Duerr and Gevorkiantz, 1938). Yield tables usually require some estimate of stand age, and thus cannot easily be applied to uneven-aged stands. Growth tables attempt to overcome this limitation by tabulating growth under various stand conditions. Variables other than age used to characterize stands for growth prediction include volume, density, height, average diameter, and crown class. These variables were often used in conjunction with estimates of site productivity and time since harvest.

Expected growth can be tabulated more compactly if expressed as a percentage rather than in absolute terms, and several formulae for growth percentages are used. Formulae indicating percentage growth of tree diameters and stand basal areas were used for many years to estimate growth in pure even-aged stands. Growth percentages can also be applied directly to estimates of stand volume to predict volume increment per hectare. Other formulae exist for predicting sustainable yields from normal forests. The simplicity of these formulae is attractive, but is also a limitation, as some formulae may not account for mortality, especially if applied to individual trees. There are many alternative formulae, all with different assumptions and limitations, but the compound interest formula may be the most satisfactory for very general and basic growth predictions (e.g. Rudolf, 1930). However, the method can be unreliable and very deceptive (e.g. Wahlenberg, 1941). Although some of these formulae may give reasonable estimates of short-term increment in stands for which they were derived, it is inevitable that percentages must decline in the longer term.

Conceptually, the distinction between tables and equations is unimportant since equations can be evaluated and presented as tables (Figure 4.1). However, the converse does not hold, and equations are a concise and convenient way to express growth and yield relationships. MacKinney and Chaiken (1939), and Schumacher (1939) were some of the first authors to publish yield equations fitted by linear regression. The MacKinney and Chaiken (1939) equation was based on the hypothesis that relative growth rate varies inversely with the square of time ($\partial V/V \propto t^{-2}$), such that $\log(V) = \beta_0 - \beta_1 t^{-1}$. This equation contains two parameters: one (β_0) that defines the upper asymptote (i.e. volume at end of the rotation), and one (β_1) that determines growth rate. They assumed that these parameters were simple linear functions of site index and stand density. The equation was subsequently used in many other models.

Desirable features of the MacKinney and Chaiken (1939) equation include: (1) the mathematical form of the covariates implies relationships which agree with biological concepts of even-aged stand development; (2) the use of log(V) as the response variable rather than V will generally be more compatible with the statistical assumptions customarily made in linear regression analysis; and (3) the use of log(V) as the response variable is a convenient way to express mathematically the interactions of the explanatory variables and how they affect V (Clutter, 1963). However, yield equations assume that the forest management treatments in a projected stand will remain the same as the stands used to construct them, which significantly limits their applicability.

4.2.2 Compatible growth and yield equations

Growth equations have an advantage over simple yield equations in that harvesting or any other forest management activity may be simulated at any time during the projection. Many stand growth equations are available (Sun *et al.*, 2007). However, some analysts have failed to exploit the relationship existing between growth and yield, which may lead to instances where estimates of yield (e.g. from yield tables) differ from successive summations of growth estimates based on the same data. Buckman (1962) and Clutter (1963) are generally credited as the first to publish forest growth and yield equations explicitly formulated to give consistent estimates of growth and yield, but many researchers were aware of this relationship for a long time (e.g. von Bertalanffy, 1949). Although Buckman (1962) and Clutter (1963) were both concerned with compatibility, the two approaches were quite different, as the approach of Buckman (1962) was based more on a state-space system of differential equations.

To achieve a compatible growth and yield equation, Clutter (1963) expressed basal area as a differentiable yield function in age and site index, so that basal area increment (dBA/dt) could be estimated from stand basal area, age, and site index. This led to five compatible equations which could be used to derive estimates of present standing volume, basal area increment, volume growth, predicted future basal area, and predicted volume yield (Figure 4.2). Sullivan and Clutter (1972) refined this equation to provide numerically consistent estimates of growth and yield from initial stand age, basal area, site index, and future age. Burkhart and Sprinz (1984) presented a method for simultaneously estimating the parameters in the Sullivan–Clutter (1972) approach. Moser and Hall (1969) extended the idea of compatible growth and yield equations to uneven-aged stands by providing

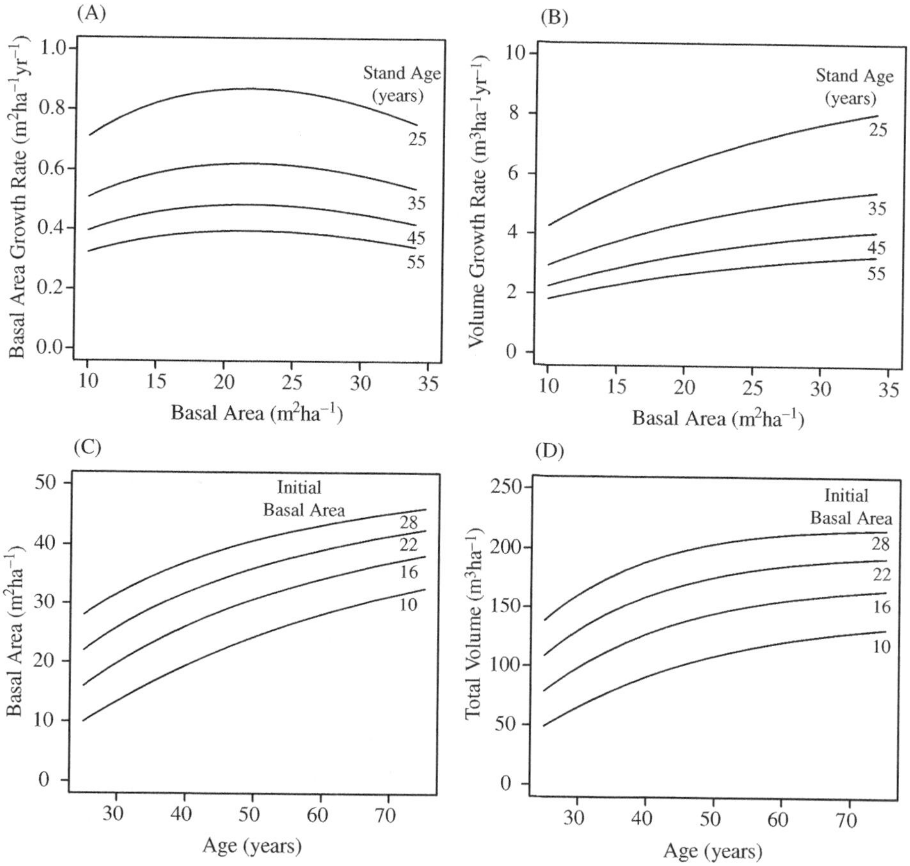

Figure 4.2 Compatible whole-stand growth and yield functions: (A) basal area growth rate ($m^2\,ha^{-1}\,yr^{-1}$) by initial basal area ($m^2\,ha^{-1}$) and stand age; (B) volume growth rate ($m^3\,ha^{-1}\,yr^{-1}$) by initial basal area and stand age; (C) basal area ($m^2\,ha^{-1}$) yield by stand age for various levels of initial basal area; and (D) volume yield ($m^3\,ha^{-1}$) by stand age and initial stand basal area. Based on equations for loblolly pine in Clutter (1963) for site index 21 m (70 ft), base age 50.

equations independent of stand age. More recently, Castedo-Dorado *et al.* (2007a) developed a compatible stand basal area growth system using the generalized algebraic difference approach (GADA: Cieszewski and Bailey, 2000; see Chapter 14), but found that enforcing compatibility reduced accuracy.

Stand growth models are effective because they have minimal data requirements and can represent a range of forest silvicultural treatments. For example, the DFSIM growth model just requires site index, total age, number of trees, and quadratic mean diameter for simulations, but can represent pre-commercial and commercial thinnings as well as fertilization in natural and planted coastal Douglas-fir stands (Curtis *et al.*, 1981; Figure 4.3; see Chapter 11).

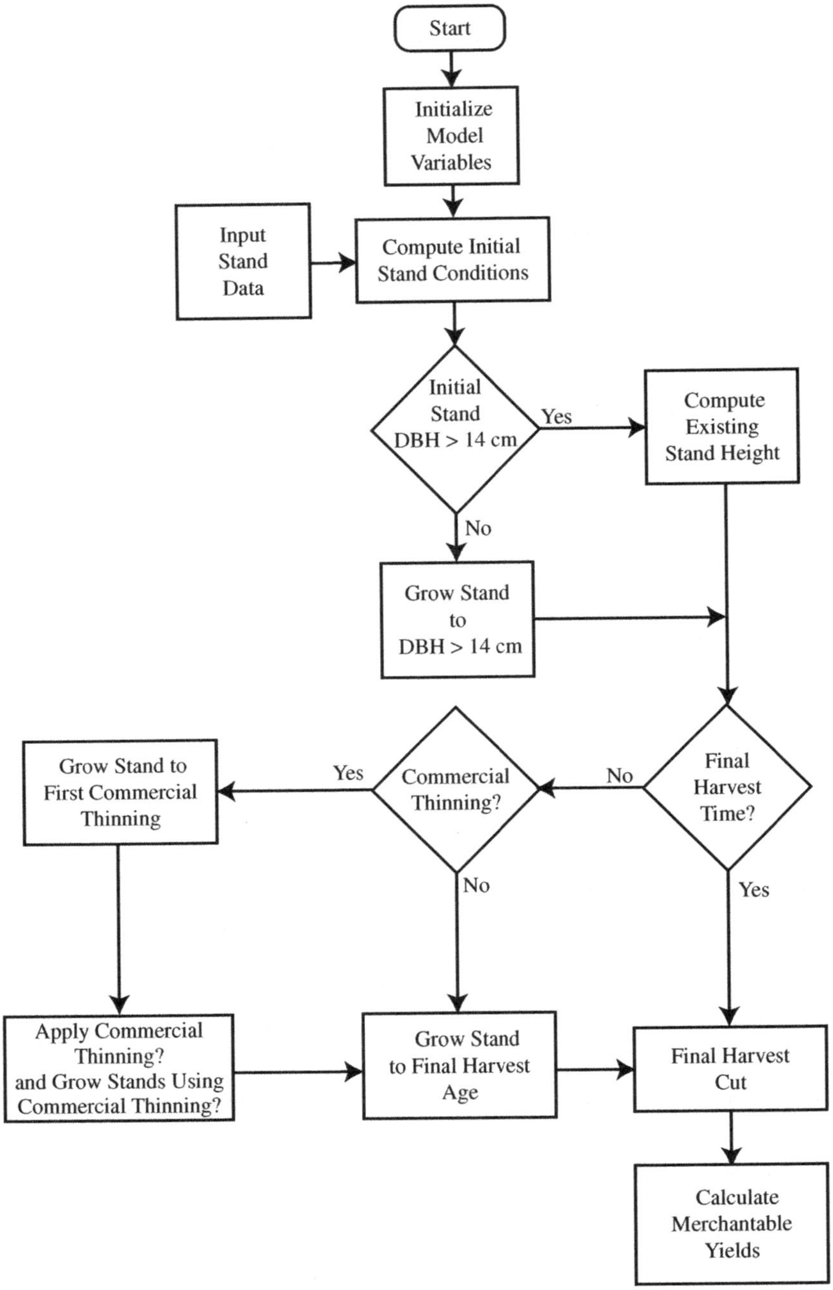

Figure 4.3 Flowchart of key model processes for the DFSIM stand-level growth model (Curtis *et al.*, 1981).

4.2.3 Systems of equations

A better understanding of growth in even-aged stands may be obtained if components of growth are individually identified and expressed collectively as a system of equations to predict stand growth. Furnival and Wilson (1971) developed a growth and yield model as a system of equations, and solved all the coefficients using simultaneous estimation. They formulated their yield model for white pine stands as the following system:

$$
\begin{aligned}
\log(\overline{\mathrm{HT}}) &= b_1 + b_2 \cdot \log(\mathrm{AGE_{BH}}) \\
\log(\mathrm{TPH}) &= b_3 + b_4 \cdot \log(\mathrm{QMD}) \\
\log(\mathrm{BA}) &= b_5 + b_6 \cdot \log(\overline{\mathrm{HT}}) \\
\log(f) &= b_7 + b_8 \cdot \log(\mathrm{QMD}) \\
\log(\mathrm{BA}) &= \log(k) + \log(\mathrm{TPH}) + 2 \cdot \log(\mathrm{QMD}) \\
\log(V) &= \log(f) + \log(\overline{\mathrm{HT}})) + \log(\mathrm{BA})
\end{aligned}
\qquad [4.1]
$$

where V is volume ($\mathrm{m^3\,ha^{-1}}$), $\overline{\mathrm{HT}}$ is mean tree height (m), $\mathrm{AGE_{BH}}$ is average stand age at breast height, BA is basal area ($\mathrm{m^2\,ha^{-1}}$), QMD is the quadratic mean diameter, TPH is number of trees ($\mathrm{ha^{-1}}$), f is the form factor, and k is a constant ($\pi/40\,000$).

This "systems formulation" allows a variance–covariance matrix to be produced and can be used to estimate standard errors of predicted growth and yields. The system of Furnival and Wilson (1971) is limited, however, by the use of simple model forms and the omission of competition-induced mortality. Regardless, it remains one of the few examples of the simultaneous estimation of a system of equations for a whole-stand model. Several other examples can be found in Borders (1989).

4.2.4 State-space models

State-space models describe the state of the stand at a point in time, and predict the rate of change in the state using first-order differential equations. García (1984; 1994) used a state-space approach to model plantations (Figure 4.4), and it has also been used by several other researchers (Zhang and Borders, 2001; Castedo-Dorado *et al.*, 2007b; Nord-Larsen and Johannsen, 2007). The stand is represented by a few state variables, usually stand basal area, number of trees per hectare, and top height. These state variables must adequately describe stand structure as well as summarize the historical events affecting future development of the stand (García, 2005b).

Only the relevant aspects of composition, structure, and past management are needed for making projections using state-space models. Future states are then determined by the current state and future actions, and other variables of interest, such as volume, are derived from these state variables. This assumption is critical to this and several other modeling approaches, and requires that two conditions be satisfied. The state variables must

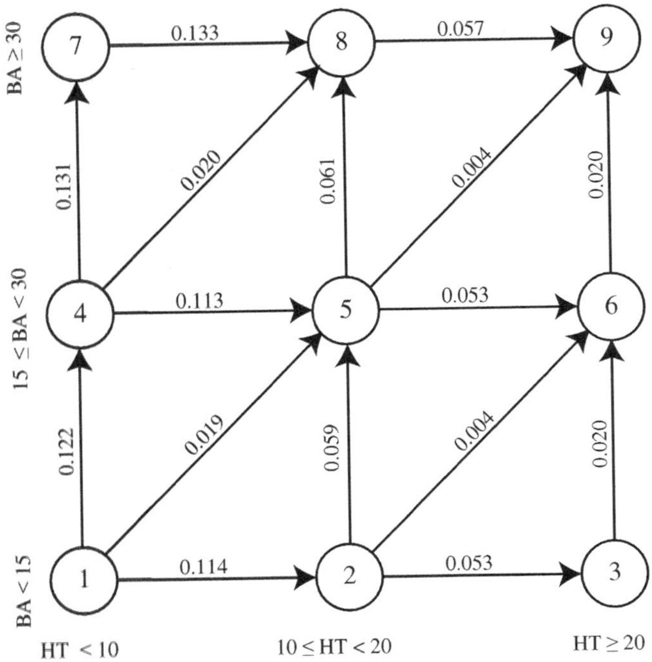

Figure 4.4 Markov chain representation of a model based on basal area (BA; $m^2\ ha^{-1}$) and top height (HT; m) developed by García (1994). The numbers in the circles identify the state (ie, 1 = BA < 15 and HT < 10) and the numbers along the lines represent transition probabilities (ie, probability of moving for state i to state j).

adequately describe composition and structure of a forest stand, and they should reflect all past forest silvicultural treatments so that growth predictions do not need estimates of stand age, time since thinning, etc. These assumptions imply that growth predictions can be made simply by updating these few state variables.

These assumptions allow the system to be described by the specified state variables and a few simple transition functions (García, 1994). In particular, the state-space approach avoids the need to directly model the complex relationships over time between inputs and outputs, by describing the state of the system at a point in time and modeling the rate of state change. García (1984) used a multivariate generalization of the von Bertalanffy (1949) function in his state-space model for *Pinus radiata* plantations, while Nord-Larsen and Johannsen (2007) used a generalized model form proposed by Zeide (1993). García (2011) recently utilized the state-space approach to develop a dynamic stand model for interior spruce in British Columbia, which used a relative occupancy variable to account for incomplete canopy closure in young stands or those recently thinned.

The state-space approach can produce accurate predictions and it can provide an effective framework for both managed, single species plantation, and natural stand growth models. However, it offers less utility for mixed species forests. Adapting the approach for these may be complex, since it is unrealistic to assume that mixed species forests could be described adequately with only three state variables. However, this has yet to be tested.

4.2.5 Transition matrix models

Transition matrix models offer two desirable features: (1) future states are easy to determine (but only for multiples of the time interval); and (2) the steady state (i.e. the long-term equilibrium) of the system can be determined by repeatedly squaring the matrix. Transition matrix models are similar to the state-space approach, but the set of possible states is finite rather than infinite. The idea of using matrices to predict stand-level growth was first used by Hool (1966), who used a two-year transition interval, and identified 36 states based on silvicultural history (thinned/unthinned), standing volume (six volume classes, each $40\,m^3\,ha^{-1}$), and stocking (three classes, each 170 stems ha^{-1}). Hool (1966) also considered three activities: undisturbed growth (over the two-year transition interval), thinning, and harvesting. The transition from state S_i to state S_j was simulated based on the probability of these activities. Likewise, Binkley (1980) used whole-stand transition matrices to examine natural succession in forest stands. The states of his matrix indicated the dominant species on each plot. His analysis suggested that assumptions inherent in the method were untenable and that transition matrices were not a reliable way to predict forest stand dynamics.

4.3 Size-class models

The size-class approach divides the stand into two or more classes, but with fewer classes than the total number of sample trees measured in the stand. This creates a relatively simple model that can portray more complex stand structures compared to whole-stand models. The types of size-class models discussed in this section are stand table projection methods, matrix models, diameter-class models, and cohort models.

4.3.1 Stand table projection

A stand table is a tabular summary showing number of trees per unit area in each of several size classes. In mixed stands, there may be rows for each species or species group. Size classes are usually diameter classes of equal width. Stand table projection is one of the oldest techniques used to determine future composition of uneven-aged forests. Stand table projections are relatively simple and can be accurate for short time periods ($<$10 years), but have limited ability to project long-term forest dynamics due to the greater accumulation of ingrowth and mortality.

Three methods are used to forecast the future stand using this technique. The first assumes that all trees in each diameter class are located at the class midpoint, and that all trees will grow at the same average rate, irrespective of their present size and vigor. This essentially involves projecting the class boundaries so that future classes contain the same trees (if there is no mortality), but may have different class boundaries (and widths), which may be inconvenient for some applications. It does not take into account the variable nature of diameter increment, so that the forecasted stand will span the same number of size classes as did the initial stand.

The second alternative assumes that trees in each diameter class are uniformly distributed through the class and each tree grows at the average rate (Table 4.1). For each class, a "movement ratio" is determined from the class width and average increment, and this indicates the proportion of trees moving to the next class (upgrowth). Projections involving stands with only a few trees require a decision regarding rounding fractions of trees. In projections for large forest areas where fractions may be considered acceptable, the method may lead to bias in predicted growth of the largest trees, since the distributions of stems in the largest classes may not be uniform, particularly in broad diameter classes. Both of these methods ignore dispersion of individual increments.

The third option attempts to account for variation in growth rate within diameter classes by using the actual movement of trees rather than average movement ratios of the size classes (Wahlenberg, 1941). Thus, this method should provide more reliable estimates of timber yields from uneven-aged stands than the previous two methods.

One problem with the last two approaches to stand table projection is the proliferation of classes with a fractional numbers of stems. This difficulty is avoided by the first approach. A related problem is the ability of some stems to move *n* classes in *n* projection periods, which may introduce bias. The first approach (updating class boundaries rather than moving stems between classes) does not have this problem either, because the approach allows no variability in increments in the projected stand, but this may underestimate future yields for the harvested component of the stand. There are three ways that the proliferation of fractions can be controlled.

1. Use longer projection intervals or narrower size classes. The optimal approach is to choose the smallest class width and longest time step (consistent with user requirements) that allows upgrowth to progress only one class in any single step.

2. Allow a non-uniform distribution of stems within each class by smoothing the stand table with a single curve (e.g. Weibull function) or with a series of curves (e.g. splines). The use of the Weibull function for improving stand table projection was demonstrated by Nepal and Somers (1992), as well as Cao and Baldwin (1999).

3. Accumulate small probabilities of upgrowth until some stems can be projected, either by
   a. specifically nominating the probability or stem number required to initiate a new class; or by
   b. explicitly modeling the growth of the largest tree in the stand, and initiating a new class only when a tree attains that size.

Table 4.1 Example 10-year stand table projection for a white pine stand in New Brunswick, Canada.

DBH class (cm)	10-year DBH increment (cm)	Movement ratio	Present stand table (No. ha^{-1})	Volume per tree (m^3)	Present stock table (m^3 ha^{-1})	Number of trees moving: 2 classes	Number of trees moving: 1 class	Number of trees moving: 0 classes	Ingrowth rate (No. ha^{-1})	Mortality rate (No. ha^{-1})	Future stand table (No. ha^{-1})	Future stock table (m^3 ha^{-1})	Present stock table (m^3 ha^{-1})	Volume production (m^3 ha^{-1})
10	1.72	0.86	273	0.029	7.9	0.0	234.8	38.2	121.9	32.6	128	3.7	7.9	−4.2
12	1.85	0.93	262	0.049	12.8	0.0	242.4	19.7		16.1	238	11.7	12.8	−1.1
14	1.95	0.98	227	0.075	17.0	0.0	221.3	5.7		4.4	244	18.3	17.0	1.3
16	2.01	1.01	179	0.109	19.5	0.9	178.1	0.0		0.0	221	24.1	19.5	4.6
18	2.04	1.02	129	0.150	19.4	2.6	126.4	0.0		0.0	178	26.7	19.4	7.3
20	2.05	1.03	85	0.199	16.9	2.1	82.9	0.0		0.0	127	25.3	16.9	8.4
22	2.02	1.01	51	0.257	13.1	0.5	50.5	0.0		0.0	85	21.8	13.1	8.7
24	1.98	0.99	28	0.324	9.1	0.0	27.7	0.3		0.1	53	17.2	9.1	8.1
26	1.92	0.96	14	0.399	5.6	0.0	13.4	0.6		0.2	29	11.6	5.6	6.0
28	1.84	0.92	6	0.484	2.9	0.0	5.5	0.5		0.1	14	6.8	2.9	3.9
30	1.74	0.87	3	0.579	1.7	0.0	2.6	0.4		0.1	6	3.5	1.7	1.8
32	1.64	0.82	1	0.683	0.7	0.0	0.8	0.2		0.0	3	2.0	0.7	1.3
34	1.53	0.77	0	0.797	0.0	0.0	0.0	0.0		0.0	1	0.8	0.0	0.8
36	1.41	0.71	0	0.921	0.0	0.0	0.0	0.0		0.0	0	0.0	0.0	0.0
38	1.29	0.65	0	1.054	0.0	0.0	0.0	0.0		0.0	0	0.0	0.0	0.0
40+	1.18	0.59	0	1.198	0.0	0.0	0.0	0.0		0.0	0	0.0	0.0	0.0
Total			1258		126.6						1327.0	173.5		46.9

Although stand table projection methods are generally becoming less common, the approach does provide consistency between stand- and tree-level growth projections as well as being able to represent multimodal diameter distributions (Borders and Patterson, 1990; Knowe *et al.*, 1997). Cao (2007) showed that the method can be used to project long-term growth (15–17 years) when mortality is sufficiently represented in the model. Trincado *et al.* (2003) demonstrated the usefulness of stand table projections for fast growing plantations of *Eucalyptus*. However, a fundamental limitation with most stand table projection techniques is their assumed independence of stand density (García, 1974).

4.3.2 Matrix models

Matrix models are a logical and formalized extension of the stand table projection method. They allow movement ratios and other details to be summarized in a matrix, so that growth projections are made with a single matrix multiplication. Three variations of this approach exist, and differ in their underlying assumptions: (1) Markov chains; (2) Usher and Leslie matrices; and (3) generalized matrices. Although Markov chains were used in the past for size-class predictions (e.g. Bruner and Moser, 1973; Cassell and Moser, 1974), the approach is generally too restrictive to represent forest dynamics. For example, predictions become unreliable if stand conditions (e.g. stand basal area) depart greatly from those conditions prevailing in the data from which the matrix was developed, and they generally require a large number of parameters.

Usher matrices attempt to reduce the large number of parameters required to form a Markov matrix. Choosing the time interval and class width so that no tree can grow more than one class during the period (for convenience, called the Usher assumption here) allows a substantial reduction in the number of parameters estimated. Unlike some ways of obtaining movement ratios for stand table projection, the Usher assumption does not introduce bias because it requires that a single time step must be sufficiently short so that a tree cannot grow more than one class. A matrix model can be reduced to four vectors for upgrowth (either a tree moves into the next class, or it does not), fecundity (recruitment), harvesting, and mortality.

Typical applications may contain 20–25 states and require the estimation of 80–100 parameters, but more than that is often required in many equation approaches. Usher (1966, 1969) demonstrated the use of this technique for forest growth projections. Bosch (1971) used a more traditional Leslie (1945) matrix approach to model redwood population dynamics. The key distinction between the approaches of Leslie (1945) and Usher (1966, 1969) is that all surviving trees grow into the next size class in Leslie matrices, while only some of the surviving trees grow into the next class in Usher matrices.

Generalized matrix models use features of both Markov chains and Usher matrices to represent forest size-class dynamics. This approach was first introduced by Buongiorno and Michie (1980), and is widely used in a variety of forest types

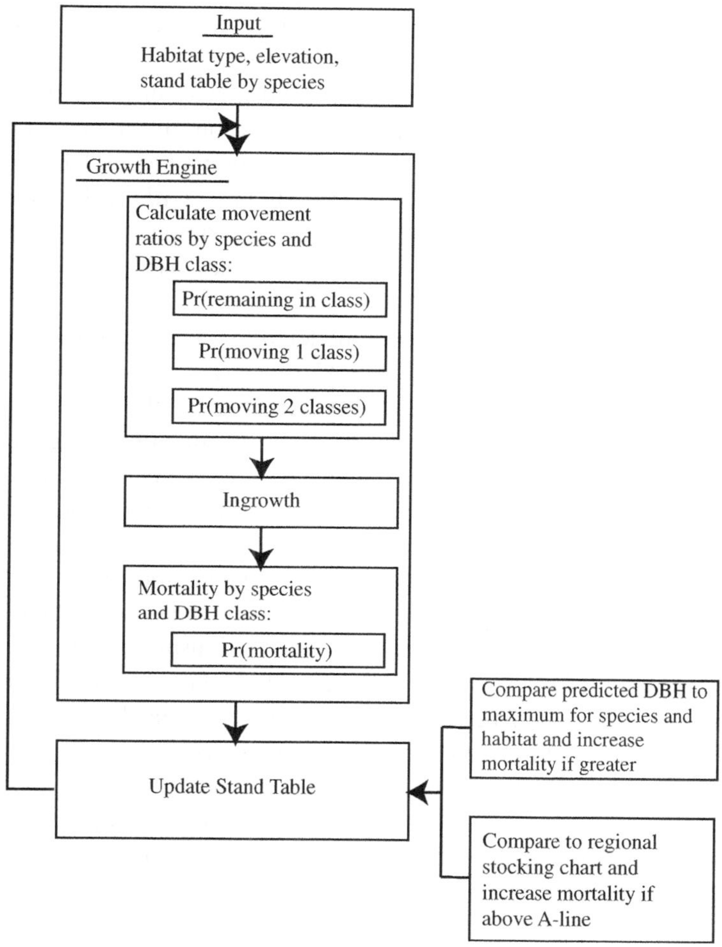

Figure 4.5 Flowchart of key model processes for the FIBER size class growth model (Solomon *et al.*, 1995).

(e.g. Lin *et al.*, 1996; Liang *et al.*, 2005; Bollandsårings *et al.*, 2008; Liang, 2010; Picard *et al.*, 2009). For example, FIBER is a generalized matrix model still widely used in the northeastern United States (Solomon *et al.*, 1995; Figure 4.5). Essentially, the approach uses stationary transition probabilities (i.e. the probabilities do not change over time) and Usher assumptions (no tree can grow more than one class during the period), but not the Markov assumption (that the transition probability depends only on the initial state). To avoid the simplicity implicit in the stationary assumption, some researchers change the transition matrix at each projection interval (Solomon *et al.*, 1986; Mengel and Roise, 1990), but this complication makes optimization difficult and may not be entirely necessary (Lin and Buongiorno, 1997). This type of matrix model is referred to as a variable-parameter approach (Lin *et al.*, 1996), which is in contrast to the constant-parameter approach of Buongiorno and Michie (1980).

One limitation of transition matrix models is the difficulty of predicting recruitment reliably. Although some earlier matrix models assumed a constant amount of regeneration at each cycle (e.g. Mendoza and Setyarso, 1986), recent matrix models have predicted it as a function of number of trees and basal area (Liang *et al.*, 2005; Liang, 2010). However, Yang and Kant (2008) found that their deterministic equations predicted very little of the variation in hardwood ingrowth, and they used just the mean species ingrowth instead.

Matrix models also generally rely on linear model forms, as more complex models have not improved predictions (Buongiorno *et al.*, 1995; Liang *et al.*, 2005). However, the third-order polynomial model form of Liang (2010) suggests high nonlinearity and may greatly limit the reliability of model extrapolation. In addition, equations to predict transition probabilities are usually fitted separately by diameter class when they should instead be fit as a system of equations constrained to be less than 1 (Yang and Kant, 2008).

Liang *et al.* (2005) found that the matrix approach was more accurate and stable, particularly for long-term projections, than a regional yield table and an individual-tree growth model (i.e. Forest Vegetation Simulator). However, the various alternative models evaluated by Liang *et al.* (2005) were parameterized using different, independent modeling data, and both Dhrymes *et al.* (1972) and Hocking (1976) have pointed out that the mean square error (MSE) calculated from an independent dataset must exceed the MSE from the original dataset. As a result, the findings of Liang *et al.* (2005) are problematic. Picard *et al.* (2002) developed a combined model that used a matrix approach for trees below a certain threshold (<40 cm) and an individual-tree approach for trees above the threshold value, which was found to have several important advantages over a traditional matrix approach. In addition, Picard *et al.* (2010) recently presented a methodology for optimally selecting the number and width of classes in a matrix model, which also improved predictions.

4.3.3 Diameter-class models

The lack of stand density measures in early stand table projection techniques (García, 1974) was addressed by Ek (1974a) and Hann (1980). Ek (1974a) developed a diameter-class model for a mix of even- and uneven-aged northern hardwood stands in Wisconsin, USA. The diameter classes were 5.08 cm in size, and the model predicted the mortality rate in each class, the upgrowth rate between classes, and the ingrowth rate into the smallest class, using the stand density measures of basal area and the number of trees per hectare. The equations also incorporated the site index of the stand and the basal area and number of trees per hectare in each diameter class. A modified version of this model was used by Adams and Ek (1974) to demonstrate how optimal management strategies could be determined for uneven-aged stands.

The diameter-class model of Hann (1980) was created to characterize the development of even- and uneven-aged ponderosa in Arizona, USA. The model used 2.54 cm diameter classes and it incorporated dynamic equations for predicting the diameter increment and mortality rate in each class and the amount of ingrowth into the first two diameter classes. Diameter increment was estimated from a basal area increment equation and used to calculate the amount of upgrowth into larger diameter classes under the assumption that the number of trees within a diameter class is uniformly distributed. Hann (1980) tested the appropriateness of the uniformity assumption for 714 diameter classes and rejected the assumption only 7 times. The model also incorporated a Monte Carlo swindle using a record tripling process (Stage, 1973), to introduce within-class variability in predicted increment. All of the dynamic equations included stand density through the usage of total stand basal area and the basal area in various diameter-class groupings. Other predictor variables included site index and the amount of time since the last cutting. The architecture of the Hann (1980) model was chosen in order to determine optimal stand management prescriptions using the very limited computing power available at that time.

The sudden development of the personal computer and its exponential improvement in speed and memory capacity made the use of such a limited architecture no longer necessary, and, as a result, research on growth and yield modeling shifted from the approach of Hann (1980) to the development of the ORGANON individual-tree model (Hann, 2011). Nevertheless, the model of Hann (1980) demonstrated the evolution of the architecture of size-class models towards the architecture of individual-tree models (see Chapters 5, 6, and 7).

4.3.4 Cohort models

The distinction made here between the previous types of size-class models and cohort models is based on how trees are assigned to size classes. Stand table, matrix models, and diameter-class models generally use "metric" size classes of equal and predetermined size. Cohort models allow more flexibility so that empty classes can be eliminated, and cohorts can be formed so that each has approximately equal importance in the model.

For example, Alder (1979) uses deciles, and models the development of the 10 median trees corresponding to the 5th, 15th, ..., 95th percentiles of the cumulative tree size distribution. In mixed stands, the formation of cohorts may be determined by species and other attributes, instead of computational niceties such as equal numbers of trees. Vanclay (1989a) presented a cohort model that grouped species according habit, size at maturity, and harvesting guidelines. A similar approach was used by Alder and Silva (2000). Cohort models are rather efficient, logical, and eliminate many limitations inherent in alternative modeling strategies, but the modern computer's power may have removed the need for this approach.

4.4 Summary

Stand and size-class models offer numerous advantages including simplicity, computational efficiency, and robustness. Although stand-level models are generally limited to single species, even-aged situations, size-class models are used in many forest structures ranging from pure even-aged plantations to tropical rainforests. However, size-class models often need detailed individual-tree data for proper parameterization (e.g. Michie and Buongiorno, 1984). Given this need and the fact that individual-tree models are the preferred means for projecting uneven-aged (Peng, 2000), multi-cohort (Groot *et al.*, 2004), and mixed species stands (Porté and Bartelink, 2002), whole-stand and size-class models should only be preferred when the study's objectives dictate their use (e.g. optimization) or detailed individual-tree data are not available.

On the other hand, combining stand-level models with size-class or individual-tree models is an effective approach for taking full advantage of their strengths and, often, improving predictions (see Chapter 10). In particular, stand-level models are generally found to be most effective for predicting mortality (e.g. Cao, 2006), while size-class models can be advantageous when representing the dynamics of small trees (e.g. Picard *et al.*, 2002).

5

Tree-level models

5.1 Introduction

Tree-level models are the new standard for modeling growth and yield in many regions of the world. They offer several advantages compared to other approaches. Individual-tree models are flexible, accurately characterize growth under a range of stand conditions, and provide high-resolution and detailed output that is used to assess such things as wood quality (e.g. Maguire *et al.*, 1991) or wildlife habitat (Hayes *et al.*, 1997). Tree-level models are the preferred means for simulating growth in mixed species (Porté and Bartelink, 2002) and uneven-aged stands (Peng, 2000), because competition between different species or size classes is better represented.

Likewise, tree-based models are better able to characterize impacts of many types of damaging agents and stand responses to management practices, particularly thinning and pruning, compared to stand-level modeling approaches, because influence of these types of events usually varies within a stand. The primary disadvantages of tree-level models include: (1) they are more expensive to develop; (2) more detailed data are required for their use; and (3) error compounding is potentially greater. Also, some tree-level models require a significant amount of computing power and time if updating a large forest inventory.

Tree-level models operate on representative trees in a stand and sum their attributes to produce stand-level values. Like stand-level models, tree-level models are further classified into a range of categories. First, tree-level models are either *single-tree models* or *tree-list models*, which differ with respect to how mortality is applied to each tree. Single-tree models kill whole trees in a growth period by stochastically deciding whether a tree dies or not during a particular growth period. Tree-list models kill fractions of the trees in a growth period by reducing each tree's per unit area expansion factor (EXPF) by

Forest Growth and Yield Modeling, First Edition. Aaron R. Weiskittel, David W. Hann, John A. Kershaw, Jr. and Jerome K. Vanclay.

the amount of mortality. This distinction has important implications for processing inventory data obtained from point samples.

Another important distinction of tree-level models is whether they are *distance-dependent* or *distance-independent*. Distance-dependent models require spatial coordinates of all trees, whereas distance-independent models assume an average spatial pattern of individuals found in the stands selected for model construction. As discussed in Chapter 2, distance-dependent models utilize quite different measures of competition than those used in distance-independent models. This distinction has important implications for scale of usability, as distance-dependent models are often limited to small plots and the complexity of stand structures capable of being simulated.

This chapter focuses on two primary types of individual-tree models: (1) single-tree distance-dependent; and (2) tree-list distance-independent models. For each type of model, the model framework, key variables, and example models are given (Table 5.1). A third type of individual-tree model is single-tree gap models. These type of models are not discussed in this chapter because gap models primarily focus on regeneration dynamics (see Chapter 9) and are often not well parameterized for predicting overstory tree growth, which can cause significant biases compared to other individual-tree modeling approaches (e.g. Yaussy, 2000).

5.2 Single-tree distance-dependent models

Single-tree distance-dependent models were developed under the assumption that knowing the spatial arrangement of trees in a stand would better represent competition between trees and would provide the capability to simulate complex harvesting systems like patch cutting or row thinning. This modeling approach also has the potential to be effective in characterizing uneven-aged stands managed under group selection practices. One of the first single-tree, distance-dependent models was described by Newnham and Smith (1964) for even-aged stands of either Douglas-fir or lodgepole pine in British Columbia. This work was followed by significant activity in the 1960s and early 1970s, when a number of alternative model structures were developed for various species in Canada and the United States (e.g. Fries, 1974). This activity, however, did not result in many of the models being adapted for operational decision making. Experience shows that, despite their initial appeal, single-tree distance-dependent models have several limitations.

Development and practical use of this type of model is hampered by a lack of suitable spatial data because stem mapping is an expensive and time-consuming process. Procedures are available for predicting tree coordinates in an attempt to avoid this problem when the models are applied to plots without coordinates. For example, Hanus *et al.* (1998) used a nonsimple sequential inhibition approach, based on maximum crown-widths of open-grown trees, to generate spatial coordinates for natural, even-aged Douglas-fir stands, and found the method to be quite effective. However, using predicted rather than measured tree

Table 5.1 Attributes of example tree models by model type.

Model (source)	Inputs	Region	Stand types	Variety of sampling designs	Silvicultural modifiers	Interface	Local calibration	Website
Individual tree, distance dependent								
TASS (Mitchell, 1975)	Tree species, DBH, HT, and HCB, spatial coordinates	British Columbia	Even-aged	No	No	GUI with graphical capabilities and visualization	No	www.for.gov.bc.ca/hre/gymodels/tass/
SILVA (Pretzsch *et al.*, 2002)	Tree species, DBH, HT, and HCB, spatial coordinates, and site characteristics (site index, ecoregion, elevation, slope and aspect)	Germany	All	No	No	GUI with graphical capabilities and visualization	No	www.wwk.forst.tumuenchen.de/research/methods/modelling/silva/
Tree list, distance independent								
PROGNAUS (Ledermann, 2006)	Tree-list, geographical location, site fertility, and soil type	Austria	All	Yes	Yes	GUI with graphical capabilities and visualization	Yes	www.waldwissen.net/technik/inventur/bfw_prognaus/index_EN
ORGANON (Hann, 2011)	Tree-list and site index	Oregon and Washington, USA	All	Yes	Yes	GUI with graphical capabilities and visualization; dynamic link library	Yes	www.cof.orst.edu/cof/fr/research/organon

coordinates eliminates any small advantage spatial competitive indices might have at predicting stand development. It is possible that light detection and ranging (LiDAR) procedures may make measurement of tree coordinates less expensive and time consuming in the future for at least even-aged stands with minimal numbers of understory trees (e.g. Falkowski *et al.*, 2010; Mäkinen *et al.*, 2010).

Plots that are stem mapped are generally small in scale, which makes them more susceptible to plot edge biases. Martin *et al.* (1977) assessed impacts of plot edge bias and found it to have a significant influence on estimation of distance-dependent indices, irrespective of plot size or shape. A linear expansion method was effective at reducing bias, but did not completely remove it, particularly for smaller plot sizes (<0.02 ha).

As discussed in Chapter 2, gains from including competitive influence zone (CIZ), size–distance (SD), growing space (GS), or open-sky view (OSV) measures of competition are relatively small compared to measures used in distance-independent models. These measures of competition conflate one-sided and two-sided sources of competition, which often operate at different scales. Light-interception (LI) measures of competition have potential for better characterizing the light environments usually associated with a one-sided measure of competition (Hanus, 2003). However, practical application of LI measures of competition requires elevation of each tree in addition to their usual two-dimensional coordinates for stands on sloped terrain (Hanus, 2003), and they generally require substantial computing resources. The use of a LI measure in a single-tree, distance-dependent model also requires the addition of a measure of competition for moisture and nutrients in order to fully characterize the competitive environment.

Because mortality is applied stochastically, any particular growth projection will also be stochastic. Deterministic estimation necessitates the conductance of a Monte Carlo analysis in which many stochastic runs are made and the results averaged (see Chapter 15). This requirement and the procedures for calculating the competition variables make running such models computationally burdensome.

5.2.1 Example models

To illustrate the structure of some single-tree distance-dependent growth models, two example models were selected and are described in some detail. The two models are the Tree And Stand Simulator (TASS; Mitchell, 1975) and SILVA (Pretzsch *et al.*, 2002).

5.2.1.1 TASS

TASS is a unique growth model because it represents each individual tree's crown in great detail and relies on an index of crown interference rather than CIZ or area potentially available (APA) to predict growth. TASS is also driven by height growth, whereas most other models are driven by diameter increment or, more commonly, diameter and height growth are predicted independently. TASS was originally developed for white spruce

(Mitchell, 1969) and then applied to Douglas-fir (Mitchell, 1975), where it is now the primary growth model for coastal and interior forests in British Columbia. Output from the model has been used to generate stand tables for the region (Di Lucca, 1999), compare long-term implications of alternative silvicultural regimes (Mitchell and Cameron, 1985; Di Lucca *et al.*, 2003), and represent growth response to thinning in three dimensions (O'Hara and Oliver, 1988).

TASS makes several important assumptions, as follows.

1. The ratio of foliar volume of a subject tree (FV) and that of a similar-sized, open-grown tree (FV_{max}) is an effective measure of tree vigor and degree of suppression.

2. Height growth of any tree can be predicted from a randomly assigned genetic potential that is a function of dominant tree height and FV/FV_{max}.

3. Longitudinal growth of the main axis of all open branches is related to concurrent height growth, regardless of suppression, and thus the radial growth of tree crowns can be predicted from height growth subject to space limitations imposed by competing trees.

4. Bole volume increment can be predicted from a nonlinear statistical function of FV and FV/FV_{max}.

5. Mortality can be predicted from the distance a subject tree is below the crown surface of competing neighbors.

6. Distribution of radial bole increment approximates Pressler's hypothesis, which states that the area increment on any part of the stem is proportional to the foliage above it.

7. Stochastical modeling of mortality, the height growth potential, and the rate of branch extension relative to height growth propagates through other model components and accounts sufficiently for all natural variation in the stand.

A simplified flow diagram of TASS is given in Figure 5.1 (refer to Mitchell (1975) for a more detailed system diagram).

Competition in TASS is represented by the ratio between a weighted foliated crown volume and the maximum foliated crown volume of the identical tree if it were open-grown (i.e. FV/FV_{max}). The foliated crown volume of stand-grown trees is determined from the projected crown area, which is then converted to a volume. This is done for each foliage age class, which is weighted based on the photosynthetic efficiency and retention of the foliage. Both height growth and bole increment are directly related to foliar volume and the degree of competition. The amount of foliated crown volume in subsequent years is determined by the amount of height and branch growth, which are nonlinearly related.

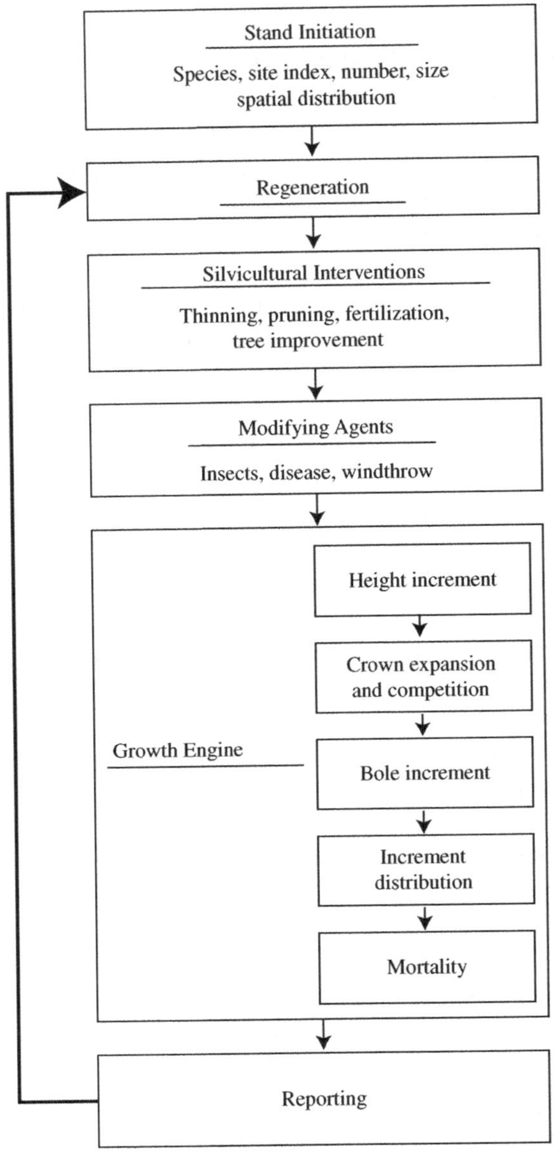

Figure 5.1 Flowchart of key model processes for the TASS distance-dependent, individual-tree growth model (Mitchell, 1975).

The distribution of annual radial increment is based on the ratio of cross-sectional area increment at the base of the bole to the cross-sectional area increment at the base of the crown, the location of the crown base, and the distance to butt swell. Mortality is dependent on the stand's dominant height and a critical overtopping distance. Trees die if

Figure 5.2 Example realistic tree depictions generated by the TASS distance-dependent, individual-tree growth model (Mitchell, 1975): (A) individual tree species; (B) two species, even-aged stand; (C) multiple age classes; and (D) complex stand structure.

they are less than the critical overtopping distance during the early stages of stand development (dominant height < 26 m). In later stages, trees with crown lengths shorter than the minimum length will die.

When compared to other models, TASS has been shown to perform well across a range of stand structures (Mitchell, 1986; O'Hara and Oliver, 1988; Johnson, 2005). The model has been recently extended to hardwood species like red alder and trembling aspen, and Suárez-Mínguez (2010) linked LiDAR data from Sitka spruce stands in Scotland to TASS to evaluate wood quality and wind firmness. A future release will enable the model to handle more complex stands by incorporating the radiation transfer model of Brunner (1998). Amongst its strongest attributes is realistic visualization of simulated stands, an important component in model verification and communication (Figure 5.2).

TASS requires a considerable amount of training to run, but many user needs are met by the distributed software, Table Interpolation Program for Stand Yield (TIPSY; Mitchell *et al.*, 2004). TIPSY quickly accesses and interpolates a TASS-generated database that represents a subset of the possible prediction scenarios. Simulated stand size has generally been limited to approximately 5 ha due to computational limits, but stands up to 20 ha have been simulated. TASS has influenced several other models such as PLATIPUS in Queensland (Vanclay, 1988), and other similar models in France (Ottorini, 1991; Ottorini *et al.*, 1996).

5.2.1.2 SILVA

Single-tree, distance-dependent models are primarily developed and used in North America, but they are rare in Europe (Pretzsch *et al.*, 2008). One exception to this trend is the SILVA growth and yield model (Pretzsch *et al.*, 2002). SILVA was built to elucidate potential benefits of structurally diverse forests (Hanewinkel and Pretzsch, 2000), assess long-term consequences of forest management (Pretzsch *et al.*, 2002), and assess implications of climate change (Pretzsch, 2002). The model was originally parameterized for the Bavarian region of Germany using a long-term dataset (Pretzsch *et al.*, 2002), but was successfully extended to other regions like southwestern Germany (Mette *et al.*, 2009) and parts of Switzerland (Schmid *et al.*, 2006a).

Like TASS, SILVA represents tree crowns in three dimensions based on empirical species-specific crown shape models. Competition is represented by an index that represents both the type and spatial distribution of competitor trees, which are defined by a "virtual-reverse-cone method" (Pretzsch *et al.*, 2006; Figure 5.3). Essentially, a cone with its base located in the crown (relative location and vertex angle vary by species), and with an axis of equal length to the tree axis, is estimated for each tree, and all surrounding trees whose tops are intersected by the virtual cone are considered competitors. For each competitor, the angle between the base of the cone and the top of the competitor tree is calculated, which is weighted by the ratio of the crown cross-sectional areas and a species-specific light transmission coefficient. Two additional distance-dependent measures of competition are also calculated. One is based on the ratio of crown surface area for coniferous and all competitor trees, while the other estimates the "horizontal distance of the center of gravity of competition from the stem center of the subject tree" (Pretzsch *et al.*, 2006).

Based on these measures of competition and crown surface area, both potential diameter and height growth are reduced with nonlinear multiplicative modifiers (Chapter 6) that vary by species. Potential height growth for a tree is based on a derived theoretical age and site factors, while potential diameter increment is just dependent on tree size. Once tree diameter and height are updated, height to crown base is predicted from a static equation. Tree mortality is dependent on estimated diameter increment, the ratio of tree height to diameter, and site index. The model has a five-year time step and corrects for plot edge effects using the linear expansion method of Martin *et al.* (1977). Site productivity is defined by both traditional site index and a suite of environmental variables such as ecoregion, elevation, slope, and aspect. In addition, model predictions can be altered based on long-term climate data such as mean temperature, precipitation, and atmospheric CO_2 concentration, which can be held constant or altered during a simulation.

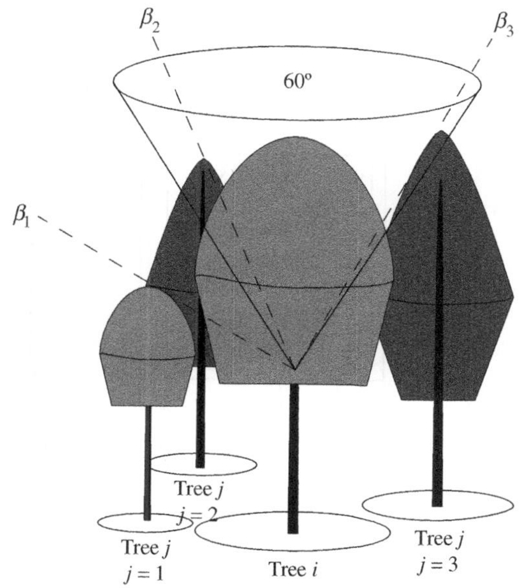

Figure 5.3 Virtual-reverse-cone method used in the SILVA distance-dependent, individual-tree growth model to calculate competition (Pretzsch *et al.*, 2002). Tree *i* is the subject tree and trees *j* (*j* = 1, 3) are potential competitors. The inverted cone, which is determined by species- and individual-specific parameters, defines a search area for competitors. Trees whose crown tips intersect the search area are competitors. Thus trees 2 and 3 are competitors of tree *i*.

The model behaves well when evaluated against long-term growth data (Pretzsch, 2002; Schmid *et al.*, 2006a), but reparameterization was required to obtain proper behavior in southwestern Germany (Mette *et al.*, 2009). Regardless, SILVA is useful for a range of practical, scientific, and educational purposes (Pretzsch, 2009, p. 515). For example, the model was recently used to analyze and compare biomass development of different stand types across a range of site productivity and thinning regimes in Germany (Rötzer *et al.*, 2010).

The model is linked to both stand- and landscape-level visualization programs (Pretzsch *et al.*, 2008). In addition, the model has the capacity to generate stem coordinates of all trees with a structure generator, STRUGEN (Pretzsch, 1997), if they are unavailable. The unique attribute of STRUGEN is that it generates stand structures based on verbal descriptions such as "row mixture" and "single tree mixture" (Pretzsch *et al.*, 2008).

5.3 Tree-list distance-independent models

Tree-list distance-independent growth models are the standard for predicting growth and yield in many regions because of their precision, high resolution, and computational

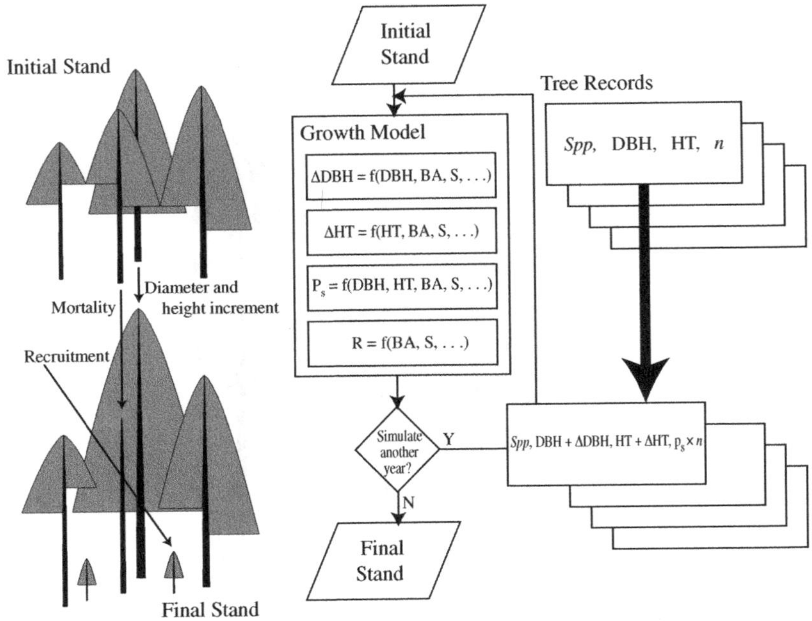

Figure 5.4 Key processes represented by individual-tree growth models. The most important include diameter increment (ΔDBH), height increment (ΔHT), probability of survival (P_s), and often recruitment (R). S = site quality; *Spp* = species; *n* = initial number of trees per unit area.

simplifications. Single-tree models maintain a list of attributes (species, DBH, etc.) for each individual tree on a plot. The tree-list approach does the same, but also simulates number of trees per unit area represented by each tree record rather than keeping track of every individual in a stand (Figure 5.4). This simplifies processing large stands, accommodates a range of sampling schemes, and allows mortality to be predicted in a variety of ways.

Like gap models, tree-list models assume that competition is relatively homogeneous in a stand, and tree social position can be represented without knowledge of tree spatial coordinates. Unlike gap models, tree-list models are relatively insensitive to plot size and can be initialized from inventory data or tree-list generators. Trees can be sampled using either fixed- or variable-radius plots or even a combination of the two. It is just important that all trees be measured on a plot because measurement of only crop trees or a partial subsample results in biased forecasts. In addition, it is recommended to use a plot type and size similar to the one used to construct the tree-list model (Hann and Zumrawi, 1991), since the measures of competition used in the predictions can be sensitive to plot size (Stage and Wykoff, 1998).

Tree-list models are often comprised of a series of equations that predict diameter increment, height increment, and mortality, while ingrowth is sometimes, but rarely, simulated. Most tree-list models also predict changes in live crown ratio. The structures of the diameter and height increment equations, as well as crown recession equations, are described in more detail in Chapter 6, while mortality and ingrowth are discussed in Chapters 8 and 9, respectively.

The equations in tree-list models are generally species specific, but they can also be developed for species groups when multiple species are present (Vanclay, 1991a) or data for some species is scarce (McTague *et al.*, 2008). The equations are generally parameterized from permanent plot data. One of the first attempts to use a tree-list approach was by Stage (1973), who developed prediction equations for species groups in the Inland Northwest of North America. This work became the foundation for the Forest Vegetation Simulator (FVS) that currently is widely used throughout the United States (Crookston and Dixon, 2005). Other early tree-list models were also in North America and included those of Leary (1979) in the Lake States (STEMS), as well as Krumland and Wensel (1981) in northern California (CACTOS).

Most tree-list models predict periodic (e.g. 5- and 10-year) rather than annual growth because of the nature of remeasured plot data and the minimization of compounded error. Tree-list models can include age, or not. Use of age limits model applications to even-aged stands, while models without age do not have such restrictions. Therefore, most tree-list models do not use age and, as a result, make the assumption that tree size rather than chronological age is more important for predicting stand development. This assumption is supported by many studies, including Peet (1976; 1981), Ryan and Yoder (1997), Ryan *et al.* (2006), and Bond *et al.* (2007), whose research findings indicate that tree growth rate is limited by tree size rather than tree age.

Additional variables are often included to describe tree vigor, competitive stress, and site quality. Tree vigor is often represented by live crown ratio or the ratio of tree height to diameter, while competitive stress is generally characterized by either tree-level variables like basal area in larger trees and crown competition factor in larger trees or stand-level indices such as total basal area or crown competition factor (Krajicek *et al.*, 1961). Site quality is most often represented by site index, habitat type, physiographic variables like slope, aspect, and elevation, or mean climate variables. Most models rely solely on site index, but many of the FVS variants used in North America are sensitive to habitat type as well as slope and aspect.

Growth in tree-list models is predicted with equation forms that have a peaking or asymptotic behavior, which limits tree size. Some models employ a potential tree-growth function multiplied by a modifier as is widely used in gap models, while others predict realized increment directly. Diameter growth can be predicted as an increment (e.g. Weiskittel *et al.*, 2007a), squared diameter increment (e.g. Wykoff, 1990), inside bark basal area increment (e.g. Cole and Stage, 1972), and relative diameter increment (e.g. Yue *et al.*, 2008). Height growth is most often predicted as an increment, but is sometimes inferred from a static equation as is commonly done in gap models. Tree mortality is commonly predicted as probability of death for each tree in the tree list, which is used to adjust expansion factors. Hence, individual trees are often not removed from a tree list, which is the opposite of single-tree models. Likewise, thinning is often represented by modifying each tree's expansion factor rather than removing trees from a tree list.

Although most tree-list models are deterministic, a few allow stochasticity in the predictions. For example, Stage (1973) assumed that a stochastic diameter increment function was sufficient to account for enough stand variation so that other equations could be deterministic. Stochasticity was implemented by adding random noise to the predictions, with the variance being determined from the residuals of the empirical equation (Stage, 1973), which was later refined to incorporate serial correlation and better account for variance of growth rather than variance induced by measurement error (Stage and Wykoff, 1993).

Another method for inducing stochasticity is to modify the tree list by record doubling (e.g. Vanclay, 1989a) or tripling (e.g. Stage, 1973). In this approach, initial tree attributes are repeated in a tree list, and the expansion factor modified to maintain plot consistency; then different predicted increments are applied during model simulations. Vanclay (1989a) doubled tree records, gave each tree 50% of the original expansion factor, and applied an increment of 0.8- and 1.2-times the predicted diameter increment. Another approach is to model tree mortality stochastically (e.g. Belcher *et al.*, 1982). These various methods of including stochasticity in tree-list models were reviewed by Fox *et al.* (2001), who suggested that models should all characterize and include a stochastic element.

Although tree-list models all operate based on a tree list, they differ in the resolution of the simulations. For example, most models average stand density based on all plots in a stand, while other simulators simulate at the individual-plot level and stand density will vary from plot to plot. Likewise, most tree-list models simulate on a per-unit-area basis, but some models adjust predictions based on plot size. For example, Stage and Wykoff (1998) developed a method that modified model parameters based on plot size, which had a significant influence on model projections. In addition, some models employ algorithms to compress tree lists in order to minimize computing time and power (e.g. Stage *et al.*, 1993).

Comparisons between tree-list models and other modeling approaches primarily focus on differences in predictions between distance-independent and distance-dependent approaches. Most studies find little gain with the distance-dependent approach (e.g. Wimberly and Bare, 1996). In comparison to a whole-stand model, Qin and Cao (2006) found a tree-list model to give poorer predictions of number of trees, basal area, and volume across a range of projection lengths in loblolly pine plantations. For example, the mean absolute difference for stand basal area was 21, 37, and 43% lower for the whole-stand model compared to the tree-list model for short- (4–7 years), medium- (10–12 years), and long-term (15–17 years) projections, respectively (Qin and Cao, 2006). This illustrates one drawback to tree-list models: prediction errors accumulate as a result of the number of equations employed, which results in a lack of accuracy and precision despite their higher resolution. On the other hand, Härkönen *et al.* (2010a) found that a distance-independent individual-tree model provided better predictions of mean height and stand volume after 10 years of projection for the National Forest Inventory in southern Finland when compared to stand-level and combined models.

Although Fox *et al.* (2001) concluded that growth models should include stochasticity in their simulations, comparisons between deterministic and stochastic tree-list models suggest a minimal difference between long-term predictions. Weber *et al.* (1986) found no practical difference between a deterministic and stochastic model in terms of mean stand values of density, basal area, volume, and diameter distribution after 100 years of simulation. Finally, a comparison between a tree-list model and a gap model showed the tree-list model to be much more accurate (Yaussy, 2000).

5.3.1 Example models

Despite many commonalities, tree-list models can differ in their structure and application. Two models were selected to illustrate this: PROGNAUS (Monserud *et al.*, 1997) and ORGANON (Hann, 2011).

5.3.1.1 PROGNAUS

Following development of Prognosis in the Inland Empire of the United States (Stage, 1973; Wykoff *et al.*, 1982), a similar approach was taken to develop a model for the mixed species forests of Austria. The resulting model was PROGNosis for AUStria (PROGNAUS; Monserud *et al.*, 1997), which continues to be refined and developed today (Ledermann, 2006). PROGNAUS uses species-specific diameter and height increment equations that do not rely on the potential growth concept, tree age, or site index (Monserud and Sterba, 1996; Hasenauer and Monserud, 1997; Nachtmann, 2006). Tree mortality is also independent of site index (Monserud and Sterba, 1999). Instead, PROGNAUS relies on elevation, slope, aspect, relief, soil depth, humus thickness, soil type, soil humidity, vegetation type, and growth district to represent forest site quality. One- and two-sided competition in these equations is represented by basal area in larger trees and crown competition factor, respectively.

In addition, record tripling is used to more effectively estimate basal area in larger trees (Ledermann and Eckmüllner, 2004). Crown ratio is statistically predicted from a static equation (Hasenauer and Monserud, 1996). The minimum DBH for overstory trees is 5 cm, and the model has a five-year time step. PROGNAUS estimates the number, tree species, and dimensions of ingrowth (Ledermann, 2002a). The model represents different thinning and forest harvest regimes (Ledermann, 2002b; Söderbergh and Ledermann, 2003). Log grade and quality can be estimated for all trees (Vospernik, 2004; Ledermann, 2006).

PROGNAUS is primarily used for research and education (Ledermann, 2006), but also for determining sustainable amounts and structures of harvests in forest enterprises that are shifting away from even-aged towards uneven-aged forest management (Sterba, 2002). The model was extensively tested and validated (Sterba and

Monserud, 1997; Sterba *et al.*, 2001; Monserud *et al.*, 2004). The model has a Microsoft Windows graphical user interface (GUI) (Ledermann, 2006). The model is used to assess a number of forest management issues including: long-term changes in species composition (Huber and Sterba, 2009; Huber, 2010); the need for self-thinning constraints (Monserud *et al.*, 2004); evaluation of growth in pure and mixed-species stands (Sterba *et al.*, 2002); comparisons of silvicultural systems (Sterba and Ledermann, 2005); prediction of deer habitat availability (Vospernik *et al.*, 2007); and estimation of timber and wood biomass supplies in Austria.

5.3.1.2 ORGANON

ORGANON (ORegon Growth ANalysis and projectiON) was developed at Oregon State University for even- and uneven-aged stands of pure or mixed species (Hann, 2011). The program incorporates three versions of the model, namely southwest Oregon (SWO-ORGANON), northwest Oregon (NWO-ORGANON), and a version intended for intensively managed plantations of Douglas-fir or western hemlock in northwest Oregon, western Washington, and southwest British Columbia (SMC-ORGANON). The most recent version has just been completed for red alder plantations (Hann *et al.*, 2011).

The forests that ORGANON simulates represent some of the longest lived, largest sized, most structurally and environmentally complex natural conifer forests in the world (Franklin and Dyrness, 1973). The model simulates influences of thinning, pruning, fertilization, and combinations of silvicultural treatments on growth and mortality (Hann *et al.*, 2003), and recently was modified to incorporate impacts of planting with genetically improved seedlings (Gould *et al.*, 2008) and impacts of the Swiss needle cast (SNC) disease (Maguire *et al.*, 2002).

ORGANON uses a five-year growth period and relies on a user-supplied tree list and site index (Figure 5.5, Appendix 2). Missing heights and height to crown base are imputed from regional static equations, which are calibrated to each plot if a subsample of measurements is present, by the use of simple linear regression through the origin. The model represents individual crowns in detail through the estimation of stand-grown crown width and vertical profile. The growth and mortality equations have a five-year time step, are insensitive to tree age, and are primarily deterministic, but stochasticity can be implemented in both the diameter and height growth predictions if desired by the user. The diameter increment equation determines diameter increment directly, while the height increment equation uses the approach of multiplying potential increment by a modifier (Chapter 6). Each primary species has its own parameter estimates and they are not grouped. Like the static equations, diameter increment also can be calibrated if local observations are available.

Diameter increment is sensitive to tree size, crown ratio, basal area in larger trees, site index, and stand basal area. Height increment is determined from potential height growth (estimated from site index and tree-growth effective age), crown closure in taller trees, and

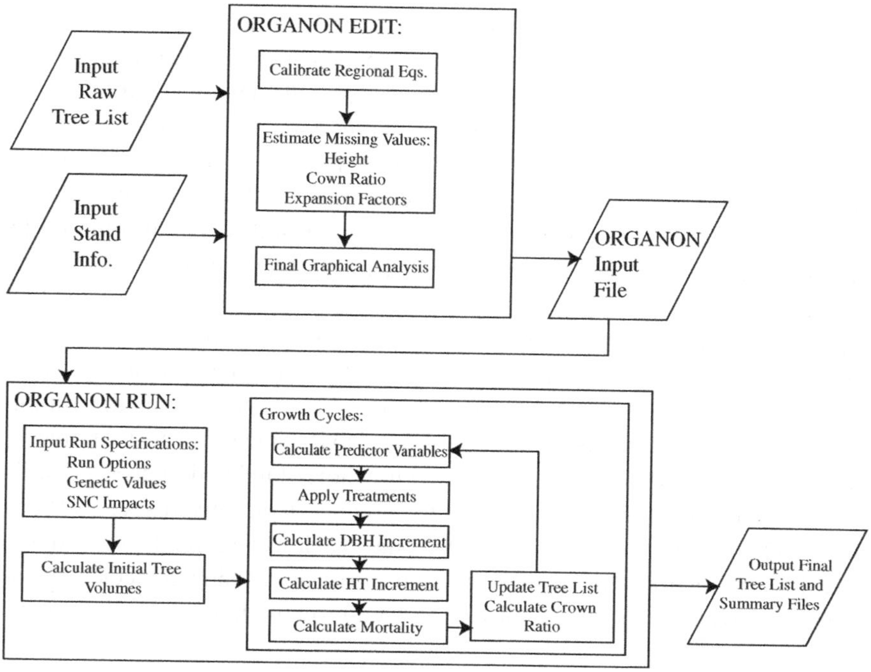

Figure 5.5 Flowchart of key model processes for the ORGANON distance-independent, individual-tree growth model (Hann, 2011). A detailed description of ORGANON EDIT and RUN steps is given in Appendix 2.

crown ratio. Probability of tree mortality is determined from tree size, crown ratio, site index, and basal area in larger trees. Crown recession is predicted from a static height-to-crown-base model that is a function of tree size, tree form, site index, crown competition in larger trees, and stand basal area (Hann and Hanus, 2004). A maximum size–density equation is used to constrain stand densities, but its need for Douglas-fir in the latest version of the SMC-ORGANON is questioned (Hann *et al.*, 2006). Small trees are added to the model by interfaces with the SYSTUM-1 (Ritchie and Powers, 1993), RVMM (Shula *et al.*, 1998), or CONIFERS (Ritchie, 2009a) young-stand models, which predict growth of planted seedlings, competing vegetation, and their interactions.

ORGANON has a flexible user-interface, and the model is available as a dynamic link library, which allows it to be called from a variety of software platforms. The model offers a wood quality module and is linked with a stand visualization program (Hanus and Hann, 1997).

5.4 Summary

Tree-level models provide the highest level of predictions, but come with the costs of higher data requirements, longer computing times, and potential for compounding errors. Both distance-dependent and distance-independent models are used for applied situations like comparison of silvicultural systems, while distance-independent gap models are primarily utilized to answer ecological research questions. Initially, distance-independent and distance-dependent models were equally represented in number (Munro, 1974; Dudek and Ek, 1980), but distance-independent, tree-list models like FVS and ORGANON now predominate. For example, the Prognosis model of Stage (1973) has been expanded to cover most of the United States (Crookston and Dixon, 2005), British Columbia (Temesgen and LeMay, 1999), Ontario (Lacerte *et al.*, 2004), and Austria (Ledermann, 2006).

A few reasons for this transition are because distance-independent models do not require tree spatial information and often give predictions as good as distance-dependent models. The number of distance-dependent models may increase in the future with advances in remote sensing technologies like LiDAR, since the cost of tree spatial information will be greatly reduced. Compared to whole-stand and diameter distribution models, tree-level models are the best tools for predictions in mixed species (Porté and Bartelink, 2002) and uneven-aged (Peng, 2000) stands. However, tree-level models might result in a loss of accuracy in predicting stand attributes of single-species, even-aged stands compared to other modeling approaches. Tree-level models are the focus of the next two chapters.

6

Components of tree-list models

6.1 Introduction

Tree-list models are developed and utilized in many regions because they are flexible, provide the highest resolution of predictions, and simulate a range of stand compositions and structures. Although they can be sensitive to error compounding, a properly specified tree-list model emulates many of the important stand-level principles. In addition, tree-list models can be coupled with other modeling approaches, which can significantly improve their predictions (see Chapter 10). A tree-list model is driven by four primary equations that are often species specific: diameter increment, height increment, crown recession, and mortality. A variety of approaches and model forms are used. This chapter explores these various approaches and discusses advantages and disadvantages associated with each approach for modeling diameter, height, and crown increment. Mortality is the focus of another chapter (see Chapter 8).

Growth is the increase in dimensions of each individual tree in a forest stand through time, while increment is the rate of change in a specified period of time. Growth consists of both elongation and thickening of the various parts of a tree through the actions of meristematic zones located throughout the tree. Apical meristems cause the elongation of the main stem, branches, and roots of the tree, and secondary meristems cause the thickening of the main stem, branches, and roots. As a result of meristematic activity, growth adds the xylem and phloem necessary to transport moisture, nutrients, and photosynthate. Although growth occurs throughout a tree, foresters are primarily concerned with changes in both tree diameter at breast height (DBH) and tree height (HT) because of their ease of measurement and strong correlation with total tree volume. Changes in tree volume are often inferred from DBH and HT increment, which can have important implications for the accuracy of predictions (Hann and Weiskittel, 2010).

Forest Growth and Yield Modeling, First Edition. Aaron R. Weiskittel, David W. Hann, John A. Kershaw, Jr. and Jerome K. Vanclay.

Tree growth has multiple inter- and intra-annual stages that must be considered by tree-list models. For example, a cumulative growth curve of height over age shows three primary stages: (1) juvenile period where growth is rapid and often exponential; (2) a long period of maturation where the trend is nearly linear; and (3) old age, where growth is nearly asymptotic. A diameter growth curve would show much the same trend, except there is a tendency toward more sustained growth rate as the tree matures (Hann and Hanus, 2002b). While height increment may nearly cease in maturity, diameter increment must continue in order to produce the xylem and phloem needed for tree survival. In any given season, height increment generally begins a few days before diameter increment, but the length of the growing season is much longer for diameter increment (Emmingham, 1977; Jayawickrama *et al.*, 1998). Height increment is more dependent on differences in genetics and environmental conditions, while diameter increment is more closely related to current amount of foliage and tree competitive status (Tappeiner II *et al.*, 2007). A variety of factors influences these trends and must be considered in any growth equation.

The growth of many stand and tree attributes exhibits a sigmoidal form over age or size. Increment at any age or size can be determined by the derivative of the sigmoidal form evaluated at that age or size. A graph of the resulting increments will produce a peaking model form that is often skewed towards the older or larger values of the stand or tree attribute of interest, and will never predict negative values. Various theoretical sigmoid model forms are used to predict growth in forestry (Zeide, 1993), but most of them can be generalized with a single equation form (García, 2005c). The most common sigmoid model forms include the Gompertz (1825), Bertalanffy (1949), and Richards (1959) equations.

Although these theoretical models offer some biological interpretability (e.g. Zeide, 2004), it is easily shown that well-formulated empirical sigmoid equations can be just as accurate or even more accurate for a wide range of data (e.g. Martin and Ek, 1984). For this reason, the peaking model forms derived from empirical sigmoid equations are more widely used in predicting increment and will be the primary focus of this chapter.

Both tree diameter and height increment are influenced by tree size and vigor, competition (one- and two-sided), and site quality. Due to the complexity of the interactions between these factors, one way to examine expected behavior is to assess tree growth without competition (i.e. an open-grown tree) and then evaluate influences of changes in vigor due to competition. A peaking model form over size is frequently used because increment is more physiologically related to size than chronological age (e.g. Bond *et al.*, 2007). An appropriate peaking model form is particularly important for diameter increment because it can ensure that predictions will never go to zero over the biologically reasonable age or size of the species. The influence of changes in site quality are to change the height of the peak, but not the position, as would be the case if the relationship was plotted over chronological age (Figure 6.1). Tree vigor is often represented by crown ratio in that increment increases as crown ratio does. Increases in both one- and two-sided competition reduce maximum increment.

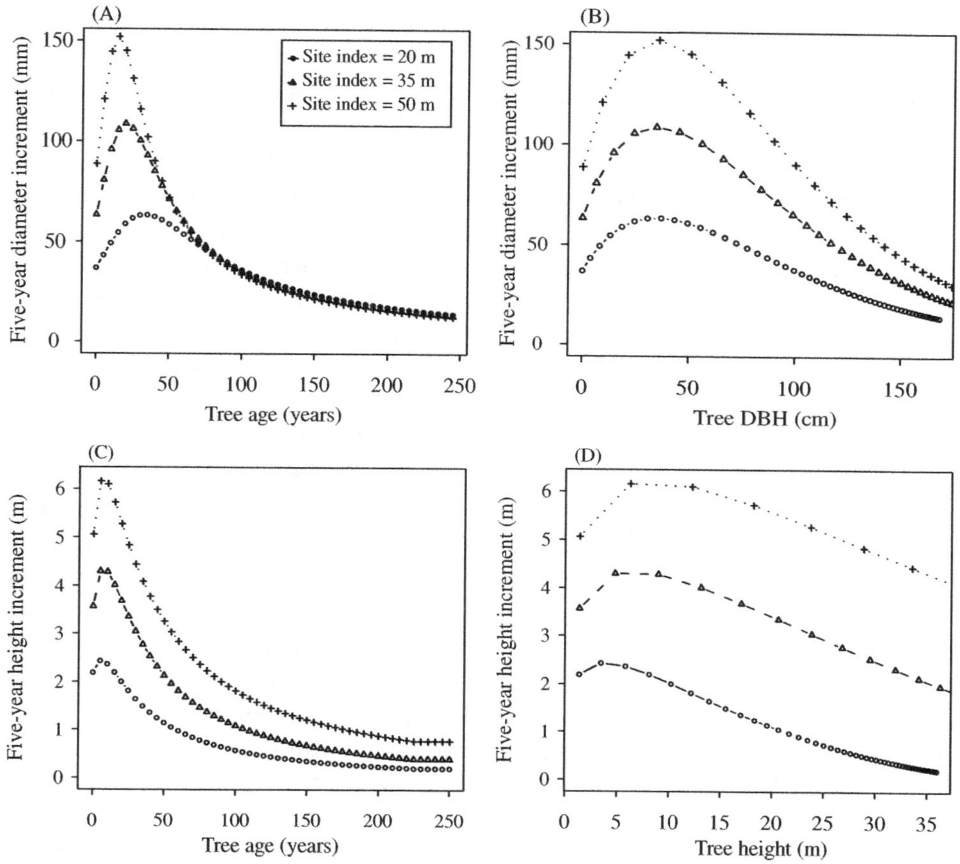

Figure 6.1 Predicted five-year diameter (top) and height (bottom) increment for coastal Douglas-fir for different site indices over total tree age (left) and tree size (right). Based on the equations of Hann *et al.* (2003).

6.2 Diameter increment

Prediction of diameter increment is particularly critical in tree-list models since diameter is the primary determinant of stem volume. One of the first examples of an empirical equation used to predict periodic diameter increment was in the work of Lemmon and Schumacher (1962b), who estimated ponderosa pine growth as a function of tree age, competition, and site quality. In comparison to predicting height increment, a wider array of approaches is used because of differences in the dependent variable, model form, and key covariates. The dependent variables for updating individual-tree DBH have included diameter increment (Hann *et al.*, 2006; Weiskittel *et al.*, 2007a), squared diameter inside bark (Cole and Stage, 1972), relative diameter increment (Yue *et al.*, 2008), and inside bark basal area increment (Monserud and Sterba, 1996).

While the optimal dependent variable is often debated, West (1980) found no difference between using diameter or basal area to predict short-term increment (one- to six-year) in *Eucalyptus* and even-aged hardwood stands. In a study of lodgepole pine, Cole and Stage (1972) compared inside bark radial increment and inside bark basal area increment, with the latter being considered superior. This has led others to use squared diameter inside bark as the dependent variable because it is directly proportional to basal area increment and can be transformed to an estimate of diameter increment if diameter is known (Stage, 1973; Wykoff, 1990).

Basal area increment is a function of DBH at the start of the growth period, diameter increment, and squared diameter increment, and many diameter and basal area increment equations use transformations of DBH at the start of the growth period as predictor variables. Therefore, models of basal area increment will generally have a higher correlation with DBH than models of diameter increment (West, 1980). Given that most studies measure stem circumference and convert it to a diameter under the assumption of a circle, the use of basal area increment as a dependent variable further extends the assumption of a circle, resulting in more mensuration error in the response variable and more unexplainable variation about the equation (Kmenta, 1997). The influence of using ΔDBH versus ΔBA was recently demonstrated by Russell *et al.* (in press), who found essentially no difference between the two approaches for short-term simulations (<10 years). However, the prediction root mean square error was reduced by over 20% when ΔDBH was used instead of ΔBA in forecasting long-term, individual-tree growth (26–30 years).

In contrast, some studies used relative diameter increment (Yue *et al.*, 2008) and relative basal area increment (Larocque, 2002) successfully, while other studies found relative dependent variables to be inferior to an absolute dependent variable for growth predictions (MacFarlane and Kobe, 2006). Although Yue *et al.* (2008) suggested that relative growth measures were less sensitive to differences in site productivity, tree size is generally included as a covariate in predicting absolute increment and provides predictions nearly equivalent to a relative-based dependent variable (MacFarlane and Kobe, 2006), which indicates that the gain by using a relative measure is likely to be small.

Two general conceptual approaches to model formulation are used to predict diameter or basal area increment: (1) a maximum potential increment multiplied by a modifier; and (2) a unified equation that predicts realized increment directly. Although the differences between the two are mostly semantic because they both give reasonable behavior (Wykoff and Monserud, 1988), they do illustrate a key philosophical decision in modeling increment and are discussed separately. For brevity, we will often use the term "diameter increment" in the remaining discussion as a synonym for both diameter increment and basal area increment.

6.2.1 Potential diameter increment equations with multiplicative modifiers

The approach of using potential diameter increment equations with multiplicative modifiers is adopted in most ecological gap models (Bugmann, 2001) and several statistical tree-list models (Belcher *et al.*, 1982; Wensel *et al.*, 1987; Soares and Tomé, 2003). The advantage of this approach is that it is constrained and robust because predicted increment can never exceed the maximum or potential increment. The disadvantage is that it can be difficult to estimate potential diameter increment and, consequently, a variety of approaches are used (Table 6.1).

Hahn and Leary (1979) estimated potential diameter increment by first developing a regression equation between diameter increment, tree size, crown ratio, and site index for dominant and codominant trees. Potential diameter increment was then calculated as the predicted mean diameter increment plus 1.65 standard deviations, or essentially the 95th percentile of predictions. Moore (1989) used open-grown trees to parameterize a potential diameter increment equation and found the approach to increase reliability in estimating Virginia pine growth, which is similar to the approach used by Amateis *et al.* (1989). Bragg (2001a) predicted potential relative increment as a function of DBH using a multi-criteria iterative approach to estimate three species-specific parameters. The potential relative increment was found to be unrelated to species shade tolerance, and differed appreciably from optimal values made by the approach used in the JABOWA gap model (Bragg, 2001a). Finally, Wensel *et al.* (1987) estimated potential diameter increment as a function of live crown ratio based on the largest 33% of the trees in each stand with crown ratios greater than 0.5.

Once a potential diameter increment equation is developed, the parameters are fixed and a multiplicative modifier function is calibrated to the remaining modeling data. The modifiers can include environmental factors like those typically used in gap models (Bugmann, 2001) or site and tree-level factors like site index and basal area in larger trees (Figure 6.2). The modifiers generally try to characterize both site factors and competition not accounted for in the potential increment equation. The modifier can be the product of several growth multipliers (e.g. Monserud, 1975), the minimum of several growth multipliers, or a single asymptotic function of several factors (e.g. Wensel *et al.*, 1987). The multiplicative approach assumes that the factors influence increment simultaneously and independently. The minimum approach is additive and conforms to Liebig's law of the minimum, which states that increment is controlled by the factor of lowest availability and not by the total of resources available.

A fundamental problem with the multiplicative approach is that it can produce artificially low values, especially if the number of multipliers increases, which means that the maximum has to be parameterized to be much higher than measured (Bugmann, 2001). In the Monserud (1975) model for northern hardwoods, the competition modifier was adjusted by two additional multipliers, which reflected shade tolerance and the ability to respond to release. Although the minimum or additive approach is adopted in several

Table 6.1 Equation forms and covariates[a] for predicting potential/average diameter increment, and associated multiplicative modifier equations used in several individual-tree models.

Model	Potential/average
CACTOS	$\dfrac{\left(\left(\left(1-e^{(-5\cdot b_2)}b_0\right)^{\left(\frac{1}{b_3}\right)}\right)\cdot SI^{\left(\frac{1}{b_3}\right)} + \left(e^{(-5\cdot b_2)}\right)\cdot (DBH^2)^{\left(\frac{1}{b_3}\right)}\right)^{\left(\frac{1}{b_3^{-1}}\right)} - DBH^2}{\left(1+e^{(4-b_4\cdot CR)}\right)}$
STEMS (stand-level, pure species)	$b_0 + b_1 \cdot DBH^{b_2} + b_3 \cdot (SI \cdot CR \cdot DBH)^{b_4}$
STEMS (individual tree)	$b_0 + b_1 \cdot DBH^{b_2} + b_3 \cdot (SI \cdot CR \cdot DBH)^{b_4}$
FVS-NE	$b_0 SI^{b_1} \cdot \left(1 - e^{(-b_2 \cdot BAL)}\right) \cdot 0.7$
—	$b_0 \cdot e^{(-b_1 \cdot DBH)} \cdot DBH^{b_2}$
—	$(b_0 + b_1 \cdot HT_{DOM}) \cdot DBH^{b_2} \cdot e^{(-b_3 \cdot DBH)}$
NIRM	$b_0 \cdot DBH^{b_1} \cdot b_3^{DBH}$

[a] Where DBH is diameter at breast height, RDBH is relative diameter at breast height, CR is crown ratio, CC_{66} is crown closure at 66% of tree height, BAL is the basal area in larger trees, CCLS is tree crown class, SI is site index, $\overline{DBH}$ is average stand diameter, BA is total stand basal area, BA_{max} is maximum species area, HT_{DOM} is dominant height, SDI_{sum} is a modified stand density index, PC is physiographic class, LAT is latitude, LNG is longitude, CI is a competition index, DI is a soil drainage index, NI is an available nitrogen index, and HSI is a heat sum index; the w_ks are weighting factors, and the b_is and c_is are parameters.

gap models (Bugmann, 2001), it is not widely adopted in statistical tree-list models because the asymptotic approach is preferred. In the Wensel *et al.* (1987) asymptotic model, the modifier is an exponential function of a one-sided measure of competition: crown closure at 66% of a tree's height (CC_{66}).

Modifier	Source
$e^{\left(c_0 CC_{66}^{c1} \cdot BA^{c2}\right)}$	Wensel *et al.* (1987)
$1-e^{\left(\frac{\left(-c_0\left(1-e^{\left(c_1 \cdot \overline{DBH}\right)}\right)\right)}{BA}\right)}$	Leary and Holdaway (1979)
$1-e^{\left(-\left(\left(c_0\cdot\left(1-e^{\left(c_1 \cdot RDBH\right)}\right)\cdot c_2+c_3\right)\right)\cdot\left(c_4\cdot(\overline{DBH}+1)^{c_5}\right)\cdot\left(\frac{BA_{max}-BA}{BA}\right)\right)^{0.5}}$	Belcher *et al.* (1982)
$e^{(-c_0 \cdot BAL)}$	Dixon *et al.* (2007)
$e^{\left(\begin{array}{l} c_0 \cdot (CR-4)+c_1 \cdot (BAL-11.5)+c_2 \cdot (BA-23)+c_3 \cdot (CCLS-3) \\ +c_4 \cdot (PC-5)+c_5 \cdot (PC-5)^2+c_6 \cdot (LNG+93)+c_7 \cdot (LAT-47) \end{array}\right)}$	Lessard *et al.* (2001)
$c_0+c_1 \cdot SDI_{sum}+c_2 \cdot (SDI_{sum} \cdot HT_{DOM})$	Pretzsch and Biber (2010)
$\frac{\sum_{k=1}^{4} w_k}{\left(\left(\frac{w_1}{CI}\right)+\left(\frac{w_2}{DI}\right)+\left(\frac{w_3}{NI}\right)+\left(\frac{w_4}{HSI}\right)\right)}$	Bragg *et al.* (2004)

In the STEMS model, Leary and Holdaway (1979) developed stand-level modifier functions that differed for pure and mixed stands. The modifier for pure-species stands was a function of average diameter, while an interaction term that differed by forest type was incorporated in the modifier for mixed stands (Leary and Holdaway, 1979). This modifier

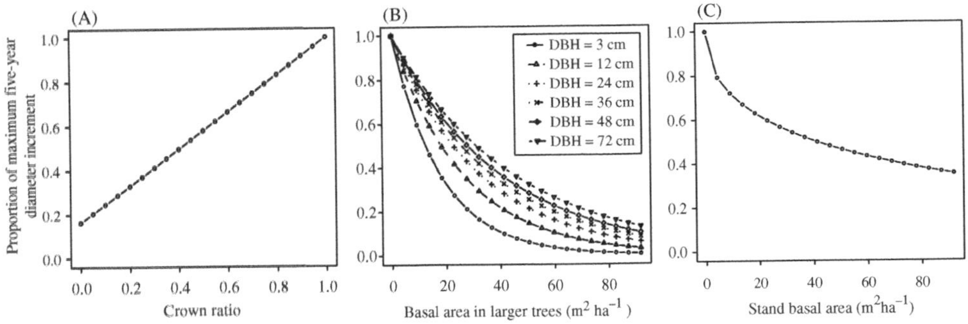

Figure 6.2 Influence of crown ratio (A), basal area in larger trees (B; $m^2\,ha^{-1}$), and total stand basal area (C; $m^2\,ha^{-1}$) on five-year diameter increment in coastal Douglas-fir. Based on the equations of Hann *et al.* (2003).

was later revised by Holdaway (1984) to operate on an individual-tree basis through the use of relative tree size, along with average stand DBH, current basal area, and a maximum basal area expected for the stand. In contrast, the northeastern variant of the Forest Vegetation Simulator (FVS) simply uses basal area in larger trees to modify potential basal area increment (Dixon *et al.*, 1995).

The potential-multiplied-by-modifiers approach has multiple limitations. The most important include: (1) potential increment is not easily observed; (2) procedures for estimating potential increment are complex, cumbersome, and often subjective; and (3) fitting equations in two stages does not allow proper estimation of uncertainty and may cause bias (Lessard *et al.*, 2001). In addition, most potential increment equations use site index, which can be difficult to estimate in uneven-aged, mixed-species stands (e.g. Wykoff, 1990).

In contrast to the potential-multiplied-by-modifiers approach, Lessard *et al.* (2001) used an equation predicting average increment rate times a modifier equation, and found it worked well. Newer statistical methods such as quantile regression may provide a more objective method for estimating potential increment (e.g. Pretzsch and Biber, 2010), but the inability to estimate parameters simultaneously for both the potential and modifier is an important limitation of this approach. For example, attempts to simultaneously estimate model parameters for both the potential and modifier were found to result in biased model parameters (e.g. Wensel *et al.*, 1987; Wykoff and Monserud, 1988).

6.2.2 Realized diameter increment equations

The realized diameter increment approach predicts increment directly based on a single equation and an array of covariates, which avoids some of the limitations with the potential-multiplied-by-modifier approach. Furthermore, a diameter increment equation properly constructed using the realized approach should also provide reasonable predictions of potential increment for free-to-grow trees. Like the potential-multiplied-by-modifier approach, a variety of equation forms and covariates have been used in the realized approach (Table 6.2). Wykoff (1990) indicated that three classes of covariates need to be included in a diameter increment model: tree size, competition, and site.

Although equation forms differ due to differences in the dependent variable (i.e. diameter increment, basal area increment, or some transformation), the primary distinction for realized equations is the use of tree age as a covariate. For example, some studies show age-dependent diameter increment equations to be superior (e.g. Nyström and Kexi, 1997), while other studies found just the opposite (e.g. Tomé *et al.*, 2006). One shortcoming of equations that incorporate age is their limitation to even-aged stands. Furthermore, diameter increment studies that include age and DBH often produce parameter estimates for the two variables which result in them behaving like the ratio of DBH to age (e.g. Quicke *et al.*, 1994), which essentially is the average past diameter growth rate of the tree. While average past diameter growth rate might be highly correlated with diameter increment if the stand remains untreated, it can be poorly correlated with diameter increment after application of a treatment or occurrence of natural disturbance. In addition, age can only be used in species with well-defined rings, is often not collected in large-scale inventories, and is of limited value in multi-cohort or uneven-aged stands (Tomé *et al.*, 2006). Finally, several physiological studies suggest that tree growth is limited by size rather than age (Ryan *et al.*, 2006; Bond *et al.*, 2007).

Tree size is generally represented by DBH in most diameter increment models. Since the shape of diameter increment versus diameter is unimodal and has a maximum that is dependent on species, models have accounted for this peaking behavior in a variety of ways. Equations including DBH and DBH^2 (Monserud and Sterba, 1996), log(DBH) and DBH^2 (Wykoff, 1990; Hann *et al.*, 2006), $\log(DBH + k)$ and DBH^2 (Hann and Larsen, 1991), $\log(DBH + k)$ and DBH (Wykoff, 1990; Hann *et al.*, 2006), and DBH, DBH^2, and $1/DBH$ (Zhao *et al.*, 2004) all induce a peaking behavior. Although a logarithmic transformation of DBH is most commonly used in the relationship between tree size and increment, a square-root transformation might work better for species with slow initial increment (Hökkä *et al.*, 1997; Hökkä and Groot, 1999).

Crown attributes are measures of tree size commonly included in diameter increment equations. The ratio of crown length to total height (crown ratio) is the most commonly used variable because it reflects tree size and vigor, social position within the stand, and stand density (Wykoff, 1990). Crown ratio was shown to be an effective predictor of diameter increment for both deciduous and conifer species in southwestern Oregon (Hann and Hanus, 2002a), coastal Oregon and Washington (Hann *et al.*, 2003), California Sierra Nevadas (Dolph, 1988), Austria (Monserud and Sterba, 1996), and northern Rocky Mountains (Wykoff, 1990). Other studies have used alternative measures of crown, like crown spread ratio (ratio of crown width to tree height; Schröder *et al.*, 2002), crown class (Zhao *et al.*, 2004), and crown width (Ritchie and Hamann, 2008), but these variables are

Table 6.2 Equation forms and covariates for predicting realized diameter increment, used in several individual-tree models.

Model	Dependent variable	Equation form[a]	Source
—	BAI	$b_0 \cdot \left(e^{(b_1 \cdot \sqrt{BA})} \cdot e^{(b_2 \cdot BAL)} \cdot e^{\left(\left(b_3 \cdot \left(1 - e^{(b_4 \cdot DBH)}\right)\right) - b_5\right) \cdot AGE} \right)$	Quicke *et al.* (1994)
MGM	ΔDBH	$\left(\frac{b_0 + b_1 \cdot BA + b_2 \left(\frac{BA_{SP}}{BA}\right) + b_3 \cdot HT + b_4 \cdot \left(\frac{DBH}{\overline{DBH}}\right) + b_5 \cdot PI}{\left(b_0 + b_1 \cdot BA + b_2 \cdot \left(\frac{BA_{SP}}{BA}\right) + b_3 \cdot HT + b_4 \cdot \left(\frac{DBH}{\overline{DBH}}\right) + b_5 \cdot PI\right) - \left(b_6 - b_7 \cdot \sqrt{TPH}\right)} \right) \cdot \left(e^{\left(-\left(b_6 - b_7 \cdot \sqrt{TPH}\right) \cdot DBH\right)} - e^{\left(-\left(b_0 + b_1 \cdot BA + b_2 \cdot \left(\frac{BA_{SP}}{BA}\right) + b_3 \cdot HT + b_4 \cdot \left(\frac{DBH}{\overline{DBH}}\right) + b_5 \cdot PI\right) \cdot DBH\right)} \right)$	Huang and Titus (1995)
Prognosis	ΔDBH^2	$e^{\left(b_0 + b_1 \cdot \log(SI) + b_2 \cdot ELEV + b_3 \cdot \log(CCF) + b_4 \cdot \log(CR) + b_5 \cdot \log(DBH) + b_6 \cdot BA_{PCT}\right)}$	Stage (1973)

Prognosis	$\log(\Delta \text{DBH}^2)$	$b_0 + b_1 \cdot \log(\text{DBH}) + b_2 \cdot \text{DBH}^2 + b_3 \cdot (\text{SL} \cdot \cos(\text{ASP})) + b_4 \cdot (\text{SL} \cdot \sin(\text{ASP})) + b_5 \cdot \text{SL} + b_6 \cdot \text{SL}^2 + b_7 \cdot \text{ELEV} + b_8 \cdot \text{ELEV}^2 + b_9 \cdot \text{CR} + b_{10} \cdot \text{CR}^2 + b_{11} \cdot \frac{\text{BAL}}{\log(\text{DBH} + 1.0)} + b_{12} \cdot \text{CCF}$	Wykoff (1990)
SPS	$\Delta\text{DBH}/\Delta\text{HT}_{\text{DOM}}$	$b_0 \cdot \left(\frac{\text{CCF}}{100}\right)^{b_1} \cdot \left(1.0 - e^{\left(b_2 \cdot \left(\frac{\overline{\text{DBH}}}{\text{HT}_{\text{DOM}}}\right)\right)^{b_3}}\right)$	Arney (1985)

[a] Where BAI is basal area increment, DBH is tree diameter at breast height, BA is stand basal area, BAL is basal area in larger trees, AGE is total stand age, BA_{SP} is the basal area of a particular species, HT is total tree height, CR is crown ratio, BA_{PCT} is tree percentile, TPH is trees per hectare, PI is the productivity index of Huang and Titus (1993), $\overline{\text{DBH}}$ is the mean DBH in the stand, HT_{DOM} is stand dominant height, SI is site index, ELEV is elevation, CCF is crown competition factor, SL is percentage slope, ASP is site aspect, the b_is are parameters, and log is the natural logarithm.

generally not available from routine forest inventories. Tree size variables generally account for most of the variation explained by a diameter increment model. Monserud and Sterba (1996) found that DBH and crown ratio accounted for 14 to 47% of the variation in basal area increment for deciduous and conifer species in Austria, while the entire equation explained 20 to 63% of the variation.

Although DBH and crown ratio reflect effects of past competition to some degree, additional measures of both one- and two-sided competition (see Chapter 2) often are included in diameter increment equations. One-sided competition in the stand has been quantified with a range of measures, while a more limited range of variables has been used to represent two-sided competition. Stage (1973) originally accounted for one-sided competition by estimating a tree's percentile in the basal area distribution of trees, but this approach is only applicable in unmanaged or undisturbed stands (Wykoff, 1983; Schröder and von Gadow, 1999). Despite similar limitations, tree relative height (Arney, 1985; Yang *et al.*, 2009a) or relative diameter were used in the past.

Basal area in larger trees (BAL) is the most common means of quantifying one-sided competition since it is an absolute measure, easy to calculate, and well correlated with growth. Basal area in larger trees was shown to be an effective predictor of diameter increment for tropical species (Vanclay, 1991a), as well as conifer and deciduous species in southwestern Oregon (Hann and Hanus, 2002a), coastal Oregon and Washington (Hann *et al.*, 2003), California Sierra Nevadas (Dolph, 1988), Finland (Hökkä *et al.*, 1997), Austria (Monserud and Sterba, 1996), and northern Rocky Mountains (Wykoff, 1990). Various expressions of BAL are found to be more effective than BAL itself. For example, Hann and Hanus (2002a) and Wykoff (1990) found the ratio of BAL or BAL^2 to $\ln(DBH + k)$ to be a more effective predictor. However, when BAL is combined with stand-level covariates, mixed results are obtained. Kiernan *et al.* (2008) found that the ratio of BAL / BA, which is analogous to Stage's (1973) percentile in the basal area distribution, was inferior to BAL, while Schröder and von Gadow (1999), as well as Corral Rivas *et al.* (2005), indicate that multiplying BAL / BA by relative spacing was more effective than BAL.

One limitation of BAL is that it is a DBH-indexed variable that assumes that all species of a given DBH exert equal competitive influence. However, it is likely that some species with different crown structures or different values of HT for a given DBH would exert a greater influence than others with the same DBH. Consequently, alternative measures of one-sided competition have been explored for mixed-species stands. Two examples of measures found to work well in mixed-species stands are (1) crown competition factor in larger trees (CCFL), which is indexed by DBH; and (2) crown closure in higher trees (CCH), which is indexed by HT. Ritchie and Hann (1985) used CCFL, which accounts for species differences in maximum crown width, to model growth of Douglas-fir and grand fir in Oregon, and found it to be more effective than BAL; while Hann and Larsen (1991) and Biging and Dobbertin (1995) found relatively little gain using CCFL compared to BAL. Although not compared to CCFL, Bravo *et al.* (2001) found the HT-indexed variable CCH to be superior to BAL for predicting Douglas-fir diameter increment across a range of stand types.

Another HT-indexed variable used is CC_{66} (crown closure at 66% of subject tree height). However, comparisons between CCH and CC_{66} have varied, as Biging and Dobbertin (1995) found CC_{66} to be a more effective predictor of diameter increment than CCH for white fir, and just the opposite was found, by a small margin, for ponderosa pine. The advantage of CCH and CC_{66} is that they account for differences in vertical canopy profiles, which makes them effective measures of one-sided competition in mixed-species as well as uneven-aged stands. The primary disadvantages of these measures are that they require crown profile models for each species, and their application requires measurement of tree height and height to crown base on every sample tree (Bravo *et al.*, 2001). One alternative to these crown-based covariates is to separate BAL by species. Pukkala *et al.* (2009) used this approach in uneven-aged stands in Finland, and found it to be an effective technique, while Ritchie and Hann (1985) did not. However, additional work on various schemes of weighting species by their competitive ability is warranted.

Two-sided competition is commonly expressed by total stand basal area (BA) (Quicke *et al.*, 1994; Cao, 2000; Hann and Hanus, 2002a; Hann *et al.*, 2003; Zhao *et al.*, 2004). A square-root transformation of BA was found to work better than a logarithmic or untransformed expression (e.g. Hann and Hanus, 2002a). It is hypothesized that this transformation is more closely related with total sapwood basal area at breast height in the stand. For example, Gould and Harrington (2008) have shown that a direct connection exists between the amount of sapwood area in the stem and in the roots. Therefore, total sapwood basal area at breast height should be closely related to the amount of fine root biomass in the stand. BA is a commonly used measure of two-sided competition because it is easy to obtain and combines both tree size and density into a single measure, but it is often confounded with stand age and site quality (Curtis, 1970).

Using BA to represent stand species composition in models is often accomplished by including proportions of basal area occupied by selected species (Sterba *et al.*, 2002; Andreassen and Tomter, 2003; Zhao *et al.*, 2004). This approach for modeling the degree of species mixture is relatively simplistic, and its utility would likely be limited in stands of four or more species of varying shade tolerances. Although stand-level growth is related to stand structural indices (e.g. Lei *et al.*, 2009), relatively little work attempting to incorporate these types of measures into tree-level equations exists, and likely warrants further investigation.

Another widely used two-sided measure of competition is crown competition factor (CCF; see Chapter 2) because it is generally assumed to be independent of site and age. The measure is also effective because it can account for differences in species, which BA does not accomplish. CCF was used in diameter increment models for mixed-species stands in Austria (Monserud and Sterba, 1996) and the northern Rocky Mountains (Wykoff, 1990). Interestingly, Bravo *et al.* (2001) found that BA was superior to a crown-based indicator of two-sided competition (total crown surface area) for predicting Douglas-fir diameter increment across a range of conifer-dominated mixed-species stand types. Biging and Dobbertin (1995) found similar results, as more complex measures of stand density, such as stand density index, were better predictors of white fir diameter increment, while BA was just as effective as other predictors for ponderosa pine. Additional measures of stand competition used in diameter increment models include: number of trees per hectare (Andreassen and Tomter, 2003), basal area weighted mean diameter (Nyström and Kexi, 1997), and stand density index (Uzoh and Oliver, 2008).

Measures of site productivity often explain a limited amount of variation in diameter increment, but are generally included to give models some site specificity. For example, Monserud and Sterba (1996) found that various attributes of site did not explain more than 6% of the variation for several species in Austria. Site index is the most commonly used measure of site productivity in diameter increment models (Dolph, 1988; Hann and Hanus, 2002a; Hann *et al.*, 2003). For species missing a site index estimate, equations are developed to predict site index from other species in the stand (e.g. Dixon *et al.*, 2007). However, site index has multiple limitations (see Chapter 3) and is not appropriate in all regions, particularly those dominated by uneven-aged or mixed-species stands. Consequently, various alternatives to site index are commonly used.

Quicke *et al.* (1994) developed a regional diameter increment model for naturally regenerated, even-aged longleaf stands in the southern USA that did not explicitly characterize site differences and performed well across a range of stands. This finding is similar to the results of Deleuze *et al.* (2004) for Douglas-fir and Norway spruce in France, where site index was not a significant predictor of diameter increment. However, the Quicke *et al.* (1994) and Deleuze *et al.* (2004) models included a DBH-age modifier and dominant height, respectively, which would theoretically account for some site to site differences. Several studies show that additional site variables can be just as effective as site index (e.g. Wykoff and Monserud, 1988; Nyström and Kexi, 1997; Pokharel and Froese, 2009), while Hann and Larsen (1991) found site index to be the strongest productivity variable of the many they examined. Hann and Hanus (2002a) reported procedures for obtaining reasonable site index estimates in uneven-aged stands containing Douglas-fir, which Hann (1998) found could recover from early suppression.

Variables used instead of site index are generally of four types: habitat or forest type; physiographic; geographic; and soil based. Wykoff (1990) allowed the intercept of a diameter increment model to vary by six different habitat types common in the northern Rocky Mountains, which was just as effective as including site index (Wykoff and Monserud, 1988). This is similar to the findings of Pokharel and Froese (2009), where an indicator for forest type was included in their model. The limitation of this approach is that a consistent basis for determining habitat type is required, and interpretation can be difficult (e.g. Wykoff, 1990). Physiographic and topographic variables like elevation, slope, and aspect also are widely used (Stage, 1973; Wykoff, 1990; Monserud and Sterba, 1996). Most often the transformations and interactions for slope and aspect suggested by Stage (1976) are employed, as they provide more flexibility and biological interpretability. Stage and Salas (2007) presented an approach that accounts for interactions between slope, aspect, and elevation, but it has yet to be used in an individual-tree diameter increment model.

The limitation of topographic factors is that multiple parameters (e.g. three to eight) are needed to define proper behavior, their influences vary by species, and they are surrogates for climatic factors, which can create a complex pattern in prediction bias (e.g. Froese and Robinson, 2007). Geographic variables like latitude are used in several diameter increment models (Dolph, 1988; Andreassen and Tomter, 2003), both as a continuous and as a categorical variable. The limitation of this approach is that the model cannot be extrapolated beyond the range of the dataset used for parameterization. Rather than using latitude, Wykoff (1990) included indicators for different National Forests in his model to represent effects of geographic location, which was found to be ineffective for reducing prediction biases (e.g. Froese and Robinson, 2007).

These previous measures are almost all indirect measures of site productivity since they are proxies for other environmental factors that actually influence growth. Several attempts to include direct measures of site productivity in diameter increment models have been made, with varying degrees of success. Given that climate data are rarely available, most measures focus on soil attributes such as depth of the humus horizon (Monserud and Sterba, 1996), peat thickness (Hökkä and Groot, 1999), soil type (Sterba *et al.*, 2002), soil water-holding capacity (Ritchie and Hamann, 2008), effective soil depth (Schröder *et al.*, 2002), and the ratio of soil carbon to nitrogen (Laubhann *et al.*, 2009). However, Dolph (1988) found that several soil variables were ineffective at explaining variation in basal area growth for six mixed-conifer species in the Sierra Nevadas. The uneven success of soil-based covariates is likely due to their high variability within the stand.

Climatic variables, like mean temperature (Pokharel and Froese, 2009) and temperature sum (Nyström and Kexi, 1997), can be effective at explaining variation in diameter increment. However, the use of both soil and climatic variables assumes that this information is available to model users when applied to new areas of the population, which can be an important limitation. Finally, Hann and Larsen (1991) conducted an extensive comparison of alternative productivity measures, including elevation, latitude, annual rainfall, rainfall during the growing season, bedrock type, amount and size of coarse fragments in each soil horizon, abundance of roots in each soil horizon, water-holding capacity of each soil horizon, slope, aspect, vertical angles to the tops of ridges that might block the sun, predicted average monthly minimum and maximum temperatures, predicted solar irradiation (Kaufmann and Weatherred, 1982), and predicted net photosynthesis (Emmingham and Waring, 1977) at each site. They found that some of the alternatives were statistically significant, but that no combination of them explained more than three to four percent of the variation in the logarithm of diameter increment, and no more than two percent of the variation when used with site index (Hann and Larsen, 1991).

Additional factors that influence diameter increment and are included in varying degrees are damage and the influence of diseases. Hann and Hanus (2002a) examined the impacts of various damaging agents on predicted diameter increment in several conifer species of southwest Oregon, and found significant impacts from 14 of the 29 damaging agents considered. For example, they found that a dead or missing top caused reductions in five-year diameter increment that ranged from nearly 15% for Pacific madrone to more than 35% for grand/white firs (Hann and Hanus, 2002a). Long-term and extensive defoliation by foliar diseases or insects has significant influence on diameter increment. For example, Weiskittel and Maguire (2004) developed an individual-tree diameter increment model for Douglas-fir trees with varying degrees of Swiss needle cast disease, and found that five-year diameter increment was reduced up to 18% by the disease. Most individual-tree diameter increment models assume that trees are perfect specimens; whereas significant deviations from expected values occur when trees deviate from normal conditions.

A variety of approaches to predicting individual-tree diameter increment exist. This chapter explored the two most common approaches and the range of covariates generally considered important. Alternatives to modeling diameter increment directly include modeling it as a simple function of predicted height increment (Nunifu, 2009), or as a probability (Lowell and Mitchell, 1987; Vanclay, 1991b). The potential-multiplied-by-modifier approach to modeling diameter increment has a long history, but suffers from the inability to estimate parameters simultaneously, and estimating potential increment is challenging. Consequently, empirical model forms that predict realized diameter increment are more common and differ primarily in the covariates considered. The majority of equations include two expressions of DBH to induce a peaking behavior, BAL, a measure of two-sided competition, and site index. In general, these equations are just as effective as ones that are distance- or age-dependent. Although the dependent variable of diameter increment equations differs, including basal area increment or a relative measure of growth, modeling diameter increment directly is recommended because it avoids circularity, the potential of compounded error, and often performs just as well as other measures.

6.3 Height increment

Modeling height increment is generally much more difficult than diameter increment because of higher within-stand variability, a more limited number of remeasurements, and a closer connection to environmental factors rather than stand-level ones. Like diameter increment, two approaches have been most commonly used to directly model height increment: (1) potential multiplied by modifier and (2) realized.

Height increment can also be indirectly estimated by first predicting diameter increment and the resulting DBH at the end of the growth period. The starting and ending DBH values are then used in a static height-to-diameter equation to calculate the trees' heights at the start and end of the growth period. Height increment is calculated from the difference in the estimated heights. This indirect approach is used in several gap models (Bugmann, 2001), certain variants of the Forest Vegetation Simulator (Crookston and Dixon, 2005), and for some minor conifer and hardwood species in ORGANON (Hann, 2011), but has some major limitations, especially in multi-cohort stands, and stands with management interventions such as thinning.

6.3.1 Potential height increment equations with multiplicative modifiers

The potential-multiplied-by-modifier approach is more commonly used for predicting height increment than diameter increment (Hegyi, 1974; Arney, 1985; Burkhart *et al.*, 1987; Wensel *et al.*, 1987; Hann and Ritchie, 1988; Hann and Hanus, 2002b; Hann *et al.*, 2003; Weiskittel *et al.*, 2007a; Nunifu, 2009; Table 6.3). One reason for this is that commonly available dominant height equations can be used as estimates of potential height. Both age-dependent (e.g. Arney, 1985; Burkhart *et al.*, 1987) and age-independent (e.g. Wensel *et al.*, 1987; Hann and Ritchie, 1988) approaches to estimating potential height increment are used. The age-dependent approach calculates dominant height for the start and end of a growth period using a dominant-height increment equation (tree- or stand-level), and the difference is used to estimate potential height increment. Because the method assumes that all trees in a stand are of the same age, the resulting potential height increment is the same for all trees in the stand (given site and age). In addition, the method also assumes that trees below dominant height are shorter solely because of competition (Hann and Ritchie, 1988). As a result, the approach is applicable to even-aged, homogeneous stands only.

In the age-independent approach, estimates of a growth effective age (GEA) for each tree in a stand are obtained by solving, either algebraically or numerically, the dominant-height growth equation used to predict potential increment, so that age is expressed as a function of tree height and site index. Therefore, GEA is the age of a dominant tree with the same height and site index as the subject tree. The length of the growth period is then added to GEA, and future potential height of the tree at the end of the growth period is predicted from the dominant-height growth equation. Finally, present height is subtracted from future potential height in order to calculate potential height increment for that tree. This approach assumes that potential height increment of a tree is defined by the rate attainable by a dominant tree of the same height and that tree size is more important than age in predicting increment. Consequently, the approach has been applied to both even- and uneven-aged stands. Hann and Ritchie (1988) found that an age-independent estimate of potential height increment was more effective in even-aged stands than an age-dependent estimate, with a reduction in mean square error of nearly 10% when the former was used.

Dominant-height increment models for some species include corrections for density, as a means of removing the impact of density when estimating a stand's site index (e.g. Flewelling *et al.*, 2001; Weiskittel *et al.*, 2009a). Whether such corrections for

Table 6.3 Equation forms and covariates[a] for predicting potential height increment and associated multiplicative modifier equations used in several individual-tree models.

Model	Potential/average
MGM	$b_0 \cdot e^{\left(b_1 \cdot SI + b_2 \cdot \left(\frac{HT_{DOM} \cdot TPH}{10\,000}\right)\right)} \cdot HT^{b_3} \cdot e^{\left(-b_4 HT^2\right)}$
CACTOS	$\frac{b_4 \cdot \left(-\left(\frac{1}{b_2}\right) \cdot \log\left(1 - \left(\frac{HT - 1.3}{b_0 \cdot SI^{b_1}}\right)\right)^{\left(\frac{1}{b_3}\right)}\right)}{\left(1 + e^{(4 - b_5 \cdot CR)}\right)}$
FVS-NE	$b_0 + b_1 \cdot SI^{b_2}$
	$\cdot \left(1 - e^{\left(b_3 \cdot \left(\left(\frac{1}{b_3}\right) \cdot \log\left(1 - \left(\frac{HT - b_0}{b_1 \cdot SI^{b_2}}\right)\right)\right)^{\frac{1}{b_4 \cdot SI^{b_5}}}\right)}\right) \cdot b_4 \cdot SI^{b_5}$
SPS	$f(SP, SI, AGE_{BH})$

Where SP is species, SI is site index, AGE_{BH} is breast-height age, HT is total tree height, HT_{DOM} is dominant height, CCF is crown competition factor, BA is total basal area, TPH is trees per hectare, BAL is basal area in larger trees, BAL_{SW} is softwood species BAL, BAL_{HW} is hardwood species BAL, $\overline{HT}_{SP}$ is average height for the species, QMD is quadratic mean diameter, QMD_{SP} is the species quadratic mean diameter, CR is crown ratio, RHT is individual-tree relative height, the b_is and c_is are parameters, and log is the natural logarithm.

density should be included in the potential component or the modifier component of the height increment model has varied between studies. Wensel *et al.* (1987) used tree crown ratio to adjust potential height increment so that trees with a crown ratio less than 0.5 had a significant reduction in potential increment. They found this relationship was similar in

Modifier	Source
$e^{\left(-\left(c_0 \cdot BAL_{SW} + c_1 \cdot BAL_{HW} + c_2 \cdot \left(\frac{HT_{DOM} \cdot TPH}{10\,000}\right)\right)\right)}$, white spruce	Nunifu (2009)
$\left(\begin{array}{c} c_0 + c_1 \cdot \overline{HT}_{SP} + e^{(-(c_2 \cdot BAL^2))} \\ + c_3 \cdot \log\left(\frac{QMD_{SP}}{QMD}\right) + c_4 \cdot \log\left(\frac{HT_{DOM} \cdot TPH}{10\,000}\right) \end{array}\right)$, lodgepole pine	
$e^{\left(\begin{array}{c} -\left(c_0 \cdot \overline{HT}^2_{SP} + c_1 \cdot BAL + c_2 \cdot BAL_{SW}\right. \\ \left. + c_3 \cdot \frac{QMD_{SP}}{QMD} + c_4 \cdot \left(\frac{HT_{DOM} \cdot TPH}{10\,000}\right)\right) \end{array}\right)}$, trembling aspen	
$e^{(c_0 \cdot CC_{66}^{c_1} \cdot BA^{c_2})}$	Wensel *et al.* (1987)
$\left(1 - \left(\left(1 - e^{(-c_0 \cdot BAL)}\right) \cdot (1 - RHT)\right)\right) \cdot 0.8$	Dixon *et al.* (2007)
$\left(\frac{HT}{HT_{DOM}}\right)^{c_0} \cdot \left(1.0 - \left(\frac{CCF}{c_1}\right)^{c_2}\right)$	Arney (1985)

form and extent for six northern California conifer species with varying levels of shade tolerance (Wensel *et al.*, 1987). Likewise, Nunifu (2009) included crown ratio as well as an index for crowding in the stand to account for influences of stand density on dominant-height growth in his potential height increment equation.

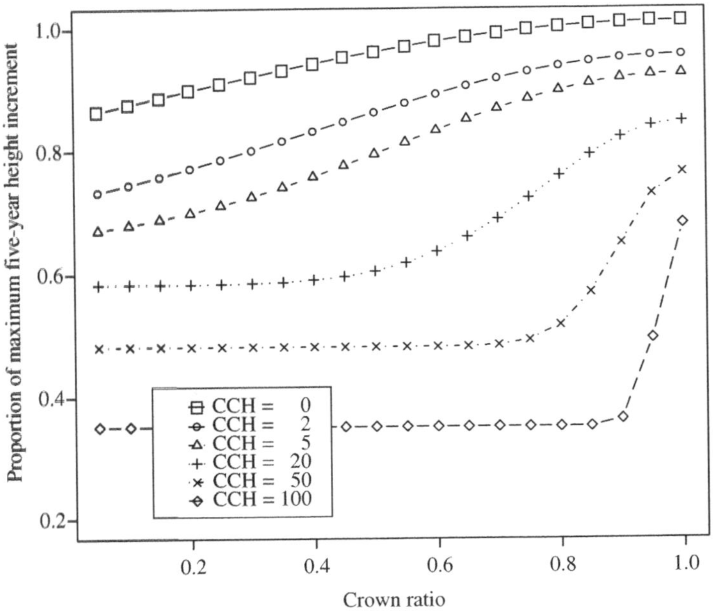

Figure 6.3 Modifier function for five-year potential height increment in coastal Douglas-fir that is dependent on crown ratio and a measure of one-sided competition, which in this example was crown closure in higher trees (CCH). Based on the equation of Hann *et al.* (2003).

Similar to potential height increment equations, both age-dependent and age-independent modifiers are used to adjust predicted potential height increment. For example, Arney (1985), as well as Ritchie and Hann (1986), used a transformation of site height, which was a function of stand age, in their modifier equations. This effectively limits the use of the modifier to even-aged stands, although the modifier of Ritchie and Hann's (1986) equation was also used in a two-storied stand with stand age being defined by the overstory. Hann and Ritchie (1988) found that a properly specified age-independent modifier performs just as well as, or in some cases better than, age-dependent modifiers. The form of the modifiers varies, but Hann and Ritchie (1988) suggested that they must include at least a measure of crown size and an index of one-sided competition like CCH (Figure 6.3).

The measure of crown size most commonly used is crown ratio (Wensel *et al.*, 1987; Hann and Hanus, 2002b; Hann *et al.*, 2006). The measures of one-sided competition have been based on relative differences in either height or DBH of the trees in the stand (see Chapter 2). The most frequently used height-based measures of one-sided competition have been CC_{66} and CCH. Ritchie and Hann (1990) and Hann and Hanus (2002b) found that CCH was a more effective predictor of height increment than CC_{66} for Douglas-fir, while Biging and Dobbertin (1995) found CCH to be superior for white fir. As a result, Hann and Hanus (2002b) used CCH for six conifer species in southwest Oregon. On the other hand, Wensel *et al.* (1987) used CC_{66} in their model for several northern

California conifers, and Biging and Dobbertin (1995) found CC_{66} was more effective than CCH for ponderosa pine.

Hann and Hanus (2002b) explored possible causes for the difference in performance between the modifier of Wensel *et al.* (1987), which used CC_{66}, and the modifier of Hann and Ritchie (1988) and Hann and Hanus (2002b), both of which used CCH. They attributed the superiority of the modifier with CCH primarily to differences in the effectiveness of the underlying model forms used with the two measures of one-sided competition. In contrast to these studies, Nunifu (2009) used the DBH-based BAL as a measure of one-sided competition for mixed-species stands in Alberta, but separated BAL by conifer and deciduous species, and he used the average height of a species in the stand as a covariate to account for vertical stratification in mixed-species stands.

Height increment is commonly assumed to be independent of stand density; therefore, measures of two-sided competition are incorporated into modifier functions to a lesser degree. Wensel *et al.* (1987) included stand basal area in their modifier equation, but found it was statistically significant for only one of the six species examined. Arney (1985) used CCF in his height increment model for Douglas-fir, but it did not have much influence except at very high densities. Nunifu (2009) included an index of crowding (i.e. dominant height times trees per hectare) in his height increment models for lodgepole pine, trembling aspen, and white spruce. In contrast, several studies have found that the inclusion of two-sided measures of competition did not improve the predictive behavior of their modifier functions (Hann and Ritchie, 1988; Hann and Hanus, 2002b; Hann *et al.*, 2003; Weiskittel *et al.*, 2007a). Perhaps the inclusion of crown ratio in these studies adequately characterized most of the impact of two-sided competition.

The potential-multiplied-by-modifier approach for height increment is easier to apply compared to diameter increment, as evidenced by its relatively wide use, but requires a species-specific dominant-height growth equation and an estimate of site index. The use of this method does facilitate in constraining predicted heights of the dominant trees in even-aged stands to be close to the values predicted by the dominant-height growth equation.

6.3.2 Realized height increment equations

Prediction of height increment with a realized approach parallels approaches used for estimating diameter increment directly (Table 6.4). For example, Hasenauer and Monserud (1997) used a height-increment model form similar to the diameter increment equation of Monserud and Sterba (1996), except that tree height squared was used instead of DBH^2. In the original Prognosis model, Stage (1975) predicted 10-year height increment as a function of both DBH and total tree height. Another key variable in the Stage (1975) model was predicted diameter increment, which was assumed to indirectly account for many site characteristics and stand conditions. Despite including

Table 6.4 Equation forms and covariates for predicting realized height increment used in several individual tree models.

Model	Dependent variable	Equation form[a]	Source
—	ΔHT	$b_0 \cdot \left(1-\left(1-\left(\frac{\mathrm{HT}}{b_0}\right)\right)\cdot e^{\left(-\left(b_1\cdot\left(b_2+\left(\frac{1}{\left(1+e^{\left(b_3+b_4\cdot \mathrm{BAL}+b_5\cdot\left(\frac{\mathrm{BAL}}{\sqrt{\mathrm{HT}}}\right)+b_6\cdot\left(\frac{\mathrm{UH}}{\mathrm{HT}}\right)\cdot(b_7\cdot \mathrm{UCOVER})\right)}\right)}\right)\right)\right)\right)}\cdot(t_2-t_1)\right)^{\left(\frac{1}{b_8}\right)}-\mathrm{HT}$	Salas *et al.* (2008)
MGM	ΔHT, aspen	$\left(\frac{b_0\cdot \mathrm{DBH}+b_1\cdot \mathrm{BA}+b_2\cdot \mathrm{PI}}{\left((b_0\cdot \mathrm{DBH}+b_1\cdot \mathrm{BA}+b_2\cdot \mathrm{PI})-\left(b_3+b_4\cdot \Delta\mathrm{DBH}+b_5\cdot\frac{\mathrm{BA_{SP}}}{\mathrm{BA}}\right)\right)}\right)\cdot\left(e^{\left(-\left(b_3+b_4\cdot\Delta\mathrm{DBH}+b_5\cdot\frac{\mathrm{BA_{SP}}}{\mathrm{BA}}\right)\cdot \mathrm{HT}\right)}-e^{(-(b_0\cdot \mathrm{DBH}+b_1\cdot \mathrm{BA}+b_2\cdot \mathrm{PI})\cdot \mathrm{HT})}\right)$	Huang and Titus (1999a)
Prognosis	log(ΔHT)	$\mathrm{HAB}+\mathrm{SP}+b_0\cdot\log(\mathrm{HT})+b_1\cdot\log(\mathrm{DBH})+b_2\cdot\log(\Delta\mathrm{DBH})+b_3\mathrm{HT}^2$	Stage (1975)
—	ΔHT	$e^{\left(b_0+b_1\cdot\log(\mathrm{HT})+b_2\cdot \mathrm{HT}+b_3\cdot\left(\frac{\mathrm{DBH}}{\mathrm{HT}}\right)+b_4\cdot \mathrm{BA}+b_5\cdot\left(\frac{\mathrm{BAL}^2}{\log(\mathrm{DBH})}\right)\right)}$	Nord-Larsen (2006)
—	ΔHT	$b_0\cdot \mathrm{HT}^{b_1}\cdot \mathrm{DBH}^{b_2}\cdot e^{\left(b_3\cdot \mathrm{AGE}+\frac{b_4\cdot \mathrm{BAL}}{\mathrm{BA}}\right)}$	Crecente-Campo *et al.* (2010)

[a] Where HT is total tree height, DBH is diameter at breast height, BAL is basal area in larger trees, AGE is total age, BA is total stand basal area, $\mathrm{BA_{SP}}$ is total basal area of a certain species, HAB is habitat type, SP is species, UH is height of the understory, UCOVER is understory percentage cover, PI is the productivity index of Huang and Titus (1993), t_i is the time of the ith measurement, the b_is are parameters, and log is the natural logarithm.

diameter increment, Dolph (1992) found it necessary to include both site index and total basal area in his equation for red fir in California and Oregon. Likewise, Uzoh and Oliver (2006) found slope, aspect, and elevation to be significant predictors in their linear model for ponderosa pine height increment, even after site index was included. These were all log-linear models, which introduces problems with log bias (Flewelling and Pienaar, 1981) and may reduce their ability to extrapolate accurately.

Huang and Titus (1999a) developed a nonlinear height increment model because they found log-linear equations inherently tended to underestimate height increment for younger trees. Their equation for mixed white spruce – aspen stands in Alberta was of a similar form to their diameter increment equation (Huang and Titus, 1995), and included tree height, tree diameter, tree diameter increment, stand density, tree social position, a measure of site productivity, and species composition (Huang and Titus, 1999a). Nonlinear height increment equations also were used by Zhang *et al.* (1996), Nothdurft *et al.* (2006), and Salas *et al.* (2008). These studies applied integrated forms of differential equations, which allowed for predictions over variable length growth periods. In addition, the nonlinear realized height increment equations offer constrained and robust predictions across a range of conditions.

Several models, like FVS (Crookston and Dixon, 2005), have separate height increment equations for small (<1.3 m) and large (> 1.3 m) trees. For example, Ritchie and Hamann (2008) predicted height increment of young Douglas-fir as a function of tree height, crown ratio, crown area of different species, and soil water-holding capacity. Crown area was a necessary covariate because young stands often have relatively small BA, and shrubs as well other types of vegetation can be significant competitors.

The use of two separate equations for modeling height increment often creates a discontinuity in predictions when trees approach the dividing point between small and large trees (Crookston and Dixon, 2005). Wykoff *et al.* (1982) used a weighted average between the small and large tree equations to smooth the transition between models. In contrast, the model of Salas *et al.* (2008) predicts growth across all tree sizes continuously by accounting for both overstory stand density and competing vegetation in the understory. The measures of competition in the Salas *et al.* (2008) model were BAL, height of the understory, and understory cover, while site productivity was represented by habitat class, elevation, slope, and aspect. Given the importance of representing juvenile growth dynamics (e.g. Golser and Hasenauer, 1997), further development of equations that work across a range of size classes is needed.

Modeling height increment is challenging because it is highly variable, due to both measurement errors and the strong influence of environmental factors. For example, Hasenauer and Monserud (1997) found measurement errors to be so high that the ability to model five-year height increment was limited. Salminen and Jalkanen (2005) observed a very strong correlation between mean July temperature of the previous year and tree height increment in Scots pine. Despites these challenges, success in modeling individual-tree height increment is achieved with a variety of approaches and covariates.

The potential-multiplied-by-modifier approach is quite effective when dominant-height growth equations are available, but the realized approach performs well. Like diameter increment equations, a common set of variables representing tree size, stand competition, and site productivity often are used in height increment equations.

6.4 Crown recession

Given the common use of crown ratio and other crown-related variables in increment models for both diameter and height, accurate representation of crown dynamics in individual-tree models is critical. The most common approach for representing the change in crown size through time is the use of a static equation (see Chapter 7; Hann and Hanus, 2004). These equations are used to estimate crown size at the start of a simulation growth period, then tree diameter, height, and expansion factors are updated with increment and mortality equations, and crown size is re-estimated with the same static equations used at the start of the simulation period. Differences in predictions provide an estimate of the amount of crown recession.

This approach is used in several growth and yield models including CACTOS (Krumland and Wensel, 1981), ORGANON (Hann, 2011), FVS (Crookston and Dixon, 2005), MOTTI (Hynynen *et al.*, 2002), and PROGNAUS (Ledermann, 2006). It has often been preferred because of the stochastic nature of crowns and the limited amount of remeasured height-to-crown base data (Hann and Hanus, 2004). The static approach generally provides unbiased estimates of crown recession (Liu *et al.*, 1995; Hann and Hanus, 2004), but equations that predict crown recession directly often significantly increase precision (Hann and Hanus, 2004). Typically, crown recession is predicted at two different resolutions: individual-tree, and branch levels.

6.4.1 Individual-tree crown recession models

Several different model forms and covariates are used to predict crown recession directly (Table 6.5). Both linear (Maguire and Hann, 1990a) and nonlinear equations (Hann and Hanus, 2004) have been used. Maguire and Hann (1990a) compared several linear and nonlinear model forms and found log-linear models to be superior, but Short and Burkhart (1992) suggested that nonlinear model forms be used because crown recession can be zero and logarithmic transformations would not be possible.

Krumland and Wensel (1981) and Hann and Hanus (2004) used a modified form of the nonlinear logistic equation that limited crown recession to between zero and the potential maximum crown recession as defined by the value of the numerator. Krumland and Wensel (1981) defined their potential maximum crown recession as coming from a part of the crown length existing at the start of the growth period and a part of the growth period's height increment. Hann and Hanus (2004) rejected the definition of Krumland and Wensel

Table 6.5 Equation forms and covariates for predicting crown recession, used in several individual-tree models.

Model	Equation form[a]	Source
CACTOS	$\frac{(b_0 \cdot \mathrm{CL} + b_1 \cdot \Delta\mathrm{HT})}{(1 + e^{(b_2 \cdot \mathrm{CC}_{\mathrm{HCB}})})}$	Krumland and Wensel (1981)
—	$e^{\left(b_0 + b_1 \cdot \log(\mathrm{HT}) + b_2 \cdot \sqrt{\mathrm{CR}} + b_3 \cdot \left(\frac{\mathrm{QMD}}{\mathrm{DBH}}\right) + b_4 \mathrm{AGE}_{\mathrm{BH}}\right)}$	Short and Burkhart (1992)
	$\frac{\mathrm{CR}^{\left(\frac{b_0}{\mathrm{HT}}\right)}}{(b_1 + b_2 \cdot \mathrm{CR} + b_3 \cdot \mathrm{CR}^2)}$, western white pine	
—	$(b_0 \cdot \mathrm{HT}) \cdot \left(1 - e^{\left(\left(\frac{b_1}{\mathrm{HT}}\right) \cdot \mathrm{CR}^{\left(\frac{b_2}{\mathrm{HT}}\right)}\right)}\right)$, ponderosa pine	Garber *et al.* (2008)
	$e^{(b_0 + b_1 \cdot \mathrm{HT} + b_2 \cdot \log(\mathrm{CR}) + b_3 \cdot (1.01 - \mathrm{CR}))}$, Douglas-fir	
—	$e^{(b_0 + b_1 \cdot \mathrm{CR} + b_2 \mathrm{AGE}_{\mathrm{BH}} + b_3 \cdot \log(\mathrm{HT}) + b_4 \cdot \log(\mathrm{CR}) + b_5 \cdot \log(\mathrm{CCF}) + b_6 \cdot \log(\Delta\mathrm{HT}))}$	Maguire and Hann (1990a)

[a] Where CL is crown length, HT is total tree height, $\mathrm{CC}_{\mathrm{HCB}}$ is crown closure at subject tree height to crown base, CR is crown ratio, QMD is quadratic mean diameter, DBH is diameter at breast height, $\mathrm{AGE}_{\mathrm{BH}}$ is age at breast height, CCF is crown competition factor, the b_is are parameters, and log is the natural logarithm.

(1981) because crown recession would first occur in the existing crown before moving into the new addition to crown length produced by height increment. Instead, they defined maximum possible crown recession as the sum of the crown length at the start of the growth period and the growth period's height increment.

Crown recession models have used both age-dependent (Maguire and Hann, 1990a; Short and Burkhart, 1992) and age-independent approaches (Hann and Hanus, 2004; Garber *et al.*, 2008), and a number of other covariates to form predictor variables. Hann and Hanus (2004) found an age-dependent model explained 51% of the variation, while an age-independent model explained 46% of the variation. The primary covariate used in most crown recession equations is crown ratio. For example, Garber *et al.* (2008) found that crown recession rates in Douglas-fir and western pine peaked between crown ratios of 0.7 to 0.9, while the rate for ponderosa pine monotonically increased with crown ratio (Figure 6.4). Other studies also found a peak in the rate of Douglas-fir crown recession (Maguire and Hann, 1990a; Hann and Hanus, 2004) and no peak for a pine species (Short and Burkhart, 1992). This may suggest that species shade tolerance may play a role in determining crown recession rates and needs to be appropriately accounted for in models. Maguire and Hann (1990a) found that height increment was an effective, positively correlated predictor of crown recession, but Garber *et al.* (2008) reported no significant relationship between crown recession and either diameter or height increment for three conifer species (Figure 6.4).

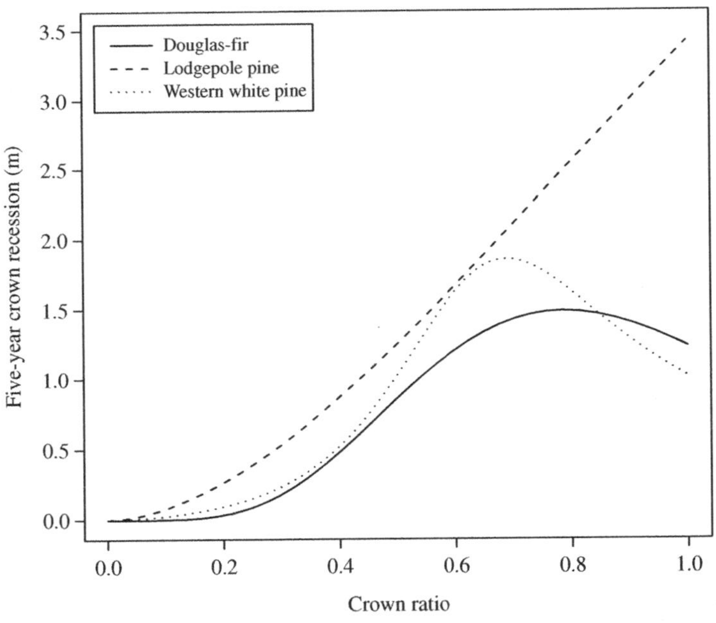

Figure 6.4 Predicted five-year crown recession over initial crown ratio for three conifer species in Idaho. Based on the equations of Garber *et al.* (2008).

Short and Burkhart (1992) reported that a distance-dependent measure of competition was superior to a distant-independent measure for predicting annual crown recession in thinned and unthinned loblolly pine, with reductions in mean square errors of nearly 16%. However, both models explained less than 20% of the original variation (Short and Burkhart, 1992). Crown competition factor was successfully used in models for Douglas-fir (Maguire and Hann, 1990a; Hann and Hanus, 2004), while Krumland and Wensel (1981) established that crown closure at the base of a subject tree's crown accounted for both density and tree social position in the stand. Hann and Hanus (2004) compared the model forms of Short and Burkhart (1992) and Krumland and Wensel (1981) to their own, and found that their model forms with CCF explained substantially more of the variation in Douglas-fir crown recession. Although originally used to model stand-level crown recession, the approach of Valentine *et al.* (1994) has been applied to individual trees in even-aged, single-species stands and shown to be effective.

Because of the predominance of years with little or no crown recession in their dataset, Garber *et al.* (2008) used a two-stage approach to improve their predictions of crown recession in three conifers. In the first stage, the probability of any crown recession occurring was predicted as a function of several tree-level covariates. If crown recession was predicted to occur, then the rate of crown recession was then predicted. Like crown recession rate, probability of crown recession occurring was strongly related to crown ratio and showed a peak between 0.7 to 0.9 for all three species (Garber *et al.*, 2008).

6.4.2 Branch-level crown recession models

Crown recession is caused by death of individual branches, and, theoretically, would be best simulated by modeling recession at this resolution. Weiskittel *et al.* (2007b) developed a mortality equation for individual branches based on branch size and location in crown, size of crown, and relative stand density. This equation was combined with several empirical crown reconstruction equations (Weiskittel *et al.*, 2007c) and annualized individual-tree increment equations (Weiskittel *et al.*, 2007a), which were used to simulate growth of intensively managed Douglas-fir plantations. The use of dynamic branch-level equations improved short-term predictions of crown recession by nearly 15%, which led to a slight improvement in stand-level volume growth predictions (Weiskittel *et al.*, 2007b). However, the results were not compared to predictions achieved with a tree-level, dynamic crown-recession model.

Additional branch-level models have been linked to individual-tree growth and yield models, but do not predict specific individual-branch mortality because it is either estimated by tree-level equations or assumed to occur randomly (Grace *et al.*, 1999; Mäkelä and Makinen, 2003; Trincado and Burkhart, 2009).

Table 6.6 Dynamic model forms and covariates used in two example distance-independent, individual-tree growth models.

Equation	Model form[a]	Reference
ORGANON		
Diameter increment (five-year)	$\Delta \mathrm{DBH} = e^{\left(b_0 + b_1 \cdot \log(\mathrm{DBH}+k_1) + b_2 \cdot \mathrm{DBH} + b_3 \cdot \log\left(\frac{(\mathrm{CR}+0.2)}{1.2}\right) + b_4 \cdot \log(\mathrm{SI}-1.37) + b_5 \cdot \left(\frac{\mathrm{BAL}}{\log(\mathrm{DBH}+k_2)}\right) + b_6 \cdot \sqrt{\mathrm{BA}}\right)}$	Hann *et al.* (2003, 2006); Hann and Hanus (2002a)
Height increment (five-year)	$\Delta \mathrm{HT} = \mathrm{POT}_{\Delta \mathrm{HT}} \cdot \left(b_0 \cdot \left(e^{(b_1 + b_2 \cdot \mathrm{CCH})} + \left(e^{(b_3 \cdot \sqrt{\mathrm{CCH}})} - e^{(b_1 + b_2 \cdot \mathrm{CCH})}\right) \cdot e^{\left(-(1.0-\mathrm{CR})^2 \cdot e^{(b_4 \cdot \sqrt{\mathrm{CCH}})}\right)}\right)\right)$	Hann *et al.* (2003, 2006); Hann and Hanus (2002b)
Crown recession (five-year)	$\Delta \mathrm{HCB} = \frac{(\mathrm{CL} + \widehat{\Delta \mathrm{HT}})^{b_0}}{\left(1.0 + e^{\left(b_1 + b_2 \cdot \log(\mathrm{CR}) + b_3 \cdot \mathrm{CR} + b_4 \cdot \mathrm{GEA} + b_5 \cdot \log(\mathrm{CCF}+1.0) + b_6 \cdot \left(\frac{\mathrm{CR}}{\mathrm{CCF}}\right)\right)}\right)}$	Hann and Hanus (2004)

PROGNAUS		
Diameter increment (five-year)	$\log(\text{BAL}) = b_0 + b_1 \cdot \log(\text{DBH}) + b_2 \cdot \text{DBH}^2 + b_3 \cdot \log(\text{CR}) + b_4 \cdot \text{BAL} + b_5 \cdot \text{CCF} + b_6 \cdot (\text{ELEV} - b_7)^2 + b_8 \cdot \text{SL}^2 + b_9 \cdot (\text{SL} \cdot \sin(\text{AZ})) + b_{10} \cdot (\text{SL} \cdot \cos(\text{AZ})) + b_{11} \cdot \text{HF} + b_{12} \cdot \text{HH} + b_{13} \cdot \text{DP} + b_{14} \cdot \text{M} + b_{15} \cdot \text{P} + b_{16} \cdot \text{S} + b_{17} \cdot \text{V} + b_{18} \cdot \text{GD}$	Monserud and Sterba (1996)
Height increment (five-year)	$\log(\Delta\text{HT}) = b_0 + b_1 \cdot \log(\text{DBH}) + b_2 \cdot \text{HT}^2 + b_3 \cdot \text{CR} + b_4 \cdot \text{CCF} + b_5 \cdot \text{BAL} + b_6 \cdot \text{ELEV} + b_7 \cdot \text{ELEV}^2 + b_8 \cdot \text{SL} + b_9 \cdot \text{SL}^2 + b_{10} \cdot (\text{SL} \cdot \cos(\text{AZ})) + b_{11} \cdot (\text{SL} \cdot \sin(\text{AZ}))$	Hasenauer and Monserud (1997)

[a] Where DBH is diameter at breast height, HT is total tree height, CR is crown ratio, CL is crown length, CCF is crown competition factor, BAL is basal area in larger trees, ELEV is elevation, SL is percentage slope, AZ is azimuth, BA is total basal area, SI is site index, the k_is are parameters that have been rounded to the number of decimal places used in measuring DBH,$\text{POT}_{\Delta\text{HT}}$ is potential height increment, HCB is height to crown base, GEA is growth effective age, CCH is crown closure in higher trees, $\Delta\hat{\text{HT}}$ is predicted height increment, HF is the depth of F humus horizon, HH is the depth of the H humus horizon, DP is an indicator variable that equals 1 if the soil depth is less than 30 cm, M is an indicator variable for soil moisture, P is an indicator characterizing slope position, S is an indicator for 26 soil groups, GD is an indicator for the 21 growth districts of Austria, V is an indicator for 21 vegetation types in Austria, and log is the natural logarithm.

Of the three attributes commonly predicted by an individual-tree model, crown recession is the most difficult to estimate accurately. This is partly because data for developing crown recession models are commonly quite limited, errors in measuring height to crown base often are high, and crown recession is inherently stochastic. However, significant gains in the accuracy of growth and yield projections are achieved with inclusion of dynamic crown-recession equations (Hann *et al.*, 2006; Weiskittel *et al.*, 2007b). Given that crown recession is also directly related to silvicultural treatments (Liu *et al.*, 1995), development of dynamic crown-recession equations, and their inclusion in more growth and yield models is warranted. In addition, more attention to crown recession patterns in hardwood species is needed since some research has indicated it to be particularly important from a physiological perspective (e.g. Ilomäki *et al.*, 2003).

6.5 Summary

Individual-tree models simulate development of trees with a set of equations that predict changes in diameter, height, and sometimes the crown. Equations are often species specific, vary greatly in form, and contain an array of covariates representing tree size, competition, and site productivity (Table 6.6). The potential-multiplied-by-modifier approach is theoretically appealing because of its constrained nature, but it often is difficult to estimate a potential growth rate for tree diameter increment. For height increment, the potential-multiplied-by-modifier approach generally requires a dominant-height growth equation that is a function of site index. However, practical differences between the potential-multiplied-by-modifier and realized approaches are likely small if the models are carefully crafted, and evaluated with quality data.

The use of age in growth equations is debated, but age-independent equations often perform just as well as age-dependent equations because of the high correlation between size and increment. Crown ratio is widely used in diameter, height, and crown increment equations emphasizing the ability of crowns to reflect both tree vigor and degree of local competition. Although few models have incorporated dynamic crown-recession equations with promising results, this finding suggests that many models would likely benefit from improved predictions of crown recession. Future work should focus on development of unified growth equations that work across species.

7

Individual-tree static equations

7.1 Introduction

In contrast to the dynamic equations discussed in Chapters 6 and 8, static equations predict the relationship between two or more attributes of an individual tree at one point in time. Static equations are used widely in growth and yield models to impute missing values prior to simulations, predict hard-to-measure attributes, and in some cases, estimate growth. For example, forest inventories rarely measure all sample trees for total height and, as a result, a static equation is needed to fill in the missing heights prior to the beginning of a growth simulation.

Static equations can take many forms. One of most widely used forms is the simple allometric or power function (Kaitaniemi and Lintunen, 2008), which expresses the relationship as $Y = aX^b$, with a and b being parameters. This form is used in ecology and several process-based growth models. More complex static equation forms also are used. This chapter outlines the important model forms and uses of individual-tree static equations in growth and yield models. The static models covered in this chapter include (1) total height; (2) crown length; (3) crown width and profile; (4) stem volume and taper; and (5) biomass. In addition, the chapter will also discuss opportunities to calibrate regional static equations to the stand-level.

7.2 Total height

Static equations to predict total tree height are quite common and take multiple forms (see Huang *et al.*, 1992). While many of the total height equations can be parameterized at the stand-level, this section will concentrate on total height equations parameterized at

---

Forest Growth and Yield Modeling, First Edition. Aaron R. Weiskittel, David W. Hann, John A. Kershaw, Jr. and Jerome K. Vanclay.

Table 7.1 Model forms commonly used for imputing total tree height.

Model	Equation form	Examples
Chapman–Richards	$1.3 + b_0 \cdot \left(1 - e^{(-b_1 \cdot \mathrm{DBH})}\right)^{b_2}$	Peng *et al.* (2001); Temesgen *et al.* (2007)
Curtis (1967)	$1.3 + b_0 \cdot e^{\left(-b_1 \cdot \mathrm{DBH}^{b_2}\right)}$	Arney (1985)
Larsen and Hann (1987)	$1.3 + e^{\left(b_0 + b_1 \cdot \mathrm{DBH}^{b_2}\right)}$	Hanus *et al.* (1999b)
Weibull	$1.3 + b_0 \cdot \left(1 - e^{\left(b_1 \cdot \mathrm{DBH}^{b_2}\right)}\right)$	Huang *et al.* (1992)
Wykoff *et al.* (1982)	$1.3 + e^{\left(b_0 + \frac{b_1}{(1+\mathrm{DBH})}\right)}$	Robinson and Wykoff (2004)

the region level. Regional equations can be calibrated at the stand-level either indirectly by incorporating appropriate stand attributes directly into the regional equation, or directly by scaling the regional equation using a subsample of trees from the targeted stand (e.g. Gilmore, 2001).

Although total tree height is strongly correlated with tree diameter at breast height (DBH), this relationship varies by species and stand conditions, so additional covariates are commonly included. Huang *et al.* (2000) found that over 27 different model forms were used to estimate height, with some of the most common model forms being the Chapman–Richards, Curtis (1967), Weibull, and Wykoff *et al.* (1982) equations (Table 7.1). Comparisons of various equations across multiple species show that most equations predict with a similar degree of precision when extensive data are available (Huang *et al.*, 1992; Temesgen and von Gadow, 2004).

Significant differences between model forms can exist, however, when limited data are available. For example, Temesgen and von Gadow (2004) found that the percentage difference in root mean square error between the best and worst equation for five commonly used model forms varied from 5 to 33%. Huang *et al.* (1992) found in their analysis that the Chapman–Richards, Weibull, and a modified logistic-type function were consistently among the best performing models because they were flexible, able to assume a variety of shapes, and extrapolated well. They did indicate that the Chapman–Richards function can approach the asymptote too quickly when there is a weak relationship between tree height and diameter (Huang *et al.*, 1992), which can be quite common in hardwood species. Both Hanus *et al.* (1999a) and Temesgen *et al.* (2007) found that the model form of Larsen and Hann (1987), which lacked pronounced asymptotic behavior, outperformed the Chapman–Richards equation when only DBH was used as a predictor variable in a region equation.

However, Temesgen *et al.* (2007) also found that the Chapman–Richards form became superior to the model form of Larsen and Hann (1987) when stand-level predictor variables were added to the equation. This would indicate that pronounced asymptotic behavior of height over DBH may be a more common feature of stands, particularly if they are even-aged. In addition to asymptotic behavior, most height-to-diameter equation forms are also constrained to predict a height of 1.3 or 1.37 (i.e. breast height) when DBH is equal to zero, but this constraint may cause poorer model performance across the full range of DBH (Newton and Amponsah, 2007).

Hanus *et al.* (1999b) found that DBH alone explained between 36 and 83% of the original variation for several conifer and hardwood species in southwestern Oregon. Hardwood heights tend to be harder to predict because of the lack of a true leader and the difficulty of measuring hardwood heights accurately (e.g. Kitahara *et al.*, 2010). In addition to DBH, several other covariates are commonly used to improve tree-height static equations. Basal area in larger trees (Temesgen and von Gadow, 2004) and crown competition factor in larger trees (Temesgen *et al.*, 2007) have been incorporated at the tree-level.

Stand-level variables are more varied and include: basal area (Huang and Titus, 1994; Temesgen *et al.*, 2007), quadratic mean diameter (Sharma and Parton, 2007; Vargas-Larreta *et al.*, 2009), number of trees (Sharma and Zhang, 2004; Saunders and Wagner, 2008), and relative stand density index (Newton and Amponsah, 2007). Other less used measures are maximum DBH on a plot (Zhang *et al.*, 1997a; Soares and Tomé, 2002), percentage of trees smaller than the basal area median diameter (Calama and Montero, 2004), and the difference between the 90th and 10th percentile in the diameter distribution (Lappi, 1997). Several studies have used stand age (Curtis, 1967; Soares and Tomé, 2002; Lopéz Sanchez *et al.*, 2003). Studies conducted in plantations indicate that inclusion of stand age and top height can adequately characterize both thinned and unthinned stands without the need to incorporate additional thinning descriptors in the equation (Zhang *et al.*, 1997a; Castedo Dorado *et al.*, 2006). Inclusion of stand age is primarily limited to even-aged stands due to the limited explanatory power of age in uneven-aged stands (e.g. Huang *et al.*, 1992).

Site index has also been included in several tree-height static equations (Ek *et al.*, 1984; Larsen and Hann, 1987; Wang and Hann, 1988), but others found that it did not improve predictions (e.g. Sharma and Zhang, 2004). Instead of site index, several researchers found stand dominant height to account for both site productivity and stand structure (Newton and Amponsah, 2007; Sharma and Parton, 2007; Kershaw *et al.*, 2008; Vargas-Larreta *et al.*, 2009; Figure 7.1). Inclusion of both average dominant diameter and height has shown significant improvements in predictions in even-aged stands (Krumland and Wensel, 1988; Hanus *et al.*, 1999a; 1999b; Castedo Dorado *et al.*, 2006).

Instead of a dominant height or diameter, Huang and Titus (1994) used a site productivity index based on the relationship between DBH and height. They also included a measure of species composition in their model (Huang and Titus, 1994). Other researchers found tree height to be significantly influenced by latitude/longitude (Russell *et al.*, 2010), elevation

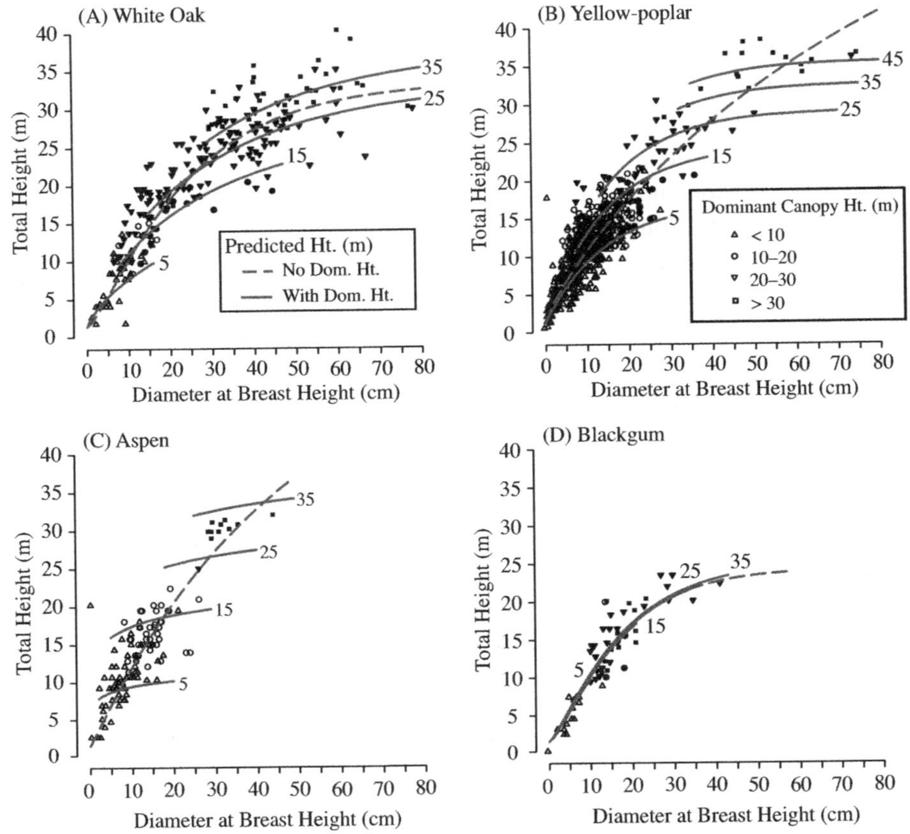

Figure 7.1 Individual-tree diameter-to-height curves, with (solid line) and without (dashed line) the inclusion of stand dominant height in the equation for white oak (A), yellow-poplar (B), aspen (C), and blackgum (D) for hardwood-dominated stands in Indiana. Closed symbols are FIA data, and open symbols are Hoosier National Forest regenerating clearcut data. Based on the equations of Kershaw *et al.* (2008).

(Hynynen *et al.*, 2002), and ecoregion (Garman *et al.*, 1995; Huang *et al.*, 2000). One attribute that most tree static equations do not generally account for, but which could have a significant influence, is tree damage. Hanus *et al.* (1999b) found that the percentage of damaged trees in their extensive dataset ranged from 33 to 64%, which resulted in biases of up to 27% for conifers and 52% for hardwoods if not corrected.

In assessing various combinations of covariates, Temesgen *et al.* (2007) found that crown competition factor in larger trees instead of basal area in larger trees, and stand basal area instead of stand crown competition factor were the best two covariates for several species in southwestern Oregon. Others found that models which included both number of trees and basal area were superior to those that just included one or the other (Temesgen and von Gadow, 2004; Sharma and Parton, 2007), while Newton and

Amponsah (2007) suggested that relative density was better because it integrated both measures into a single value and avoided issues with multi-collinearity. All of these studies clearly indicate that the incorporation of additional predictor variables into a prediction equation of total height does generally explain more variation than using just DBH alone. In addition, a number of these variables provide an indirect means of localizing a regional equation to specific stands. However, there are limitations imposed with the incorporation of a number of the additional predictor variables. For example, the predictive improvement of including density and tree position variables can lead to inaccurate predictions following silvicultural activities such as thinning.

Constructing a well-behaved tree-height static equation requires selecting an appropriate model form and an extensive dataset that covers a range of stand conditions and individual tree sizes. It has been found that including national and state champion trees to the modeling dataset significantly improved the regional equations' predictive power. For example, Bragg (2008) reported that an existing static equation for white pine consistently underpredicted heights by 20–30% for mature trees, but refitting the equation with a state champion tree removed this bias.

Various statistical methods are used to estimate parameters of tree-height static equations. Parameter estimation techniques include: (1) ordinary least squares (OLS) for either linear or linearized-nonlinear forms (Curtis, 1967; Arabatzis and Burkhart, 1992); (2) OLS nonlinear regression (Krumland and Wensel, 1988; Zhang *et al.*, 1997a; Temesgen and von Gadow, 2004); (3) weighted least squares nonlinear regression (Larsen and Hann, 1987; Huang *et al.*, 1992; Hanus *et al.*, 1999a; 1999b; Huang *et al.*, 2000); (4) mixed effects linear regression (Lappi, 1991a, 1997; Robinson and Wykoff, 2004); and (5) mixed effects nonlinear regression (Calama and Montero, 2004; Dorado *et al.*, 2006; Temesgen *et al.*, 2008). Mixed effects regression is becoming the predominant method for obtaining parameter estimates because it accounts for the nested nature of most data and can be used to localize equations (see below). Equation weighting factors vary as well. Larsen and Hann (1987) and Hanus *et al.* (1999a; 1999b) used $\frac{1}{DBH}$, while Huang *et al.* (2000) found that the inverse of predicted height raised by $\frac{3}{2}$ worked best in their analysis.

7.3 Crown length

Tree growth is strongly linked to crown size, which is often expressed as the ratio of crown length (CL) to total tree height, or crown ratio (CR). These variables are therefore commonly included in several of the key equations used in growth and yield models. However, crown measurements are less common than observations of total tree height. Regional static crown length equations are then used to fill in missing values prior to simulations, and they are quite often used to estimate crown recession during simulations. Although less attention has been given to developing crown length equations compared to total tree-height equations, a variety of approaches have been explored.

Crown length generally is estimated in one of two ways: (1) total tree height minus height to crown base (HCB); or (2) total tree height times CR. Modelers use both of these approaches. For example, the Forest Vegetation Simulator (FVS) relies on direct predictions of CR (Crookston and Dixon, 2005), while ORGANON estimates CR from predictions of HCB (Hann, 2011). Although CR is more commonly modeled (Belcher *et al.*, 1982; Wykoff *et al.*, 1982; Hynynen, 1995a; Hasenauer and Monserud, 1996; Soares and Tomé, 2001), Hann and Hanus (2004) found that static HCB equations produced more precise indirect predictions of crown recession than static CR equations. Several forest inventories, such as the US Forest Service Forest Inventory Analysis (FIA), make direct estimates of CR to the nearest 10%, which may explain why CR models are relatively more common.

A properly formulated CR model should be constrained to give predictions between 0 and 1, while a HCB equation should give predictions that do not exceed total tree height. Consequently, the most common model form used to model crown ratio and/or HCB has been the logistic equation, since it can be constrained to an asymptote of 1 or total tree height (Ritchie and Hann, 1987; Hasenauer and Monserud, 1996; Hanus *et al.*, 2000; Temesgen *et al.*, 2005; Ducey, 2009b). Other model forms used include the exponential (Hatch, 1980; Dyer and Burkhart, 1987), Chapman–Richards (Soares and Tomé, 2001), and Weibull (Dixon, 1985). In comparing different model forms for CR, Soares and Tomé (2001) found that the Chapman–Richards model form was superior, but differences in mean absolute bias between the equations were minimal.

Dell (1979) assumed that tree CR equaled average stand CR when tree DBH was equal to stand mean DBH, and, as a result, an increase in DBH implied an increase in CR. Dyer and Burkhart (1987) used long-term data to examine these assumptions and found them to be incorrect. Like static height equations, tree size is generally an effective predictor variable in either HCB or CR equations. Tree total height often is a more effective predictor (Van Deusen and Biging, 1985; Ritchie and Hann, 1987; Zumrawi and Hann, 1989; Hanus *et al.*, 2000; Hann *et al.*, 2003; Hann and Weiskittel, 2010), but tree DBH also is used (Holdaway, 1986; Soares and Tomé, 2001). Some models have included both DBH and total tree height (Hatch, 1980; Wykoff *et al.*, 1982; Temesgen *et al.*, 2005), while others have included the ratio of DBH to total tree height (Van Deusen and Biging, 1985; Dyer and Burkhart, 1987; Ritchie and Hann, 1987; Zumrawi and Hann, 1989; Hanus *et al.*, 2000; Hann *et al.*, 2003) or vice versa (Hasenauer and Monserud, 1996; Temesgen *et al.*, 2005). Both of the latter ratios are indicators of tree form (Hann *et al.*, 1987) and so is CR (Larson, 1963). Therefore, a tree with a large DBH for a given height has a greater likelihood of possessing a large CR than a tree with a small DBH for a given height. Tree age was shown to be an effective predictor in even-aged, single-species plantations (Dyer and Burkhart, 1987; Soares and Tomé, 2001), but not useful in multi-species and multi-layered stands (e.g. Temesgen *et al.*, 2005). Ducey (2009b) found that only the ratio of DBH to quadratic mean diameter (the diameter of the tree of average basal area) was significant in his crown ratio model for white pine.

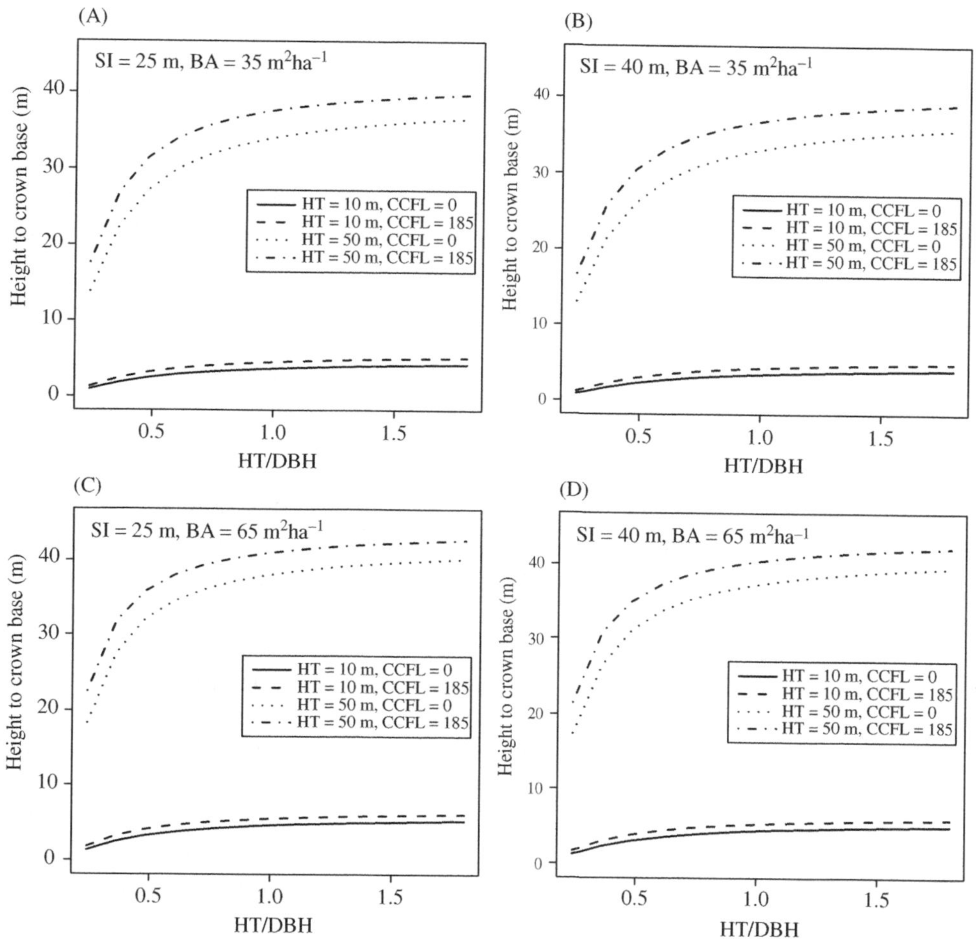

Figure 7.2 Coastal Douglas-fir height to crown base vs. the ratio of tree height to diameter for varying levels of crown competition factor in larger trees (CCFL), site index (SI) and total basal area (BA). Based on the equations of Hann *et al.* (2003).

Measures of competition are of equal importance as tree size in CR or HCB equations (Hasenauer and Monserud, 1996; Temesgen *et al.*, 2005; Figure 7.2). One-sided measures of competition used include basal area in larger trees (Temesgen *et al.*, 2005), crown competition factor in larger trees (Ritchie and Hann, 1987; Zumrawi and Hann, 1989; Hanus *et al.*, 2000; Hann *et al.*, 2003), and percentile in the basal area distribution (Hatch, 1980). Stand-level variables include crown competition factor (Hatch, 1980; Hasenauer and Monserud, 1996; Temesgen *et al.*, 2005), total basal area (Holdaway, 1986; Ritchie and Hann, 1987; Hanus *et al.*, 2000; Hann *et al.*, 2003), relative spacing (Ducey, 2009b), and number of trees (Soares and Tomé, 2001).

Hynynen *et al.* (2002) found stand origin (i.e. natural vs. planted) to have an influence, with plantations tending to have longer crowns, while Hanus *et al.* (2000) found that the product of the average height of the 12.4 largest-diameter trees per hectare multiplied by their average diameter was effective for characterizing the influence of old-growth stand structure on HCB. In addition, Hanus *et al.* (2000) found that a sampling point estimate of crown competition factor in larger trees rather than a stand-level estimate provided a modest improvement for the majority of species in their analysis, particularly for the hardwood species.

Variables representing site productivity were found to account for 5 to 23% of the original variation in CR (Temesgen *et al.*, 2005). Site index was included in several HCB equations (Ritchie and Hann, 1987; Hanus *et al.*, 2000; Hann *et al.*, 2003; Hann and Hanus, 2004). Others have avoided the use of site index by using combinations of aspect, slope, and elevation (Hasenauer and Monserud, 1996; Temesgen *et al.*, 2005) or habitat type (Hatch, 1980). Like tree-height static equations, mean dominant diameter and height are effective predictors (Hynynen, 1995a; Soares and Tomé, 2001). Temperature sum and latitude also are found to be influential predictors when the equation covers a broad geographic region (Hynynen *et al.*, 2002). Tree damaging agents, particularly suppression, can have a significant influence on HCB, even after accounting for tree size and competition (Hanus *et al.*, 2000).

In contrast to estimating CR directly for individual trees, Dixon (1985) proposed first estimating a stand-level mean CR based on relative stand density index and species-specific parameters. A within-stand distribution of CR is then calculated based on the stand mean CR, species-specific parameters, and the Weibull probability density function. A CR value is assigned to each tree based on the tree rank in the stand basal area distribution. The limitations of this approach are that it is most applicable to single-species stands and that relative stand density index can be difficult to calculate in uneven-aged stands. However, the approach can perform well (e.g. Leites *et al.*, 2009).

CR and HCB are generally much harder to predict than total tree height, particularly for hardwood species (Hasenauer and Monserud, 1996). The equations of Hanus *et al.* (2000), for example, explained between 9 and 89% of the weighted variation in HCB for undamaged trees of sixteen species in southwest Oregon, with most species having less than 50% of the weighted variation explained by the equation. In addition, significant biases in predicting CR or HCB can occur, which can have important implications for long-term growth projections (Leites *et al.*, 2009). One reason for this is the general difficulty in measuring crowns and the varying definitions used for the base of the live crown. For example, some protocols have field crews subjectively and visually rearrange the live crown branches to obtain a CR value, which can be very challenging when there is an uneven crown. Equations have been developed to convert different measurements of CR (e.g. Monleon *et al.*, 2004; Randolph, 2010), but the measurement error is still high, and a more objective measurement, like height to the lowest live branch, may be more precise.

Regional static equations can be locally calibrated with the same techniques used for total tree height when a subsample of measurements of HCB or CR exists, but most growth models do not have this option. There has been substantially more research focused on imputing missing tree heights than missing HCB or CR values. Given the greater variability and error associated with the measurement of these variables, a larger sample size than those needed for locally calibrating a tree height equation is likely required. Using imputed rather measured CR can, however, result in significant growth prediction biases. For example, Leites *et al.* (2009) found that using imputed rather than measured CR increased the root mean square error by 17 to 49% in predicting 10-year diameter growth. The prediction error was particularly high in trees with larger crowns (Leites *et al.*, 2009). Despite the difficulty and expense of making the measurement, this result underscores its importance.

7.4 Crown width and profile

Several key variables used in growth and yield models rely on estimates of crown width. For example, crown competition factor (Krajicek *et al.*, 1961) requires an estimate of maximum crown width (MCW) for all trees in a stand. Measurements of crown width are much less common than measurements of HT and CL and, as a result, estimation is based almost solely on regional static equations. At least three types of crown width are predicted in these regional equations. A MCW equation predicts the largest width reached by an open-grown, competition-free tree, while a largest crown width (LCW) equation predicts the widest crown diameter or radius for a stand-grown tree. A crown profile equation predicts the width of the crown at any point within a tree's crown.

Maximum crown width (Krajicek *et al.*, 1961; Vezina, 1962; Ek, 1974b; Paine and Hann, 1982; Farr *et al.*, 1989; Hasenauer, 1997) and LCW (Moeur, 1981; Hann, 1997; Bechtold, 2003; 2004) equations exist for several species. Maximum crown width equations are generally a linear (e.g. Krajicek *et al.*, 1961; Vezina, 1962), curvilinear (e.g. Paine and Hann, 1982), or nonlinear (e.g. Ek, 1974b; Farr *et al.*, 1989) function of DBH, which is generally more effective than total tree height (Hasenauer, 1997). Although no significant differences between linear and nonlinear model forms for predicting MCW are reported (e.g. Leech, 1984), a curvilinear or nonlinear model form is more biologically logical for trees with very large DBH.

There are two approaches generally used for collecting MCW modeling data. The first approach involves finding and measuring trees that meet the definition of an open-grown tree, and it often takes extensive travelling and effort in order to find such trees. Paine and Hann (1982) defined an open-grown tree as one that: (1) was free of present or past competition; (2) had limbs extending to the ground; (3) had lowest limbs as long or longer than limbs above; (4) had no forking in the lower 75% of the bole; and (5) had no evidence of pruning, disease, or other damages. The second approach involves measuring LCW on a large sample of trees in planting density studies and then selecting a subsample of these trees as being MCW trees. For example, Condés and Sterba (2005) used simple linear regression to fit the log of LCW as a function of log of DBH, and the log of total height as a function of log of DBH. MCW trees were then defined as trees whose residuals fell in both the upper 90th percentile of log of LCW and the lower 10th percentile of log of total height.

Due to genetics and environmental factors, significant geographic variation in the parameters of a MCW equation often exist for a single species (e.g. Paine and Hann, 1982). Hasenauer (1997) also found that the relationship between MCW and DBH was influenced by site factors such as elevation, aspect, and slope.

A variety of variables have been used in LCW equations, including predicted MCW from DBH (Hann, 1997), total tree height (Smith and Bailey, 1964), crown ratio (Hann, 1997; Bechtold, 2003; 2004), crown length (Moeur, 1981; Hann, 1997), crown class (Gill *et al.*, 2000), ratio of tree height to DBH (Hann, 1997), and stand basal area (Bragg, 2001b). Generally, DBH is effective at capturing most variation in LCW (Gill *et al.*, 2000). Bechtold (2003) found that DBH explained between 9 and 88% of the variation, with a mean of 46%, for 87 tree species in the eastern USA; while including crown ratio increased the mean to 52%. Bragg (2001b) reported that the inclusion of basal area significantly improved model fit for the majority of species examined; however, the adjusted R^2 only increased by 2.5% on average.

Like MCW, LCW also varies geographically. Bechtold (2003) discovered that a bioclimatic index was statistically significant in over one-third of the species in his analysis, but the inclusion of the index only increased the R^2 by less than 1% in most cases. Bechtold (2004) included both DBH and DBH^2 in the equation, but the resulting predictions of LCW for the majority of the species with the two predictor variables peaked and then declined within the range of DBH found in the modeling data.

It is important to note that, although LCW equations predict a mean value, crown widths often are highly eccentric because of differences in local competition (e.g. Brisson, 2001). Thus, differences in LCW parameters for a given species can arise not only from differences in geographic location, but also from whether they were systematically sampled (e.g. Gill *et al.*, 2000) or selectively chosen (e.g. Bragg, 2001b). Despite trying to sample only "idealized" crowns, Bragg (2001b) found that the majority of his sample was dominated by somewhat elliptical crowns. Hann (1997) made two measurements of LCW, one for the largest value of LCW and the other at right angles to the first, and then used the geometric average of the two in order to better characterize the LCW of elliptical crowns.

Crown profile equations are necessary for estimating crown volume, portraying trees properly in visualization programs, and calculating some crown-based measures of competition. Both a direct and an indirect approach have been used to model crown profile. Direct characterization uses deterministic or stochastic models to predict crown width (radius or area) from tree attributes; while indirect characterization predicts the size and orientation attributes of individual branches and computes crown width based on trigonometric relationships. The direct characterization has been the predominant form of predicting crown profile for a variety of species (Ritchie and Hann, 1985; Nepal *et al.*, 1996; Baldwin and Peterson, 1997; Biging and Gill, 1997; Hann, 1999; Hann and Hanus, 2001; Marshall *et al.*, 2003; Rautiainen and Stenberg, 2005; Rautiainen *et al.*, 2008; Crecente-Campo *et al.*, 2009a), but the indirect approach is used for several species (Cluzeau *et al.*, 1994; Deleuze *et al.*, 1996; Roeh and Maguire, 1997).

The indirect approach has the advantage of providing realistic depictions of crown profile and avoiding difficulties associated with making direct measurements of crown profile in the field (Roeh and Maguire, 1997). However, this approach requires data not commonly available for most species and which is very tedious to collect. The use of photographic techniques (Gill and Biging, 2002a), lasers (Biging and Gill, 1997), and a crown window based on similar triangles (Crecente-Campo *et al.*, 2009a) has made making direct measurements of crown profile much easier compared to felling a tree, sectioning it, and standing each section up for measurement.

Both continuous (Baldwin and Peterson, 1997) and segmented (Hann, 1999) approaches are used for direct characterization of crown profile. One of the first to mathematically describe crown profile was Honer (1971), who developed a polynomial equation based on depth into the crown and tree height for balsam fir and black spruce. The continuous model form of Baldwin and Peterson (1997) predicted crown profiles of loblolly pine based on relative crown height, DBH, crown ratio, and tree age. In contrast, Hann (1999) divided the crown of Douglas-fir into two sections: the portion of the crown above where LCW occurs, and the portion below that point. The upper portion was found to vary in shape and was dependent on relative location in the crown and tree form (as reflected by the ratio of tree height to DBH), which can be indicative of social position with in a stand; while the lower portion was more consistent (Hann, 1999; Figure 7.3). This approach was successfully extended to several other species including western hemlock (Marshall *et al.*, 2003), white and grand fir (Hann and Hanus, 2001), and radiata pine (Crecente-Campo *et al.*, 2009a). In comparing several deterministic crown-profile equations, Crecente-Campo *et al.* (2009a) found the Hann (1999) approach to be superior.

Crowns are highly stochastic, and predictions of mean crown profiles may not be fully representative. Consequently, Biging and Gill (1997) proposed to directly model crown profiles with a stochastic approach that depicted the intermodal reductions in crown widths (Gill and Biging, 2002a). Nepal *et al.* (1996) used stochastic frontier analysis to predict crown profiles of loblolly pine, and found it superior to techniques that relied on OLS. Doruska and Mays (1998) used nonparametric regression to model crown profiles, to increase flexibility in representing crown shape and avoid forcing trees to follow a predetermined mathematical form. All of these techniques are dependent on available data, and not widely incorporated in growth models. Another limitation of nearly all crown profile equations is that they have been primarily developed for conifer species. However, Gill and Biging (2002b) used autoregressive moving average models for two

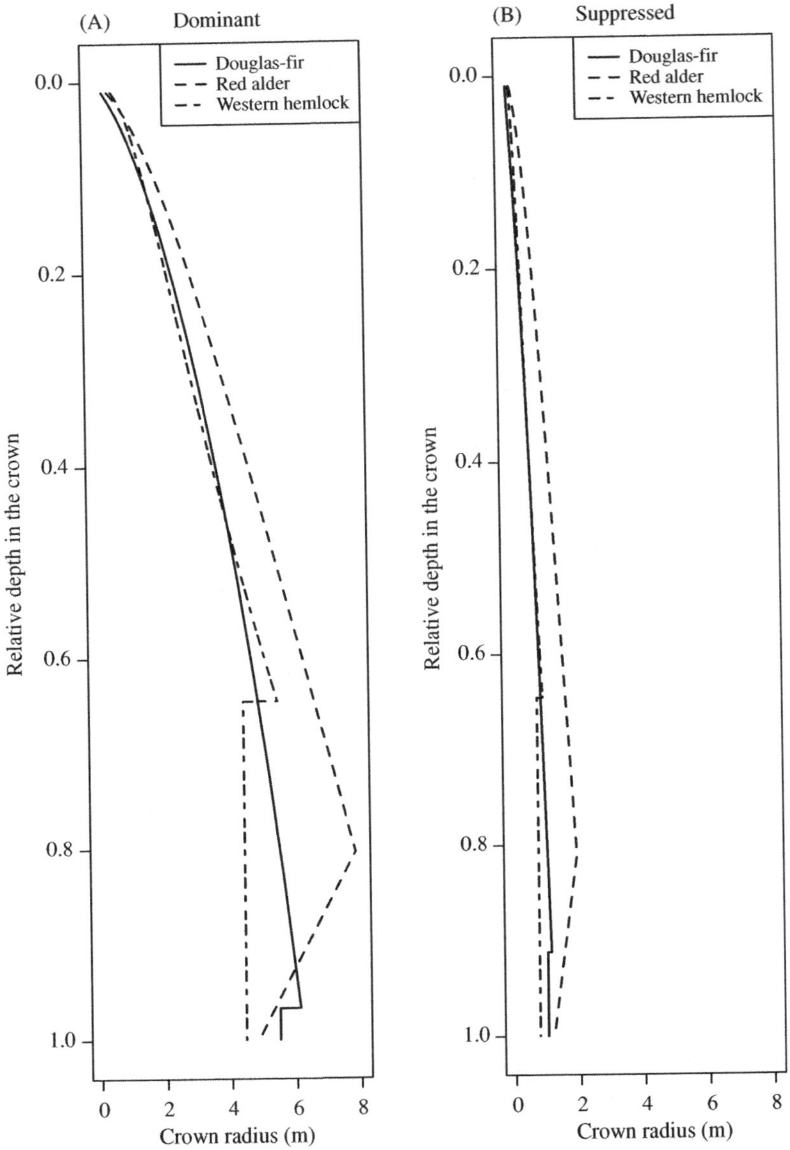

Figure 7.3 Relationship between crown radius and relative depth in the crown for (A)a dominant and (B)a suppressed tree for three species in the Pacific Northwest. The dominant-tree characteristics for Douglas-fir and western hemlock were diameter at breast height (DBH) of 80 cm, total tree height (HT) of 39 m, and a crown ratio (CR) of 0.7, while the DBH, HT, and CR for the dominant red alder were 25 cm, 14 m, and 0.7, respectively. The suppressed tree characteristics were 5 cm DBH, 8 m HT, and 0.2 CR for all three species. Based on the equations of Hann (1999), Marshall *et al.* (2003), and Hann *et al.* (2011).

hardwood species in California with promising results, and Hann *et al.* (2011) recently developed profile equations for red alder using the techniques of Hann (1999).

7.5 Stem volume and taper

Stem form and volume are the two most important tree attributes for determining value, and are the primary interest of most growth-model users. Thus, a variety of approaches for determining both attributes exists, even for a single geographic region (e.g. Hann, 1994). Much is already written about this topic (e.g. Husch *et al.*, 2003); so this section focuses more on the technical aspects of the equations. The current trend is to move away from stem volume equations and rely more on stem taper equations, which predict changes in stem diameter from tree tip to base. Taper equations are preferred because they depict stem form, provide predictions of total volume, and can be used to determine merchantable volume to any height or diameter specification. Limitations of taper equations are that they are often overly complex, which may limit their ability to extrapolate beyond the fitting dataset, and they are not optimized to give volume predictions. The first problem requires an exhaustive examination of the equation's predictive behavior before the equation is released for general usage, a practice which experience has shown has not been applied frequently enough.

Most volume equations include just DBH and total tree height, but use a variety of model forms (see Zianis *et al.*, 2005). Two of the most important aspects of a volume equation are that the equation must not exhibit abnormal behavior at any point in the modeling space, and it should extrapolate well. Several volume equations, particularly polynomial forms, can produce negative volume predictions for large trees. The vast majority of stem volume equations make predictions based on a single equation, when a two-equation approach might work better. Hann *et al.* (1987) found that developing equations to predict stem volume both above and below breast height worked better than a single equation. This approach has numerous additional benefits and was helpful in identifying CR as a key covariate (Hann *et al.*, 1987), when most other studies found it to be non-significant or only slightly improved predictions. Although CR will often explain a very little amount of variation in stem volume compared to DBH or tree height, its inclusion can be important because of its documented influence on tree form (e.g. Larson, 1963). Hann *et al.* (1987) state that the primary objective of including CR is to reallocate the variation to those factors that contribute to the variation, so that the equation can reasonably extrapolate to conditions not found in the modeling dataset.

Merchantable stem volume is often estimated from empirical ratios that predict the proportion of total volume that meets certain specifications. The limitation of this approach is that merchantable specifications often change over time and require knowledge of length. This problem can be partially avoided by including the merchantable diameter (Burkhart, 1977) or merchantable height (Cao and Burkhart, 1980) in the ratio equation. Cao *et al.* (1980) found that the ratio method was generally superior for determining merchantable volume, compared to using a taper equation.

Most stem taper equations are also functions of just DBH and total tree height, and a variety of model forms also exist. For example, Rojo *et al.* (2005) compared 31 different equation forms. Taper equations are of three primary types: (1) single (Thomas and Parresol, 1991); (2) segmented (Max and Burkhart, 1976); and (3) variable form (Kozak, 1988). Goodwin (2009) gives a list of criteria for an ideal taper equation, but most of the widely used forms do not meet all criteria (Table 7.2).

The primary limitation of most taper equations is that they all require a measurement of DBH, which limits their applicability to small trees, and may not be the best location for determining stem form. In fact, Goodwin (2009) found that relative heights between 10 and 30% provided much better estimates of tree shape than DBH, which is why some equations require an upper-stem diameter measurement (Clark III *et al.*, 1991; Kozak, 1998). Most taper equations also do not incorporate crown variables since they generally are found non-significant, or only marginally improved model fits (Burkhart and Walton, 1985; Muhairwe *et al.*, 1994), but there are studies that have found otherwise (Valenti and Cao, 1986; Walters and Hann, 1986; Farrar, 1987; Petersson, 1999; Valentine and Gregoire, 2001; Shaw *et al.*, 2003; Leites and Robinson, 2004; Jiang *et al.*, 2007; Yang *et al.*, 2009b).

Stand or site variables are also not usually included due to non-significance (Muhairwe *et al.*, 1994), but Sharma and Parton (2009) found that the inclusion of stand basal area improved their predictions of stem form. The goal of adding additional covariates to a taper equation should not necessarily be to improve model fit, but to ensure biologically reasonable extrapolations. For example, it is well established that crown length influences stem form (Larson, 1963). Consequently, applying a taper equation that includes crown ratio to two trees with the same total height and DBH but different CR should predict that the tree with the shorter crown will have: (1) wider diameters above breast height; (2) narrower diameters below breast height; and (3) the same diameter at breast height. One of the few taper equations that meets these expectations is the model of Walters and Hann (1986). This suggests that benchmarking equations based solely on their fit statistics is not fully justified, and equal focus should be placed on ensuring biologically reasonable extrapolations.

Although taper equations can provide accurate and unbiased estimates of stem volume (e.g. Filho and Schaaf 1999), most equations are unconstrained and the parameters are not optimized for predicting stem volume. Compatible taper equations provide an alternative approach because their volume integrations are constrained to match a volume equation (Bailey, 1994; Fang *et al.*, 2000), while providing accurate predictions of stem form (Diéguez-Aranda *et al.*, 2006). However, accurate predictions of stem form can be difficult to obtain for hardwood species due to their tendency to fork and have multiple stems. Rather than use a traditional taper equation, MacFarlane (2010) described the stem profile of several hardwood species using a varying centroid method.

Table 7.2 Criteria for an ideal taper equation from Goodwin (2009), and evaluation of several widely used taper model forms.

Criteria	Max and Burkhart (1976)	Clark III *et al.* (1991)	Bi (2000)	Kozak (2004)	Goodwin (2009)
Be accurate	✓	✓	✓	✓	✓
Use easy-to-obtain variables	✓	Relies on upper-stem diameter measurements	✓	✓	✓
Algebraically integrable and invertible	✓	✓	✓	Requires numerical integration	✓
Have continuity	✓	✓	✓	✓	✓
Be constrained	✓	✓	✓	✓	✓
Allow flexibility of diameter input	Requires DBH	Requires DBH	Requires DBH	Requires DBH	✓
Easily regionalized	✓	✓	✓	✓	✓
Easily localized	Relies on DBH and HT only	✓	Relies on DBH and HT only	Relies on DBH and HT only	✓
Be non-segmented	Three segments	Three segments	✓	✓	✓
Be applicable to a wide range of species	✓	✓	✓	✓	✓
Be applicable to a wide range of tree sizes	Not applicable to small trees	Not applicable to small trees	Not applicable to small trees	Not applicable to small trees	✓

Another recent improvement to the application of taper equations is the ability to localize them with additional upper-stem diameter measurements, either by direct calibration (Cao, 2009), or estimating random effects when the parameters are determined using a mixed-effects approach (Trincado and Burkhart, 2006). Given the ease of making upper-stem diameter measurements with a laser dendrometer, Cao (2009) suggested that optimum gains were obtained when the measurement was taken at the midpoint between breast height and tree tip. Sharma and Parton (2009) found that an upper-stem measurement taken at around 35% of tree height provided the best improvements, and taking a third diameter measurement was not warranted. The Kozak (2004) taper equation has often been found superior for predicting stem diameters (Rojo *et al.*, 2005) and volume (Li and Weiskittel, 2010). When multiple equations exist for the same species, taking the geometric mean of all predictions often improves estimates of stem form (Li and Weiskittel, 2010). Some researchers have found that stem taper equations fitted using the traditional parametric approach are insufficiently flexible to describe stem form, and have concluded that nonparametric techniques are superior (Lappi, 2006; Kublin *et al.*, 2008).

Most stem taper equations predict diameter inside bark, and occasionally estimates of diameter outside bark are needed. Bark thickness can also be predicted with a taper equation (Maguire and Hann, 1990b; Lassasenaho *et al.*, 2005) or with various other linear and nonlinear equations (Cao and Pepper, 1986; Johnson and Wood, 1987). Modeling diameter inside bark as a simple power function of diameter outside bark is an effective and unbiased technique across a variety of species, but predictions are improved with the inclusion of height along the stem for some species (Li and Weiskittel, 2011). However, bark thickness can vary significantly with tree age and site, even after accounting for diameter outside bark (Sonmez *et al.*, 2007), which suggests that additional covariates may be necessary when modeling bark thickness.

7.6 Biomass

Thousands of biomass equations have been developed around the world. For example, Jenkins *et al.* (2004) reported 2640 biomass equations from 177 studies in North America alone. Other extensive reviews for Europe (Zianis *et al.*, 2005), North America (Ter-Mikaelian and Korzukhin, 1997), and Australia (Eamus *et al.*, 2000; Keith *et al.*, 2000) highlight the vast amount of work completed on this topic. However, most biomass equations are simplistic, with parameters determined from relatively small sample sizes. Zianis *et al.* (2005) found that more than two-thirds of the equations they examined were functions of just DBH, and more than 75% of the studies that reported a sample size had less than 50 trees. The use of just DBH assumes that the relationship between DBH and height is static, which is often not the case, as noted above.

The application to other populations of simple model forms fitted to small datasets can produce large prediction errors (e.g. Wang *et al.*, 2002a). In addition, development of universal (Pilli *et al.*, 2006) and generalized (Muukkonen, 2007) static equations ignores significant species variability and complex relationships, particularly when the goal is to estimate regional and national biomass (Zianis and Mancuccini, 2004). Efforts have been proposed to localize static biomass equations, without requiring destructive sampling, by accounting for either the relationships between tree height and DBH as well as wood density (Ketterings *et al.*, 2001), or by incorporating the DBH distribution of the stand (Zianis, 2008). Several growth and yield models have used approaches other than static equations for estimating biomass.

Estimating stem biomass from one of the widely available stem taper or volume equations is relatively straightforward by applying a value of wood density (e.g. Smith, 1985; Heath *et al.*, 2009). Parresol and Thomas (1996) demonstrated how this technique can be extended to account for variations in within-stem density. This approach ensures compatible estimates of stem biomass and volume, which the use of two separate static equations does not provide. Foliage biomass has been shown to be well correlated with sapwood area (Grier and Waring, 1974), and sapwood area is readily predicted from a taper equation (Maguire and Batista, 1996) or an estimate of crown surface area (Maguire and Hann, 1989). Crown surface area itself can be inferred from a geometric shape, the product of LCW and crown length, or a crown profile equation if available. Unlike static equations that estimate foliage biomass solely from DBH, this approach relies more on the relationship between biomass and tree crown dimensions, which allows it be to more sensitive to changes in stand density or other factors that influence crown size.

Guiterman *et al.* (in press) found that a static equation containing both sapwood area and crown length was necessary to account for changes in foliage biomass imposed by various levels of thinning. Foliage and branch biomass can also be estimated using empirical equations that describe the frequency and structure of primary branches (Mäkinen *et al.*, 2003; Weiskittel *et al.*, 2007c). Weiskittel *et al.* (2010) found that this technique was just as effective across a range of silvicultural treatments as predicting leaf area from sapwood area, and that it was superior to several tree-level static equations.

Prediction of belowground biomass still remains problematic and difficult. Only 8% of the 607 equations reviewed in Zianis *et al.* (2005) provided predictions of total root biomass. Given the expense of collecting this information and relative lack of static equations for the majority of species, an alternative approach is often needed. A linear relationship between stand total aboveground and belowground biomass has generally been found useful (e.g. Ranger and Gelhaye, 2001). A similar relationship for fine roots and foliage biomass (e.g. Vanninen and Mäkelä, 1999) also exists. These relations would likely hold at the tree-level and could be constrained to ensure compatibility.

7.7 Use of static equations to predict missing values

Since static equations are often used to estimate missing values, various methods are used to accomplish this. The best method depends on whether missing values being estimated are from the modeling dataset or an independent dataset. The first is a common issue that growth modelers face when preparing modeling datasets, while the latter is a common issue that model users face when trying to apply a model to their datasets. Robinson and Wykoff (2004) have demonstrated that the use of mixed effects models makes solving the first problem relatively easy since estimates of the random effects allow plot-specific parameters to be determined, while making full use of the available data.

The second problem is solved by applying a regional static equation. However, if the regional static equation is nonlinear in form and parameters were estimated using mixed effects, biased predictions can be obtained if the random effects are simply set to zero (Monleon, 2003; Temesgen *et al.*, 2008). For a model including only DBH, Temesgen *et al.* (2008) found that this approach increased root mean square error by over 9% compared to a equation fitted with OLS, while the difference was only 2% when additional covariates were included in the model.

If observations of tree height or height to crown base are available in the independent dataset, a regional equation can be locally calibrated in a variety of ways. These approaches include: (1) applying a multiplier on the prediction equation; and (2) determining the random effects with a best linear unbiased predictor (BLUP) estimator if the original model was fitted with mixed effects. Multipliers often are estimated using linear regression through the origin (Hann and Hanus, 2004; Temesgen *et al.*, 2008) and by forcing the residuals to zero (Robinson and Wykoff, 2004). Robinson and Wykoff (2004) found that using the BLUP estimator was significantly superior to the use of a multiplier for a model that only included DBH. However, Temesgen *et al.* (2008) found multipliers on a fixed-effects model with DBH and several additional covariates to be just as precise as the BLUP for a mixed-effects model with just DBH when at least four subsampled heights were measured per plot. The BLUP method was significantly superior when smaller sample sizes were used (Temesgen *et al.*, 2008; Figure 7.4).

Effectiveness of the BLUP not only depends on model form, but also on the sampling scheme used to select sample trees. Calama and Montero (2004) recommended a subsample size of four trees and suggested selecting the four largest trees in a stand. Even though their model included dominant height, selection of four randomly chosen trees produced more precise predictions. Castedo Dorado *et al.* (2006) examined the impact upon the calculated BLUP of using 10 alternative sample sizes for randomly selected trees, or the options of concentrating the subsampling to the three largest, intermediate, or smallest trees. They recommended a subsample size of three trees that is concentrated in the three smallest trees, because they found that inclusion of dominant height in their equation greatly reduced any precision gained by subsampling dominant trees (Castedo Dorado *et al.*, 2006). If the equation did not incorporate dominant height, then selecting the largest

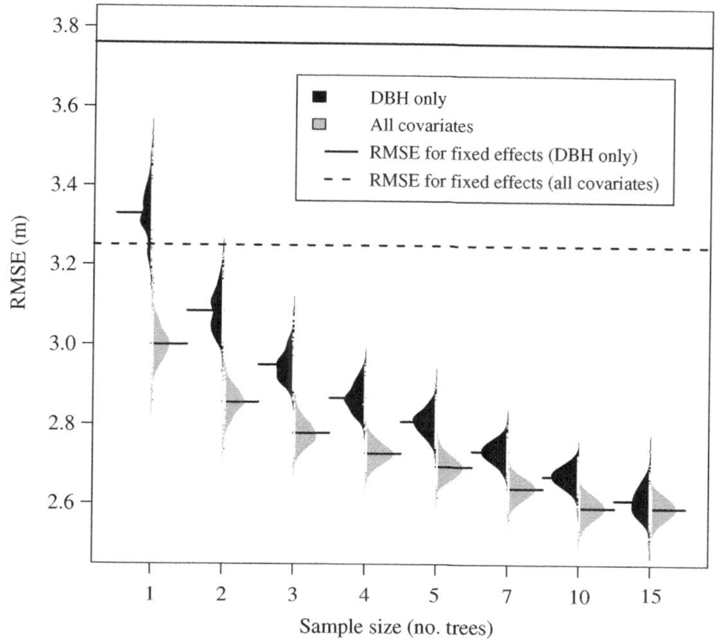

Figure 7.4 Beanplot showing the distribution and mean (short solid line) of the root mean square error (RMSE; m) after 200 Monte Carlo simulations for localizing a regional height-to-diameter static equation for coastal Douglas-fir fitted using nonlinear mixed effects with different sample sizes. The equations contained diameter at breast height (DBH) only (black beans) or DBH, crown competition factor in larger trees, and total stand basal area (gray beans). For comparison, the RMSE for the DBH only (black solid line) and all covariates (black dashed line) models fitted using traditional ordinary least squares are also given. Based on data from Temesgen *et al.* (2008).

diameter tree in a stand was more effective than randomly selecting a single tree on a plot (Temesgen *et al.*, 2008). Garber *et al.*, (2009) found that subsampling at least four trees in a plot for height measurement was the most effective method when estimating stand-level volume.

Height and height to crown base measurements are time-consuming and costly, but are important for determining stand volume and improving growth projections. Although regional static equations generally have reasonably strong predictive power, subsampling a small number of trees per plot can greatly improve predictions. However, filling in missing height and height to crown base measurements and then using these values to then predict attributes like biomass or volume may cause error propagation. For example, Fortin and DeBlois (2010) found that using predicted rather than observed heights to estimate stem volume could double and even triple the unexplained variation in the plot volume predictions.

Table 7.3 Static equation form and covariates[a] used in two example individual-tree, distance-independent growth and yield models.

Equation	Equation form	Reference
ORGANON		
Height to diameter (pure, even-aged stands of selected species)	$HT = 1.37 + (HT_{100} - 1.37) \cdot \frac{e^{\left(b_0 \cdot DBH^{(b_0 + b_1 \cdot (HT_{100} - 1.37))}\right)}}{e^{\left(b_0 \cdot DBH_{100}^{(b_0 + b_1 \cdot (HT_{100} - 1.37))}\right)}}$	Hanus *et al.* (1999b)
Height to diameter (all other species and stand structures)	$HT = 1.37 + e^{(b_0 + b_1 \cdot DBH^{b_2})}$	Larsen and Hann (1987)
Height to crown base	$HCB = \frac{HT}{\left(1.0 + e^{\left(b_0 + b_1 \cdot HT + b_2 \cdot CCFL + b_3 \cdot \log(BA) + b_4 \cdot \left(\frac{DBH}{HT}\right) + b_5 \cdot (SI - 1.37)\right)}\right)}$	Hanus *et al.* (2000); Hann *et al.* (2003)
Maximum crown width	$MCW = b_0 + b_1 \cdot DBH^{b_2}$	Paine and Hann (1982)

Largest crown width	$LCW = MCW \cdot CR^{\left(b_0 + b_1 \cdot CL + b_2 \cdot \left(\frac{DBH}{HT}\right)\right)}$	Hann (1997)
Crown profile	$CW_{ht} = I \cdot \left(LCW \cdot \left(\frac{HT - ht}{HT - HLCW}\right)^{\left(b_0 + b_1 \cdot \sqrt{\left(\frac{HT - ht}{HT - HLCW}\right)} + b_2 \cdot \left(\frac{HT}{DBH}\right)\right)}\right) + (1.0 - I) \cdot \left(LCW \cdot \left(\frac{ht - HCB}{\left(b_3 \cdot CL \cdot e^{\left(b_4 \cdot \left(1.0 - \left(\frac{HT}{HT_{max}}\right)\right)\right)}\right)}\right) + b_5 \cdot \left(1.0 - \left(\frac{ht - HCB}{\left(b_3 \cdot CL \cdot e^{\left(b_4 \cdot \left(1.0 - \left(\frac{HT}{HT_{max}}\right)\right)\right)}\right)}\right)\right)\right)$	Hann (1999); Marshall *et al.* (2003)
Stem volume (above breast height)	$b_0 \cdot \left(\frac{(HT - 1.37)}{DBH}\right)^{b_1} \cdot e^{(b_2 \cdot CR)} \cdot (DBH^2 \cdot (HT - 1.37))$	Hann *et al.* (1987)

(continued)

Table 7.3 *(Continued)*

Equation	Equation form	Reference
PROGNAUS		
Maximum crown width	$\log(\text{MCW}) = b_0 + b_1 \cdot \log(\text{DBH})$	Hasenauer (1997)
Crown ratio	$\text{CR} = \dfrac{1}{\left(1 + e^{\left(-\left(b_0 + b_1 \cdot \left(\frac{\text{HT}}{\text{DBH}}\right) + b_2 \cdot \text{HT} + b_3 \cdot \text{DBH}^2 + b_4 \cdot \text{BAL} + b_5 \cdot \log(\text{CCF}) + b_6 \cdot \text{ELEV} + b_7 \cdot \text{ELEV}^2 + b_8 \cdot \text{SL} + b_9 \cdot \text{SL}^2 + b_{10} \cdot (\text{SL} \cdot \sin(\text{AZ})) + b_{11} \cdot (\text{SL} \cdot \cos(\text{AZ}))\right)\right)}\right)}$	Hasenauer and Monserud (1996)

[a] Where DBH is diameter at breast height, HT is total tree height, HCB is height to crown base, HLCW is the height to largest crown width, CW_{ht} is the crown width at height ht, I is an indicator for position in the stem (1 if ht $\geq$ HLCW, 0 otherwise), HT_{max} is the maximum height for the species, CR is crown ratio, CCF is the crown competition factor, CCFL is the CCF in larger trees, BA is total basal area, SI is site index, HT_{100} is the average height of the 100 thickest trees per hectare, DBH_{100} is the average DBH of the 100 thickest trees per hectare, MCW is the maximum crown width, LCW is the largest crown width, CL is crown length (HT − HCB), BAL is basal area in larger trees, ELEV is elevation, AZ is aspect, SL is percentage slope, the b_is are parameters, and log is the natural logarithm.

7.8 Summary

Static equations are crucial to both the accuracy (e.g. height) and relevance (e.g. volume) of growth and yield predictions. However, growth models often differ greatly in the equation form and covariates used in their static equations (Table 7.3). Although static equations can explain a significant amount of the observed variation, there is often a large amount of variation unexplained due to differences in species or among individuals within a species (e.g. Dietze *et al.*, 2008). In addition, static equations are developed with varying degrees of statistical rigor and often need local calibration. Newer statistical techniques like mixed effects models allow local calibration to be performed quite effectively, but the effectiveness depends both on the model form and on the number of available samples.

Regardless, development of robust static models requires appropriate model forms and rigorous testing of the equations' ability to extrapolate. It is also important to note that the best equation at the tree-level may not be the best when scaled to the stand-level (Guiterman *et al.*, in press). Static biomass equations can provide accurate predictions, but also can be highly biased. Rather than using simple power functions of DBH, there are several alternative approaches for estimating biomass that likely have greater utility.

8
Mortality

8.1 Introduction

Tree mortality is a rare, yet important, event in forest stand development, and has significant implications for long-term growth and yield model projections. Although the physiological processes that occur when a tree dies are poorly understood, tree mortality is caused by a range of external factors like stand density, disease, insect defoliation, fire, and wind, to name just a few examples. Of all of the attributes predicted in growth models, mortality remains one of the most difficult, due to its at times stochastic nature and episodic occurrence. For modeling purposes, it is important to note the type of mortality, which is generally described as *regular* or *irregular.*

Regular mortality can also be expressed as density dependent, and is caused by competition-induced suppression. Irregular or catastrophic mortality is independent of stand density and is due to external factors like insects, disease, fire, or wind. However, stand density can cause slow growth rates, which can make a tree more susceptible to mortality due to insects such as bark beetles; so the distinction between the types of mortality can be easily blurred. Regular mortality often influences a small portion of stems in the stand, while irregular mortality can impact a small to large portion of the stand.

Most growth and yield models just focus on regular mortality, but some models predict both types of mortality. Previous reviews on modeling mortality have concluded that there is no best way to model it for all applications (Hawkes, 2000). The chapter focuses on both stand- and tree-level mortality equations, with emphasis on the approaches used in distance-independent, tree-list models. Although the focus of the chapter is on mortality, tree survival is also discussed interchangeably.

Forest Growth and Yield Modeling, First Edition. Aaron R. Weiskittel, David W. Hann, John A. Kershaw, Jr. and Jerome K. Vanclay.

8.2 Stand-level mortality

One of the predominant relationships describing competition-induced mortality at the stand-level is the $-\frac{3}{2}$ power law of self-thinning (Yoda *et al.*, 1963). This idea was introduced into the English forestry literature by Drew and Flewelling (1977), and has long been debated in the ecological literature (e.g. Weller, 1987; Reynolds and Ford, 2005). The theory states that a maximum size–density relationship, independent of age and site quality, exists for any stand, and the slope of the relationship is -1.5 when the logarithm of mean plant weight is regressed against the logarithm of number of trees. This slope is assumed to be independent of species and used to predict competition-induced mortality by estimating a stand's self-thinning trajectory. The concept has primarily been applied to even-aged, single-species stands (e.g. Drew and Flewelling, 1977; Drew and Flewelling, 1979; Cao *et al.*, 2000; Turnblom and Burk, 2000), but has also been used in uneven-aged (Sterba and Monserud, 1993) and mixed-species stands (Puettmann *et al.*, 1992; Solomon and Zhang, 2002; Yang and Titus, 2002). In forestry, the maximum size–density relationship is often expressed as:

$$\log(\text{TPH}) = b_0 - 1.605 \cdot \log(\text{QMD}) \qquad [8.1]$$

where TPH is the trees per hectare, QMD is the quadratic mean diameter (cm), and b_0 is a species-specific parameter (Reineke, 1933; Figure 8.1). Although there are parallels, the Reineke (1933) relationship differs from those of Drew and Flewelling (1977; 1979), as it uses QMD rather than mean tree volume.

A variety of statistical techniques are used to estimate the species-specific parameter, including ordinary least squares (Smith and Hann, 1984; Vanclay and Sands, 2009), principal components analysis (Wilson *et al.*, 1999), quantile regression (Zhang *et al.*, 2005), and stochastic frontier analysis (Bi, 2001). Although stochastic frontier analysis is recommended by some as the preferred technique (e.g. Zhang *et al.*, 2005), this method cannot overcome limitations of the data or confirm that stands are actually self-thinning. The technique of Smith and Hann (1984) simultaneously estimates both parameters of the maximum size–density line and the nature of self-thinning trajectory, while being able to test influences of additional covariates. Segmented regression also is used (Cao and Dean, 2008; VanderSchaaf and Burkhart, 2008), but requires that the proper model forms be specified in each segment.

Irrespective of the statistical technique used, an extensive dataset is required to estimate maximum size–density lines properly. Although data from yield tables (e.g. Zeide, 1987) and single measurements on a large number of plots (e.g. Wilson *et al.*, 1999) are used to determine self-thinning relationships, the best data are from remeasured permanent plots, as this allows confirmation that stands are self-thinning, and self-thinning trajectories can be assessed (e.g. Puettmann *et al.*, 1993). Spurious relationships can be detected if not all stands are self-thinning. Some researchers have found the maximum size–density line to be sensitive to stand origin (natural vs. planted; Weiskittel *et al.*, 2009b), site productivity

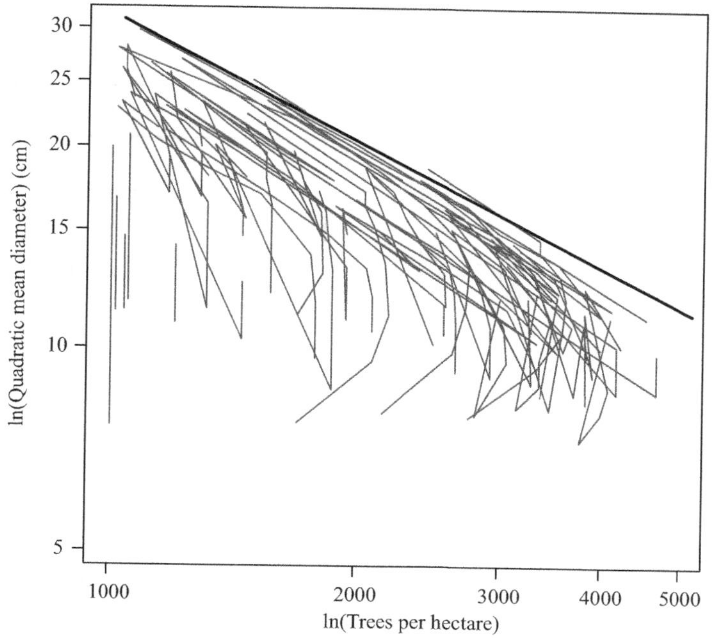

Figure 8.1 Natural development of number of trees and size for unmanaged, individual pure stands of coastal Douglas-fir in the Pacific Northwest USA, and the maximum size–density line. Based on data from US Forest Service Levels-of-Growing Stock (LOGS) study (Marshall and Curtis, 2002).

(Bi, 2001), and species composition (Puettmann *et al.*, 1992; Woodall *et al.*, 2005), while others have found it to be insensitive to stand origin and site productivity (Puettmann *et al.*, 1992; Hann *et al.*, 2003). However, it has generally been shown that the upper limit of the maximum size–density line is higher for shade-tolerant species when compared to shade-intolerant ones (Long, 1985; Pretzsch and Biber, 2005; Weiskittel *et al.*, 2009b). Regardless, a comprehensive analysis is still needed to clarify influences of these factors and quantify consequences of data flaws. These maximum size–density lines are used to constrain individual tree mortality equations (Yang and Titus, 2002; Hann *et al.*, 2003), or as a component of whole-stand models (Smith and Hann, 1986).

Stand-level mortality equations that predict changes in the number of trees also are widely used. Although model forms have differed (Table 8.1), the most common approach is to use a difference equation (Diéguez-Aranda *et al.*, 2005):

$$\mathrm{TPH}_2 = \mathrm{f}(\mathrm{TPH}_1, t_2, t_1) \qquad [8.2]$$

where TPH_i is the live number of trees per hectare at time i and t_i is the time of measurement, with 1 being the initial and 2 being the remeasurement. Rather than time, most stand-level mortality equations rely on age (Diéguez-Aranda *et al.*, 2005;

Table 8.1 Equation forms used in modeling stand-level mortality with an algebraic difference approach.

Equation form	Reference
$TPH_2 = \left(TPH_1^{b_0} + SI \cdot \left(t_2^{b_1} - t_1^{b_1}\right)\right)^{\left(\frac{1}{b_0}\right)}$	Clutter and Jones (1980)
$TPH_2 = \left(TPH_1^{b_1} + b_0 \cdot SI \cdot \left(\left(\frac{t_2}{10}\right)^{b_2} - \left(\frac{t_1}{10}\right)^{b_2}\right)\right)^{\left(\frac{1}{b_1}\right)}$	Pienaar *et al.* (1990)
$TPH_2 = \left(TPH_1^{-0.5} + SI \cdot \left(\left(\frac{t_2}{100}\right)^2 - \left(\frac{t_1}{100}\right)^2\right)\right)^{-2}$	Woollons (1998)
$TPH_2 = TPH_1 \cdot e^{\left(SI \cdot \left(t_2^{b_0} - t_1^{b_0}\right)\right)}$	Pienaar and Shiver (1981)
$TPH_2 = TPH_1 \cdot \left(\frac{t_2}{t_1}\right)^{b_0} \cdot e^{((b_1 + b_2 \cdot SI) \cdot (t_2 - t_1))}$	Bailey *et al.* (1985)
$TPH_2 = \left(TPH_1 + (b_2 + b_3 \cdot SI) \cdot \left(b_1^{t_2} - b_1^{t_1}\right)\right)^{\left(\frac{1}{b_0}\right)}$	Zhao *et al.* (2007)

Zhao *et al.*, 2007), which limits their application to even-aged stands. Site index often is found to be an important factor (Bailey *et al.*, 1985; Diéguez-Aranda *et al.*, 2005; Zhao *et al.*, 2007). Generally, mortality increases with site index (Bailey *et al.*, 1985; Diéguez-Aranda *et al.*, 2005; Zhao *et al.*, 2007). However, a few studies have found mortality to decrease with increases in site index (e.g. Woollons, 1998; Zhao *et al.*, 2007). Silvicultural factors such as thinning (Bailey *et al.*, 1985) also influence stand-level mortality. These previous examples have primarily focused on single-species stands, but Eid and Øyen (2003) extended the technique to mixed-species stands. One limitation of this approach is that ingrowth must be assumed to be negligible or predicted with another equation.

Another important limitation of stand-level mortality equations is that they always predict some level of mortality, even if no mortality occurred. Furthermore, if the equation is only fitted to data where mortality occurred, then the mortality rate will often be over-estimated (Woollons, 1998). One method of avoiding both problems is to use a two-stage approach as suggested by Woollons (1998). In this approach, an equation that predicts the probability of any mortality occurring is constructed. Then a second equation is developed that estimates the amount of mortality given that some has occurred. The method can then be applied deterministically or stochastically, but model behavior is significantly improved with the use of two equations (Woollons, 1998; Álvarez González *et al.*, 2004).

8.3 Individual-tree-level mortality

Despite difficulties in predicting stand-level mortality, multiple equations are available to estimate mortality for individual trees. Nearly all tree-level mortality equations use logistic regression to estimate the probability of a tree dying (Hamilton, 1986; Monserud and Sterba, 1999; Hann *et al.*, 2003). Thus, the primary differences between existing individual-tree mortality equations are: (1) the type of data used; (2) the statistical methods for estimating parameters; (3) the length of prediction period; (4) usage of additional equations to constrain predictions; and (5) the tree and stand variables utilized for predictions. Each of these differences will be discussed here, except that the usage of additional equations to constrain predictions will be covered more thoroughly in Chapter 10.

As for static and increment equations, diameter at breast height (DBH) is the primary variable in most individual-tree mortality equations (Table 8.2; Figure 8.2). For example, Wykoff *et al.* (1982) found DBH to be the sole predictor for 10 of the 11 conifer species in their analysis. Various transformations of DBH also are found effective, including DBH^2 (Wykoff *et al.*, 1982; Yao *et al.*, 2001), DBH^{-1} (Hamilton, 1986; Monserud and Sterba, 1999), and $DBH^{0.5}$ (Hamilton, 1986; Wykoff, 1986). Often a combination of DBH and transformations of it are used in mortality equations to capture the expected U-shaped pattern of senescence with tree size (Monserud and Sterba, 1999). In other words, trees have a higher probability of dying when they are very small, which decreases as trees increase in size and then becomes higher again as trees reach maturity.

However, a very large sample across the full range of tree sizes is needed to observe this pattern. Consequently, Buchman *et al.* (1983) developed a flexible expression of DBH in their survival equation, which was able to characterize a range of asymmetrical forms. Although Monserud and Sterba (1999) found DBH^{-1} to be the most effective expression of DBH in their mortality equation for several Austrian species, this variable can cause problems when applied to small trees (i.e. <5.0 cm; Hann and Hanus, 2001).

Diameter increment and tree basal area increment, either predicted or measured, are used as a covariate in several mortality equations (Monserud, 1976; Buchman *et al.*, 1983; Hamilton, 1986; Wyckoff and Clark, 2000; Yao *et al.*, 2001; Pretzsch *et al.*, 2002). This approach is widely used in several individual-tree gap models (Keane *et al.*, 2001). Hamilton (1986) used predicted diameter increment divided by DBH, and Pretzsch *et al.* (2002) used predicted tree basal area increment divided by DBH to improve model performance for smaller trees. Hamilton (1990) also suggested that predicted diameter increment divided by average potential diameter increment for the site is more effective than increment by itself, which was used in the equations of Wykoff (1986). Diameter increment is assumed to represent tree vigor; therefore slower growing trees have a higher probability of dying. Wyckoff and Clark (2000) found that species can differ drastically in their ability to withstand slow growth. One drawback to using predicted DBH growth is the requirement that a DBH growth equation be available, and the equation has be refitted if the increment equation changes. The advantage of using predicted instead of measured diameter increment is that predicted increment avoids problems associated with errors in variables (Monserud, 1976).

Table 8.2 Independent variables commonly used in predicting individual-tree mortality using logistic regression and distance-independent measures of competition.

Independent variables[a]	Reference
DBH, DBH^2	Wykoff *et al.* (1982)
ΔDBH, $e^{(a \cdot (DBH-1))} \cdot (DBH-1)^b$	Buchman *et al.* (1983)
$DBH^{0.5}$, DBH^{-1}, $\frac{DBH}{MD}$, ΔDBH, $\frac{\Delta DBH}{DBH}$, $BA^{0.5}$	Hamilton (1986)
DBH, CR, SI, BAL or CCH	Hann and Wang (1990)
DBH, DBH^2, DBH^{-1}, CR, BAL	Monserud and Sterba (1999)
DBH, CR, SI, CCH	Bravo *et al.* (2001)
DBH, DBH^2, CR, SI, BAL, $BAL \cdot e^{\left(a \cdot D_5 \cdot \frac{H_5}{10\,000}\right)}$	Hann and Hanus (2001)
DBH, DBH^2, ΔDBH, BA, %BA, SPI, $\frac{DBH^2}{BA}$	Yao *et al.* (2001)
DBH, BA, BAL (conifers) DBH, RDFL, DBH · RDFL (hardwoods)	Hynynen *et al.* (2002)
DBH, PCR, BAL, SI	Hann *et al.* (2003)
DBH, DBH^2, ΔDBH, $BAL \cdot DBH^2$, $\frac{BAL_H}{BA}$	Yang *et al.* (2003)
DBH, DBH^{-1}, DBH^2, BAL, $\frac{DBH^2}{BA}$	Temesgen and Mitchell (2005)
$\frac{DBH}{QMD}$, SI · BA (maritime pine) DBH^{-1}, $BAL \cdot CV_D$, SI (Scots pine)	Bravo-Oviedo *et al.* (2006)
$DBH^{0.5}$, log(BA), BAL	Pukkala *et al.* (2009)
$\frac{DBH}{QMD}$, DBH^2, BA, HT_{DOM}, BAL_{MOD}	Crecente-Campo *et al.* (2009b)
DBH, QMD, BAL_{MOD}, $\frac{BAL_{MOD}}{BA}$	Crecente-Campo *et al.* (2010)
DBH^{-1}, $\frac{HT}{HT_{DOM}}$	Adame *et al.* (2010a)

[a] Where *a*, *b* are regression parameters, DBH is diameter at breast height, ΔDBH is diameter increment, HT is total tree height, CR is crown ratio, PCR is predicted crown ratio, BAL is basal area in larger trees, BAL_H is basal area in larger hardwood trees, BAL_{MOD} is modified basal area in larger trees, CCH is crown closure in higher trees, BA is stand basal area, D_5 is average diameter of the five largest stems per acre, H_5 is average height of the five largest stems per acre, HT_{DOM} is mean dominant height, MD is mean diameter of the stand, QMD is quadratic mean diameter, %BA is species proportion of the basal area, CV_D is coefficient of diameter within a stand, RDFL is the relative density factor of trees larger than the subject tree (RDF, relative density factor, is derived from Reineke's (1933) stand density index, but based on stump diameter rather than DBH), SI is site index, SPI is the site productivity index of Huang and Titus (1993), and log is the natural logarithm.

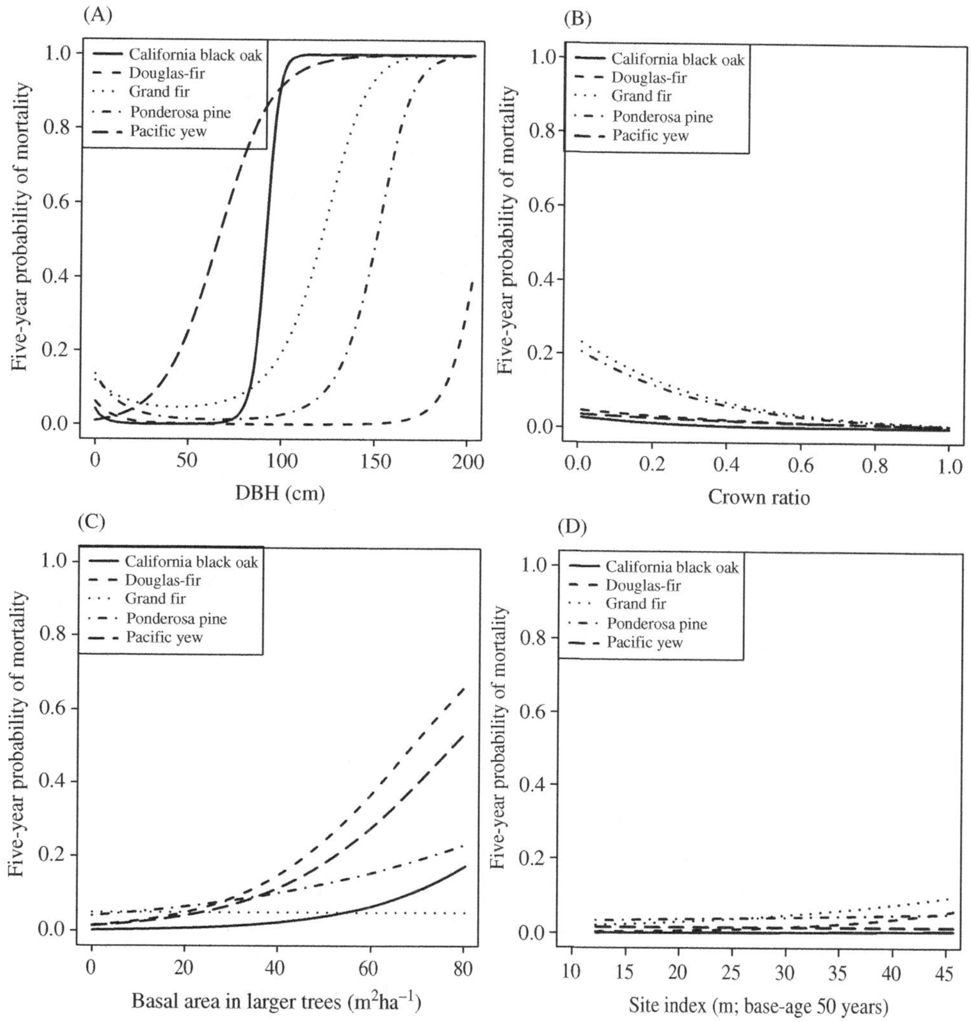

Figure 8.2 Relationship between five-year probability of mortality, diameter at breast height (A), crown ratio (B), basal area in larger trees (C), and site index (D) for several species in southwestern Oregon. Based on the equations of Hann and Hanus (2001).

One alternative to using DBH increment is to utilize crown ratio (Hann and Wang, 1990; Monserud and Sterba, 1999; Hann and Hanus, 2001). A disadvantage of crown ratio is that it is not commonly measured in forest inventories. Rather than use observed crown ratio, Hann *et al.* (2003) included crown ratio predicted from a static equation, and found it to be an effective predictor of mortality. When crown ratio is subsampled, Hann *et al.* (2006) found that that crown ratio predictions calibrated to the measured values could be effectively used along with the measured values for predicting mortality. Monserud and Sterba (1999) found that crown ratio was a more important predictor of mortality for shade-tolerant species than shade-intolerant ones. In addition, most mortality equations use either DBH growth or crown ratio, which suggests that only one measure of tree vigor is needed for representing mortality accurately (Monserud and Sterba, 1999). The ratio of tree height over DBH, used by Pretzsch *et al.* (2002), can be viewed as a surrogate for crown ratio because of the strong negative correlation between the two variables in untreated stands.

Within a stand, trees in lower social positions often have a higher probability of dying than ones that are more dominant. Hamilton (1986) described tree social position with relative DBH (DBH divided by stand mean DBH). However, the use of relative DBH is undesirable if the equation will be applied to thinned stands, since it would predict an increased mortality rate for residual trees if thinning from below, and a decrease in residual mortality rate if thinning from above. Instead, Hann and Wang (1990) accounted for tree social position by including one-sided measures of competition in their mortality equation: basal area in larger trees (BAL) or crown closure in higher trees (CCH). Basal area in larger trees was included in several other mortality equations (Monserud and Sterba, 1999; Hann and Hanus, 2001; Hann *et al.*, 2003; Temesgen and Mitchell, 2005; Hann *et al.*, 2006). Bravo *et al.* (2001) compared the effectiveness of BAL and CCH in pure Douglas-fir, mixed conifer, and mixed hardwood stands in southwestern Oregon. CCH was found to be a better predictor of tree survival (1 – mortality), regardless of the stand species composition (Bravo *et al.*, 2001). However, the use of CCH requires all heights on a plot to be measured and an existing crown profile equation.

To represent mortality patterns in old-growth Douglas-fir stands, Hann and Hanus (2001) found that BAL needed to be multiplied by the product of the average diameter and height of the five largest trees per acre. This variable was found to better characterize the more open, less shady environment that occurs in the understory of old-growth stands, and was found to be significant only for Douglas-fir (the targeted old-growth species; Hann and Hanus, 2001). Crecente-Campo *et al.* (2009b) also indicated that a modified estimate of BAL suggested by Schröder and von Gadow (1999) was more effective than BAL. Rather than BAL, Peltoniemi and Mäkipää (2011) found that the sum of $DBH^{0.5}$ larger than the subject tree was more effective than DBH raised to other exponents, as it captured the influence of smaller trees better. In contrast, Hynynen *et al.* (2002) found that relative stand density in larger trees was more effective in predicting mortality in deciduous species. Relative stand density is calculated by summing the minimum growth space of each larger tree, where minimum growth space is calculated as a power transformation of the tree's predicted stump diameter (used instead of DBH to extend the density measure to seedlings).

Several individual tree mortality equations also suggest that a measure of two-sided competition must be included (Chapter 3). This frequently is represented by total basal area (Hamilton, 1986; Yao *et al.*, 2001; Hynynen *et al.*, 2002; Pukkala *et al.*, 2009). The ratio of DBH to stand basal area is used in some mortality equations (Yao *et al.*, 2001; Temesgen and Mitchell, 2005). This variable is assumed to account for the high rate of mortality in older stands composed of very large trees. However, Bravo *et al.* (2001) found that neither stand basal area nor total crown surface area were effective predictors of Douglas-fir mortality across a range of stands, and Monserud and Sterba (1999) found BAL to be a superior predictor for several species in Austria when compared to the two-sided crown competition factor (CCF). Instead of basal area, Rathbun *et al.* (2010) found Curtis's (1982) relative density to be an effective predictor of mortality.

Stand species composition, expressed as a percentage of the total basal area, was an effective predictor in mixed-species stands (Eid and Tuhus, 2001; Yao *et al.*, 2001; Zhao *et al.*, 2004). Rather than using a species basal area over total basal area, Yang *et al.* (2003) used BAL of hardwood trees over total basal area to model white spruce mortality in Alberta. It was found that white spruce mortality rate increased as the components of hardwood BAL increased (Yang *et al.*, 2003). However, hardwood BAL divided by total basal area, or total BAL divided by total basal area, suffer from the same problems noted earlier for relative DBH when they are applied after thinning.

As with stand-level equations, site index has been used to predict individual-tree mortality (Hann and Wang, 1990; Bravo *et al.*, 2001; Pretzsch *et al.*, 2002; Hann *et al.*, 2003, 2006). Hann and Hanus (2001) found site index to have a significant influence on mortality in 8 of the 15 species they analyzed in southwestern Oregon. When it is significant, a higher site index is related to an increased probability of mortality (Hann and Wang, 1990; Bravo *et al.*, 2001; Eid and Tuhus, 2001; Hann *et al.*, 2003; Jutras *et al.*, 2003; Bravo-Oviedo *et al.*, 2006; Hann *et al.*, 2006). This suggests that the level of competition over a given growth/mortality period increases faster on better sites due to accelerated stand development caused by greater tree-growth rates. Other studies have used alternative measures of productivity when modeling tree-level mortality. Yao *et al.* (2001) used the mean height of dominant and codominant trees at 20 cm as a measure of site productivity, and found that mortality increased with an increase in estimated productivity. Jutras *et al.* (2003) used three fertility classes to characterize productivity, and found that increased site productivity was related to higher mortality rates for Scots pine, but a lower mortality rate in pubescent birch. Fridman and Ståhl (2001) found elevation, latitude, and soil moisture to be significant predictors of mortality, while Crecente-Campo *et al.* (2009b) used mean dominant height. As suggested by Keane *et al.* (2001), the use of climatic variables might be important for representing potential effects of climate change on tree mortality.

The prediction of either random or irregular mortality has been limited to a fewer number of studies. Buchman *et al.* (1983) included a separate parameter external to the statistical logistic equation with the intent of characterizing random, background tree mortality. Several tree-list and gap models try to represent irregular tree mortality to some extent (Keane *et al.*, 2001). For example, Keane *et al.* (1996) predicted the probability of a tree dying after a fire, from DBH, species bark thickness, and percentage crown volume scorched. Defoliation caused by foliar diseases and insects is found to improve predictions of mortality when this information is available (e.g. Dobbertin and Brang, 2001). For some insects, like western spruce budworm, equations to predict defoliation and tree probability of mortality are available and incorporated into growth models (Crookston *et al.*, 1990). The influences of several other disturbances are also incorporated in the Forest Vegetation Simulator (FVS; Crookston and Dixon, 2005).

8.4 Mechanistic models of mortality

Most mechanistic equations attempt to predict tree mortality from carbon balances (Hawkes, 2000). Essentially, a tree is predicted to die when it has a negative carbon balance. At least four approaches for predicting tree mortality from carbon balance are used, including: (1) tree foliage and roots cannot be renewed (e.g. Bossel, 1986); (2) the tree has no more foliage to photosynthesize (e.g. Weinstein *et al.*, 1991); (3) no carbon is available to produce foliage (e.g. Friend *et al.*, 1997); and (4) annual net carbon gain is less than zero (e.g. Bugmann *et al.*, 1998).

The effectiveness of these approaches varies (Hawkes, 2000). For example, Mäkelä and Hari (1986) found that their approach caused a decline in accuracy of tree mortality prediction compared to a statistical approach, while Korol *et al.* (1996) found that carbon-based mortality prediction performed reasonably, and the approach emulated a maximum size–density relationship consistent with empirical observations. The mechanistic approach to tree mortality is difficult to apply because of limitations due to: (1) a lack of understanding on the complex array of physiological processes that occur when a tree dies; (2) predicting respiration accurately (see Chapter 12); and (3) estimating the ability of trees to store carbon.

An underutilized mechanistic approach suggested in the review by Hawkes (2000) is to predict mortality based on the growth efficiency concept (Waring, 1983). The advantages of this approach are: (1) growth efficiency is not site- or species-specific; (2) low growth efficiency is a sign of a stressed tree; (3) growth efficiency inherently reflects climatic variation; and (4) growth efficiency decreases with time, so age or size parameters are not explicitly needed (Hawkes, 2000). In an analysis of black spruce mortality, Mailly *et al.* (2004) found growth efficiency to be an effective predictor. However, one potential drawback is that growth efficiency of some species can be low in dominant trees with long crowns (Maguire *et al.*, 1998; Mainwaring and Maguire, 2004).

8.5 Development and application of mortality equations

Modeling mortality is difficult because many different factors influence it, extensive datasets are required for proper model calibration, and it is a relatively rare, episodic event. However, growth simulations are very sensitive to mortality predictions, particularly as projection length increases (Gertner, 1989). Consequently, different methods are used to develop and apply mortality equations. As previously mentioned, these differences include: (1) the type of data used; (2) the statistical methods for estimating parameters; (3) the length of the prediction period; and (4) usage of additional equations to constrain predictions. Finally, mortality equations can be applied stochastically or deterministically.

Data for calibrating mortality equations are derived from: (1) yield tables (e.g. Lee, 1971); (2) temporary plots (e.g. Hann and Wang, 1990; Hann and Hanus, 2001); (3) permanent forest inventory plots (e.g. Monserud and Sterba, 1999); (4) permanent forest research plots (e.g. Hann *et al.*, 2003; 2006); (5) a combination of permanent research and forest inventory plots (e.g. Buchman *et al.*, 1983); and (6) remeasurements from forest inventories and data from mortality surveys collected using large-scale aerial photography (e.g. Wykoff *et al.*, 1982). In temporary plots, mortality has to be reconstructed from tree cores (Bigler and Bugmann, 2004a) and/or published tree deterioration characteristics (Hann and Wang, 1990). Consequently, the exact time of death is never precisely known. In addition, measurements of height to crown base and perhaps total height for dead trees on temporary plots may require adjustment to avoid bias (e.g. Hann and Hanus, 2001), particularly on plots with longer growth periods.

Permanent plots provide a more accurate description of past mortality patterns, but determining the exact time of death is still a challenge if the plots are periodically measured. Regardless of the type of plot used, the most important consideration in collecting mortality data is the utilization of a large enough plot size. A plot size that is too small can result in estimates of mortality rates that exceed growth and will likely cause mortality to be overpredicted. Curtis and Marshall (2005) suggest that the plot size be large enough to differentiate between regular and irregular mortality.

Individual tree mortality is a binary event, which often causes problems when estimating model parameters. Alenius *et al.* (2003) reviewed several of the statistical techniques used in modeling individual-tree mortality, which will be further described in Chapter 14. These have included logistic regression (e.g. Monserud, 1976), artificial neural networks (e.g. Guan and Gertner, 1991a), support vector methods (King *et al.*, 2000), classification and regression trees (CART; Dobbertin and Biging, 1998), Poisson mixture models (Affleck, 2006), generalized estimating equations (Kiernan *et al.*, 2009), semiparametric regression (Vieilledent *et al.*, 2009), and nonparametric Bayesian techniques (Wyckoff and Clark, 2000; Metcalf *et al.*, 2009a).

Comparisons between these techniques vary. For example, Guan and Gertner (1991b) suggested that modeling individual-tree mortality with a logistic function as indicated by the results of Monserud (1976) was effective, while other studies suggest that the alternative techniques are superior to the logistic approach (King *et al.*, 2000; Metcalf *et al.*, 2009a). However, differences between logistic regression and other fitting techniques may be less apparent when the models are applied to an independent dataset (e.g. Hasenauer *et al.*, 2001). Consequently, for most applied purposes, logistic regression is effective for modeling mortality due to its simplicity, robustness, and ease of application.

Flewelling and Monserud (2002) compared methods for fitting logistic tree mortality equations and found that almost all of them produced similar results when data were collected on fixed period lengths. However, weighted nonlinear least squares or maximum likelihood estimation were necessary when data had unequal period lengths (Flewelling and Monserud, 2002). Commonly, observations are weighted by the variable tree expansion factors determined from the design of the plot (Monserud and Sterba, 1999; Bravo *et al.*, 2001; Hann *et al.*, 2006). Such weighting is required if variable probability sampling or fixed area plots of different sizes are used to collect the data.

Evaluating and/or validating mortality models is difficult due to its discrete nature. Hamilton (1974) demonstrated that the mean square error of a dichotomous variable approaches one and not zero as sample size increases, and Flewelling and Monserud (2002) state that the R^2 statistic is "...useless for dichotomous variables." Bravo *et al.* (2001) used Akaike's information criterion (AIC) to evaluate alternative mortality models, which was also recommended by Flewelling and Monserud (2002). Commonly used goodness-of-fit criteria include: chi-square tests (e.g. Hann *et al.*, 2003), classification tables (Hein and Weiskittel, 2010), and receiver operator curves (ROC; King, 2003). The chi-square test takes into account the size of the differences between predicted and actual values but not the signs of those differences.

The runs statistic and associated test (Draper and Smith, 1998) can be used to evaluate whether the signs of the average residuals for the predictor variable classes formed to calculate the chi-square statistics are random over the predictor variable classes. Therefore, the runs test is another aspect to testing whether the model fits the dataset well. A disturbing result from using the chi-square test is its inconsistency of conclusions when applied to the prediction of mortality versus survival, the models of which differ from each other by only a change in sign on the predictor variables. Alenius *et al.* (2003) compared the different goodness-of-fit methods, and concluded that each needs to be applied due to the limitations of using just one approach.

The temporal resolution of a mortality equation depends on the how the mortality data were collected and the desired projection interval. If all observations are made with the same length of growth period and it matches the desired projection interval, then fitting the mortality equation to the data is straightforward and the resulting predictions are the probability of mortality or survival for that length of growth period. However, if the observations are made with variable-length growth periods, or the desired projection interval differs from the growth period used in measurement, then the model form has to be modified.

The usual method is to assume that the Markov property is appropriate, and, therefore, the compound interest formula is applied to adjust for differences in the length of growth periods (Monserud, 1976). This technique is commonly used (Buchman *et al.*, 1983; Hamilton, 1986; Monserud and Sterba, 1999; Yao *et al.*, 2001; Hann *et al.*, 2003; 2006). Flewelling and Monserud (2002) illustrate how this approach is applied to produce annual predictions of mortality or survival rates; however, the same procedure can be used to convert variable-length growth periods to any standard growth period (e.g. Hann *et al.*, 2003; 2006). Given that survival and not mortality is a Markov process, the modified logistic equation must be parameterized using survival data. Once the parameters are determined, then changing their sign results in a predictor of mortality (Flewelling and Monserud, 2002).

Due to the difficulty in accurately predicting mortality and its importance for projection accuracy, various methods are used to constrain both stand- and tree-level mortality predictions. The two-stage approach suggested by Woollons (1998) is effective at improving predictions at both the stand (e.g. Álvarez González *et al.*, 2004) and individual-tree-levels (e.g. Eid and Øyen, 2003). Fridman and Ståhl (2001) proposed a three-step approach to modeling tree mortality: (1) estimating the probability of mortality occurring; (2) quantifying the amount of mortality occurring in terms of proportion of basal area; and (3) distributing the mortality among individual trees. This method was found to provide relatively robust predictions of mortality for mixed-species stands in Sweden (Fridman and Ståhl, 2001).

Rather than using a two- or three-stage approach, Wykoff *et al.* (1982) computed three correction factors, one based on normality, the second being an increase in mortality for trees of advanced age, and the third based on maximum stand basal area. The final predicted probability of mortality is a weighted combination of these three corrections factors and the rate predicted from the tree-level mortality equation (Wykoff *et al.*, 1982). This was later simplified to a correction factor based just on the maximum stand basal area approach (Wykoff, 1986), which is still used in several FVS variants (Crookston and Dixon, 2005).

Hann and Wang (1990) demonstrated how a maximum size–density trajectory could be incorporated into a tree-level mortality equation to guarantee reasonable behavior on long projections. Hann *et al.* (2003) modified this approach to provide more reasonable mortality predictions for trees with a large likelihood of dying. Yang and Titus (2002) presented a similar approach for constraining predictions in mixed-species stands. Finally, Hynynen *et al.* (2002) used two multipliers to adjust predictions from a tree-level equation. The first was a subjectively parameterized sigmoid equation that increased mortality rates for old trees, while the second function increased tree mortality rates to keep the stand on a maximum density–size line (Hynynen *et al.*, 2002).

Monserud *et al.* (2004) raised the question of whether stand-level constraints were necessary for tree-level mortality models. They concluded that the constraints were unnecessary if the mortality modeling dataset was adequate, but necessary if the modeling dataset was inadequate (Monserud *et al.*, 2004). Hann *et al.* (2006) found that a revised mortality equation for Douglas-fir kept predicted stands below the default maximum size–density trajectory for the species, which supported the findings of Monserud *et al.* (2004). However, having an adequate mortality modeling dataset requires it to sufficiently characterize the extremes of the predictor variables and their corresponding response value, which generally requires very large sample sizes spread across the entire population of interest. For example, the species in Monserud *et al.* (2004) that did not require a

Table 8.3 Mortality equation forms used in the example growth models.

Model	Resolution	Equation form[a]	Reference
DFSIM	Stand-level	$\frac{\Delta BA_{Net}}{\Delta BA_{Gross}} = b_0 + b_1 \cdot AGE_{BH} + b_2 \cdot \left(AGE_{BH} \cdot RD \cdot \frac{HT_{DOM}}{QMD}\right) + b_3(THIN \cdot RD)$	Curtis *et al.* (1981)
Scube	Stand-level	$\Delta TPH = \Delta HT_{DOM} \cdot (-b_0 \cdot TPH^{b_1} \cdot HT_{DOM}^{b_2})$	García (2009)
FIBER	Size-class	$PM = b_0 + b_1 \cdot DBH + b_2 \cdot DBH^2 + b_3 \cdot BA_a + b_4 \cdot BA_b + b_5 \cdot BA_b^2 + b_6 \cdot \%BA_{HW} + b_7 \cdot ELEV + b_8 \cdot HAB$	Solomon *et al.* (1995)
—	Size-class	$PM_1 = \left(1.0 + e^{\left(-\left(b_0 + \frac{b_1}{DBH} + b_2 \cdot BA\right)\right)}\right)^{-1}$ $PM_2 = \left(1.0 + e^{(-(b_0 + b_1 \cdot DBH + b_2 \cdot DBH^2 + b_2 \cdot BA))}\right)^{-1}$ $PM_3 = \left(1.0 + e^{\left(-\left(b_0 + \frac{b_1}{DBH}\right)\right)}\right)^{-1}$ $PM_4 = \left(1.0 + e^{\left(-\left(b_0 + \frac{b_1}{DBH} + b_2 \cdot BA\right)\right)}\right)^{-1}$ $PM_5 = \left(1.0 + e^{\left(-\left(b_0 + \frac{b_1}{DBH}\right)\right)}\right)^{-1}$	Vanclay (1989a)

SILVA	Tree-level	$PM = \left(1.0 + e^{\left(-\left(b_0 + b_1 \cdot DBH + b_2 \cdot \left(\frac{\Delta ba}{DBH}\right) + b_5 \cdot \left(\frac{HT}{DBH}\right) + b_4 \cdot SI\right)\right)}\right)^{-1}$	Pretzsch *et al.* (2002)
ORGANON	Stand-level	$\log(QMD_i) = b_0 + b_1 \cdot \log(TPH_i) - (b_0 \cdot b_3) \cdot \left(\frac{TPH_i}{TPH_0}\right)^{b_2}$	Hann *et al.* (2003, 2006); Hann and Hanus (2001)
ORGANON	Tree-level	$PM = \left(1.0 + e^{\left(-(b_0 + b_1 \cdot DBH + b_2 \cdot CR + b_3 \cdot BAL + b_4 \cdot SI + KR)\right)}\right)^{-1}$	Hann *et al.* (2003, 2006); Hann and Hanus (2001)
PROGNAUS	Tree-level	$PM = \left(1.0 + e^{\left(-\left(b_0 + \frac{b_1}{DBH} + b_2 \cdot CR + b_3 \cdot BAL + b_4 \cdot DBH + b_5 \cdot DBH^2\right)\right)}\right)^{-1}$	Monserud and Sterba (1999)

[a] Where DBH is tree diameter at breast height, BAL is basal area in larger trees, HT_{DOM} is stand dominant height, Δba is individual-tree basal area growth, HAB is an indicator for habitat type, ELEV is elevation, BA is total stand basal area, $\%BA_{HW}$ is the proportion of basal area in hardwood species, BA_a is total stand basal area before thinning, BA_b is the total stand basal area after thinning, BA_{Net} is net stand basal area, BA_{Gross} is gross stand basal area, SI is site index, QMD is the quadratic mean diameter, TPH_0 is the initial number of trees, TPH_i is the number of trees at the *i*th measurement, PM is the probability of mortality, KR is a user-optional coefficient calculated in ORGANON to adjust the tree-level mortality equation so that the QMD of the plot at the end of the growth period is on or below the maximum size–density trajectory of the plot, THIN is an indicator for thinning (1 if thinned, 0 otherwise), RD is relative density, BHAGE is breast-height age, the b_is are parameter estimates, and log is the natural logarithm.

stand-level mortality constraint had more than four-times more data when compared to the species that required a stand-level constraint.

Tree-level mortality equations often predict the probability of its occurrence. Consequently, this requires a decision on how to translate this probability into a discrete event (i.e. dead or alive) during simulations. This is often done deterministically by selecting a certain probability threshold when a tree is considered dead, with 0.5 commonly used (Crecente-Campo *et al.*, 2009b). This procedure is common in distance-dependent and individual-tree gap models (Keane *et al.*, 2001). In distance-independent models, an alternative deterministic method is to convert the mortality rate into a probability of survival and multiply the tree's expansion factor by this value. This approach is widely used in several operational growth and yield models, such as ORGANON (Hann, 2011), FVS (Crookston and Dixon, 2005), and PROGNAUS (Monserud *et al.*, 1997).

A similar technique is also utilized in several stand-level models (e.g. Woollons, 1998):

$$TPH_2 = TPH_1 + \delta \cdot (TPH_1 - TPH_2) \quad [8.3]$$

where δ is the estimated probability of survival of all trees. An alternative approach is to use the predicted probability to randomly draw a number from a binomial distribution. In comparing these three different approaches for predicting stand-level mortality, Álvarez González *et al.* (2004) found no significant differences between them.

Similar results were found at the individual-tree-level (Weber *et al.*, 1986). In contrast, Crecente-Campo *et al.* (2009b) found that using a fixed probability threshold to determine mortality provided better predictions of short-term (three to six years) radiata pine mortality, while Bravo-Oviedo *et al.* (2006) suggested that a stochastic approach was superior. Rathbun *et al.* (2010), however, found that the probability multiplier, like the one used in ORGANON (Hann, 2011), was equivalent to a stochastic approach and generally superior to using a threshold.

Regardless, the optimal threshold for classifying a tree as alive or dead is rarely equal to 0.5. For example, Álvarez González *et al.* (2004) and Crecente-Campo *et al.* (2009b) found an optimal threshold of 0.71 for both stand- and tree-level mortality equations, respectively. As demonstrated by Hein and Weiskittel (2010), cutpoint analyses can be used to determine optimal thresholds. Basically, an optimal threshold was determined at the point where the sensitivity and specificity curves cross (Hein and Weiskittel, 2010). However, Crecente-Campo *et al.* (2009b) found that this threshold overpredicted mortality rates and instead recommended that thresholds be optimized for the intended use of the growth model.

8.6 Summary

All growth and yield models struggle to predict mortality accurately due to its rare and highly episodic nature. Although data intensive, most models often explain a limited amount of variation. Consequently, Valentine and Mäkelä (2005) just assumed that a tree dies when its crown ratio was less than 10%, rather than rely on a statistical equation. However, statistical equations of mortality have tended to perform better than theoretical (Bigler and Bugmann, 2004b) and mechanistic approaches (Hawkes, 2000).

At the stand-level, maximum size–density and algebraic difference equations are the most common means for empirically predicting mortality (Table 8.3). Logistic regression is the widely used approach for predicting individual-tree mortality, but a wide variety of other semiparametric and nonparametric statistical techniques have been explored. Often, tree-level predictions need to be constrained by stand-level equations to ensure proper behavior in long-term simulations, except for tree-level equations developed with extensive datasets. Constraining predictions is accomplished by either using two- or three-stage equations, or adjusting the tree-level mortality rate to place it on the stand size–density trajectory.

Tree size and tree vigor variables are commonly used to predict tree-level mortality in both tree-list and gap models. One- and two-sided competition is commonly represented by BAL and stand basal area, respectively. At both the tree- and stand-levels, the influence of site productivity varies, with most studies finding a positive or no effect on mortality. Although generally effective for predicting stand-level mortality in even-aged, single-species stands, age often is not used as a predictor of tree-level mortality. When compared to individual trees, seedling mortality often shows a higher degree of temporal variation (Beckage *et al.*, 2005) and, consequently, fewer models are available to predict its occurrence. In addition, individual seedlings are rarely tagged and tracked, so alternative methods are commonly used to model it from count data (Lavine *et al.*, 2002).

Proper application of mortality equations during short- and long-term simulations is still debated. Mechanistic equations commonly predict mortality to occur when a tree has a negative carbon balance, which is simplistic and has multiple limitations. A better understanding of the physiological mechanisms behind tree death is needed before both mechanistic and statistical equations can be improved. Future work should concentrate on further evaluating physiological and climate-based variables like growth efficiency and annual moisture stress to improve mortality predictions.

9

Seeding, regeneration, and recruitment

9.1 Introduction

Like mortality, statistical, gap, process-based, and hybrid models all struggle to accurately predict regeneration and recruitment. In fact, several growth models do not predict regeneration or recruitment and leave it up to the user as to how they will incorporate it over time (e.g. Arseneault *et al.*, 2008). One reason for this is because seeding, regeneration, and recruitment are all highly stochastic processes and driven by an array of factors (e.g. Quero *et al.*, 2011), many of which are not captured in standard inventory or permanent sample data.

However, regeneration and recruitment processes can have a significant influence on the outcome of growth-model projections, particularly in areas dominated by naturally regenerated forests. *Seeding* is the production, dissemination, and germination of seeds. *Regeneration* is the renewal of forest stands by natural (self-sown seed, coppice, suckers, lignotubers) or artificial means (sowing and planting). *Recruitment* refers to individuals that reach a specified size (e.g. breast height or a specified DBH).

All of these concepts are related, and the distinction depends on some rather arbitrary milestones in the development of an individual from a seed to a small tree. Some of the most important factors influencing seeding are species, tree age, and climatic conditions, while site micro-environment and competition factors drive regeneration and recruitment. Simulations may begin at any stage of development, but many models start with the "established seedling", or when trees reach some specified threshold size, usually based on their height or diameter. In theory, there are some merits in simulating as much of this process as possible, but in practice, this adds considerably to the complexity of a model and may provide no measurable improvement in model predictions.

Forest Growth and Yield Modeling, First Edition. Aaron R. Weiskittel, David W. Hann, John A. Kershaw, Jr. and Jerome K. Vanclay.

Because suitable data for modeling regeneration are often difficult to obtain, frequently recruitment rather than regeneration models are used to model forest renewal. Previous reviews on modeling aspects of seeding, regeneration, and recruitment are found in Price *et al.* (2001) and Miina *et al.* (2006). Since gap models primarily concentrate on predicting changes in forest succession, they have focused, in particular, on all aspects of the regeneration process (Price *et al.*, 2001) and will be featured in this chapter.

9.2 Seeding

Of all of the regeneration processes, seeding is the most difficult to predict. This is because a large number of chance events affect the development and germination of seed; there is a general lack of empirical data to fit models, and an insufficient understanding of the physiological processes involved. The primary components of seeding are: (1) flowering and pollination; (2) seed production; (3) seed dispersal; and (4) seed germination. Various modeling approaches are used to represent each of these aspects of seeding.

9.2.1 Flowering and pollination

Flowering and pollination are controlled by numerous physiological and environmental factors, which make development of prediction models difficult (Greene *et al.*, 1999). Consequently, relatively few models actually predict flowering and pollination. Although it is generally assumed that a certain age is required before trees can start flowering, Greene *et al.* (1999) suggest that a minimum tree size is more likely and suggest a basal area threshold of 0.001 m^2 inside the bark at the base of the bole.

The other predominant factor influencing flowering is weather conditions, particularly temperature. Chuine *et al.* (1999) reviewed the various approaches for modeling the timing of flowering and found that there was no consensus on the best model since results were dependent on species. However, the best models generally used forcing and chilling rates rather than sums of degree-days and chilling days (Chuine *et al.*, 1999). Although there are various modes of pollination, including wind and insects, most models just predict flowering and assume that pollination occurs. Rather than represent the various factors that influence tree flowering and pollination, Leak (1968) simply made them a purely stochastic event.

9.2.2 Seed production

Like flowering and pollination, seed production is also difficult to predict. Price *et al.* (2001) found that only 2 of the 14 gap models they reviewed actually represented aspects of seed production. Generally, seed production is related to tree size. For example, Greene and Johnson (1994) estimated mean annual seed production of trees across a range of species using:

$$\bar{Q} = 3067 \cdot m^{-0.58} \cdot \text{ba}^{0 \cdot 92} \qquad [9.1]$$

where $\bar{Q}$ is mean annual crop size of a single tree, m is seed mass (g), and ba is tree basal area (m^2) (Figure 9.1). The equation has several limitations, but a more comprehensive one has yet to be developed (Greene *et al.*, 1999). For example, Moles *et al.* (2004) found that seed mass was not significantly related to total number of seeds produced by an individual plant throughout its lifetime; while Pairon *et al.* (2006) found that tree DBH was not an accurate predictor of seed production.

Another limitation of the Greene and Johnson (1994) equation is that it ignores the high temporal variation created by environmental conditions. On the other hand, Greene and Johnson (2004) indicated that there were no discernable endogenous cycles to seed production in 22 North American tree species, and the temporal variation across species could be modeled with a single lognormal probability distribution, which is similar to the probabilistic approach used by Rogers and Johnson (1998).

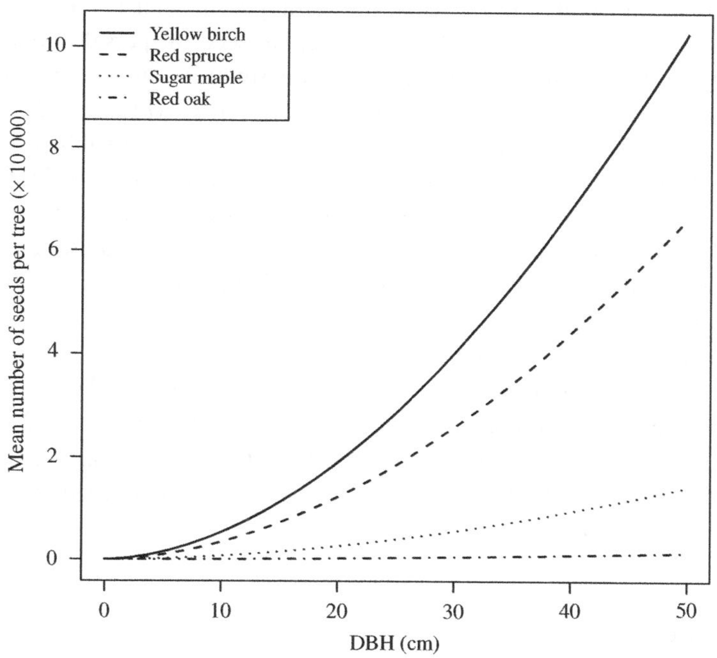

Figure 9.1 Relationship between the mean number of seeds per tree (no. × 10 000) and diameter at breast height (DBH; cm) for several hardwood species. Based on the equation of Greene and Johnson (1994).

In contrast, van der Meer *et al.* (2002) modified the gap model FORGRA so that seed production was a function of several climate variables such as the previous summer's average maximum temperature and the previous fall's radiation. Recently, Calama *et al.* (2011) used a zero-inflated statistical approach to model spatial and temporal variability in stone pine seed production from a variety of climate, stand-, and tree-level factors.

9.2.3 Seed dispersal

There are several means of seed dispersal, with wind and animals being the primary agents. Greene and Calogeropoulos (2002) reviewed the various mechanisms and modeling methods of seed dispersal. They found that empirical dispersal equations work well for small scales (<100 m), but more mechanistic approaches are needed at longer distances (Greene and Calogeropoulos, 2002). Nathan *et al.* (2001) reviewed several mechanistic models of seed dispersal and presented a model that improved upon some of the limitations noted in their review. However, most mechanistic models only predict the influence of wind on seed dispersal, and different methods are needed to properly represent the influence of animals (Pairon *et al.*, 2006). Furthermore, most mechanistic models just focus on isolated individuals, which may limit their applicability.

In contrast, Ribbens *et al.* (1994) presented a seed dispersal model based on comparisons of tree seedling and adult spatial distributions in several mixed-species hardwood stands, and found significant species differences (Figure 9.2). Clark *et al.* (1998; 1999) and Pairon *et al.* (2006) improved upon the model of Ribbens *et al.* (1994), and it is the approach used in the gap model, SORTIE (Pacala *et al.*, 1993). Although the inverse modeling approach of Ribbens *et al.* (1994) is effective, Nanos *et al.* (2010) question the need for tree size in seed dispersal models and found that an unrestricted fecundity model, which allowed fecundity to vary from tree to tree and even be zero, performed significantly better.

Overall, the use of the lognormal distribution function works well, since it is flexible enough to fit a range of observed seed dispersal curves, and provides a framework for mechanistic interpretation compared to other approaches (Greene *et al.*, 2004). Among gap models, PICUS (Lexer and Hönninger, 2001), SORTIE (Pacala *et al.*, 1993), and ZELIG (Smith and Urban, 1988) are some of the few that actually predict seed dispersal patterns (Price *et al.*, 2001).

9.2.4 Seed germination

Species have different microsite and environmental requirements needed to promote seed germination. In general, large-seeded species are purported to have higher survivorship than small-seeded species. However, Moles and Westoby (2004) found that the magnitude

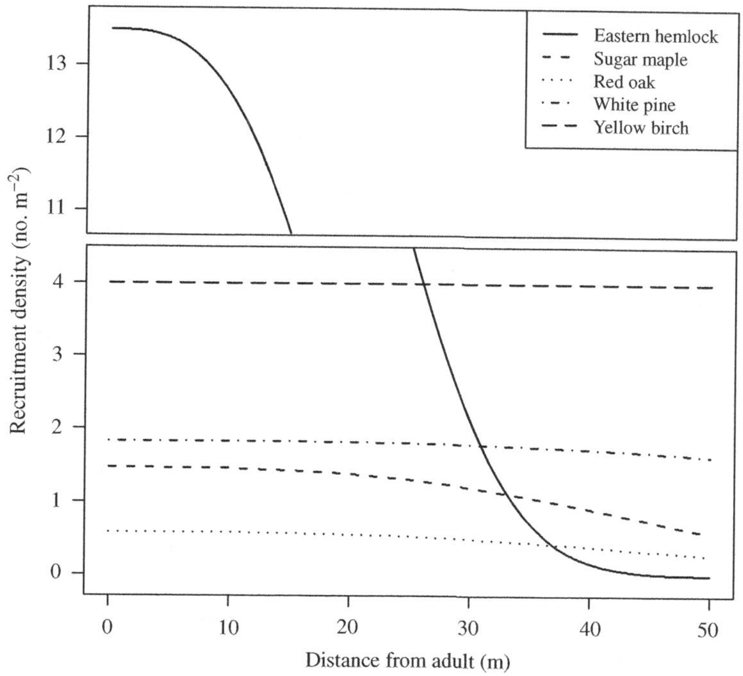

Figure 9.2 Relationship between the recruitment density (no. m^{-2}) and the distance from the adult, for several species in the northeastern USA. Based on the equations of Ribbens *et al.* (1994).

of the advantage was generally not enough to counterbalance the greater number of seeds produced by small-seeded species. Wilson and Maguire (1996) modeled germination density and survival of two conifer species in Maine, and found that both soil substrate and stand structure were highly influential.

In contrast, several gap models predict seed germination as a function of light availability, temperature, and moisture conditions in the litter layer (Price *et al.*, 2001). For example, the gap model PICUS (Lexer and Hönninger, 2001) modifies a species-specific potential germination rate by the amount of regenerable area and average environmental response of all species represented in the seed potential, which itself is a function of available light and chilling degrees. Stochastic approaches also are used to predict seed germination (e.g. Leak, 1968).

9.3 Regeneration

Rather than predict the aspects of seeding, *regeneration models* are those that simulate development of trees from seed or established seedlings, and thus attempt to take into account more of the factors influencing the process of stand renewal. Some regeneration models recruit seedlings directly into the main model, but most simulate growth of trees from seedlings to breast height within a separate regeneration model (e.g. Ek and Monserud, 1974). This approach is sufficiently flexible that almost any size may be used as the criterion for recruitment into the main stand.

One advantage of this approach is that it more realistically models the time taken for regeneration to be recruited following a reduction in stand density due to harvesting. Although germination provides a suitable starting point, many models start with an "established" seedling that has survived its first year following germination. Several regeneration models for hardwood species have been developed, particularly for oak-dominated stands (Loftis, 1990; Dey *et al.*, 1996; Rogers and Johnson, 1998; Gould *et al.*, 2006; 2007).

Growth and development of regeneration can be modeled in several ways. Hett and Loucks (1968) used age-based Leslie matrices, but one problem with this approach is that age is not a good predictor of understory growth. Ek and Monserud (1974) adopted a more sophisticated approach to predict recruitment into their stochastic, spatial, single-tree model. Their regeneration model used cohorts representing numbers of stems for each species and age in each subplot within the main plot being simulated. Each year, germinants or trees in the understory may die or survive and grow in height by an amount predicted from cover, species, and age. When trees reached breast height they were recruited into the main model. If trees did not attain this height within a specified time (e.g. 25 years for black spruce), they "died."

Monserud and Ek (1977) refined this model, improving the efficiency by reducing the number of cohorts modeled. They assumed that understory tree size was more relevant than tree age, and modeled development of trees to 7.6 m in height using five height cohorts of varying sizes and movement ratios. The height increment of the mean tree in each cohort was predicted from potential height increment (a function of height and site), overstory competition index (a relative size–distance index), shade tolerance (a function of species and height), and stand density.

Likewise, Ritchie and Hamann (2006; 2008) developed equations to predict height, basal diameter, and crown width increment for Douglas-fir and several competing shrub and hardwood species. These equations were incorporated into the CONIFERS young stand growth model (Ritchie, 2009a). In addition to using measures to describe overstory and regeneration competition, Hasenauer and Kindermann (2002; 2006) found that a variable representing the incidence of light at stand edges was needed to ensure proper prediction of regeneration growth.

Hasenauer and Merkl (2001) found artificial neural networks to be more effective at predicting regeneration establishment than regression equations such as used by Vanclay (1989a). Furthermore, Hasenauer and Kindermann (2002) indicated that artificial neural networks were effective predictors when regeneration data were not available. Nearest neighbor imputation techniques also are used to estimate regeneration characteristics. For example, Ek *et al.* (1997) used a tabular approach based primarily on stand age following harvest and site index, while Hassani *et al.* (2004) used both tabular and most similar neighbor (MSN) approaches with the latter giving better results.

One alternative to developing a statistical approach is to predict regeneration by an expert system such as REGEN, which was first presented by Loftis (1989) and later refined by Vickers *et al.* (2011). REGEN uses species-specific size classes, namely germinants, small (<61 cm tall), medium (61–122 cm tall), and large seedlings (> 122 cm tall) as well as potential stump-sprouts (trees > 122 cm tall and with DBH $\geq$ 5 cm), to represent forest regeneration. Each plot is stochastically populated with root-suckers and new seedlings. Future species composition is predicted based on an expert-determined species–size–source-specific competition ranking ranging on a scale from 1 to 8. Using these competition rankings and a series of rules, six individuals on a 40.5 m^2 plot are selected for dominance and projected forward. Vickers *et al.* (2011) improved REGEN by defining competition rankings for different site classes and found realistic predictions, but other statistical regeneration models were found to perform better.

Regardless of the approach used, stochasticity is an important element of any regeneration model given the high unexplained variance of deterministic functions. Miina and Heinonen (2008) demonstrated this importance in planted Norway spruce stands that had varying levels of natural regeneration. They used an existing model of regeneration establishment (Miina and Saksa, 2006), and incorporated the variance–covariance of the model into a simulation framework. They found that the amounts and costs of pre-commercial thinnings were properly estimated when this stochastic approach was used, which was not the case when just a deterministic function was utilized.

9.4 Recruitment

Recruitment models predict trees reaching a specified threshold size, usually based on height (e.g. breast height) or diameter (e.g. 10 cm DBH). Generally, two types of recruitment models are distinguished: (1) static and (2) dynamic. *Static* approaches take relatively little account of stand conditions and thus predict a fairly constant amount of recruitment, indicating the long-term average expectation under "typical" conditions. *Dynamic* approaches respond to stand conditions, predicting recruitment as a function of stand density, composition, and other parameters.

9.4.1 Static

Some recruitment models assume that amounts of recruitment observed during the period of data collection reflect long-term averages, and that these amounts do not vary greatly over the simulation timeframes. This assumption is common in many stand table and matrix modeling approaches. For example, the matrix model of Mendoza and

Setyarso (1986) assumed a constant amount of recruitment at each cycle. In addition, positive fecundity coefficients in transition matrices predict an exponential increase in tree numbers, with the number of recruits increasing proportionally with the number of trees in the stand table. Although this is easy to implement in a matrix model, it is unrealistic and unlikely to provide acceptable results. More satisfactory approaches are illustrated in Liang's (2010) generalized matrix model, where recruitment was negatively correlated with stand basal area, and in Bosch's (1971) Leslie matrix model, which allowed regeneration to occur only after the death of another tree.

The special case of even-aged regeneration (natural or artificial) following clearfelling is often modeled by predicting future stand structure at a specified point in time. Alder's (1979) cohort model commenced with a prediction of stand structure when dominant height was 7 m. Belcher *et al.* (1982) predicted regeneration entering the STEMS projection system 15 years after clearfelling, using subjective decision trees based on site characteristics and details of the original stand. Although these approaches are rather empirical, they may provide useful estimates of recruitment for stands that do not differ greatly from the source stands used for model development.

9.4.2 Dynamic

When proper data are available, a dynamic approach is often much more effective than a static approach. Several dynamic models are published, and range from highly empirical to those with a biological basis. Variables commonly used include site productivity, stand density (e.g. stand basal area, stem number, competition indices), species composition of the prior or residual stand, site preparation methods, and the nature of and time since harvesting. Including details of the last harvest and site preparation methods in a model may help to obtain a good fit to the data, but also may limit applications of the model if this information is not known. Where several species are involved, models may predict recruitment for each species group independently, or may predict a total and assign it to candidate species. The primary distinction in dynamic recruitment models is whether they are one- or two-stage.

In a one-stage recruitment model, frequency of ingrowth is predicted with a single equation. This approach is well illustrated by Shifley *et al.* (1993), who developed a model to predict recruitment of six tree species in mixed hardwood stands. The analysis of Shifley *et al.* (1993) also is unique in that it allowed for multiple threshold sizes. Recruitment can only be defined relative to a threshold size, and one threshold may not suit all applications equally. The smallest threshold consistent with permanent plot procedures may provide the best model, but operational inventories may adopt different standards and may use a larger threshold size for measurement of trees (Figure 9.3). Data "censorship" occurs when the threshold in the model is less than the minimum diameter recorded in operational inventory, and degrades model performance. Shifley *et al.* (1993) assumed that site occupancy is revealed by crown competition factor (CCF) and that recruitment could

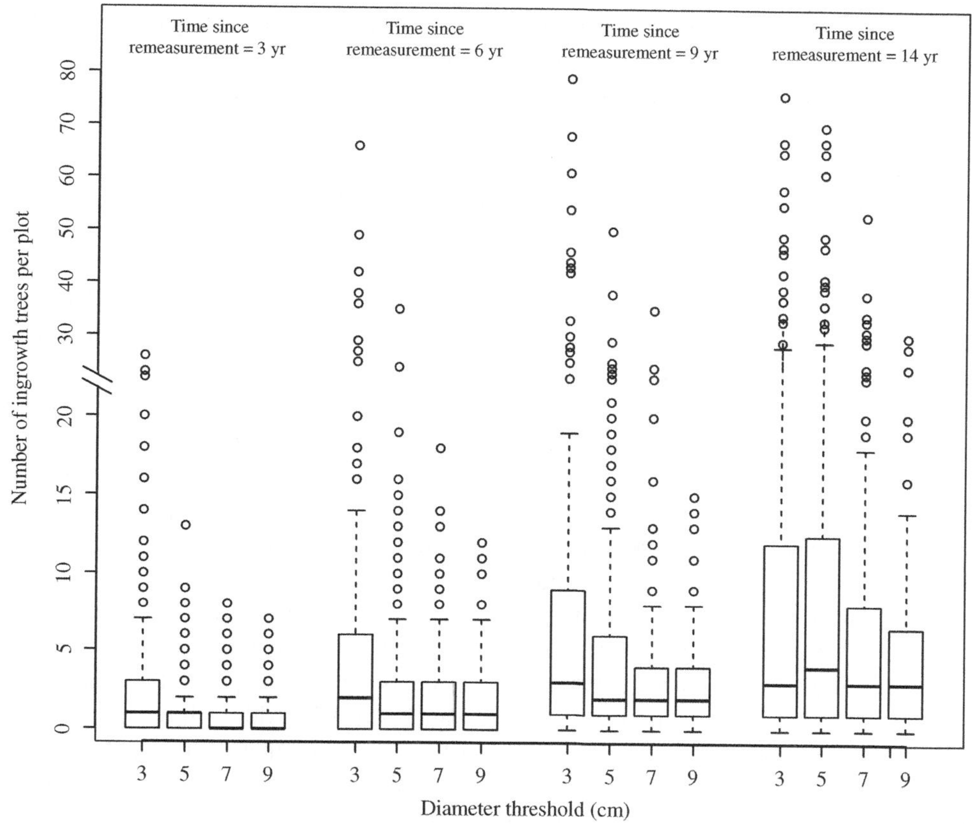

Figure 9.3 Number of ingrowth trees per plot by threshold diameter and time since first measurement for mixed-species Acadian stands in New Brunswick, Canada. Based on permanent sample plot data from the New Brunswick Department of Natural Resources.

potentially be equal to the number of trees of threshold size need to lift the CCF to the maximum for a site. Actual recruitment was predicted from potential based on forest type and ingrowth threshold size.

One difficulty in modeling recruitment is the great variability in regeneration rates among species, spatially over the stand, and over time. Much of the variability seen in recruitment data reflects the fact that during any period regeneration may or may not occur. This characteristic (none/some recruitment) becomes especially apparent when individual species or species groups are modeled. Such data are typically modeled using a two-stage approach, which helps to satisfy the usual statistical assumptions and generally improve predictions.

In a two-stage approach, the probability that some recruitment will occur is predicted using a logistic function. Then a conditional function is used to predict amount of recruitment, given that some is known to occur. The conditional function generally is estimated using ordinary linear regression. This type of approach is widely used to predict recruitment in a variety of forest types (Ferguson *et al.*, 1986; Vanclay, 1992; Solomon *et al.*, 1995; Schweiger and Sterba, 1997; Adame *et al.*, 2010b). Increased realism from application of the two-stage approach can occur when the approach is applied to each plot in a grid of small sample plots distributed over the stand to increase spatial variation, and when the approach is applied at multiple growth periods after treatments to increase temporal variation (Ferguson *et al.*, 1986).

Similar to the conditional approach, Fortin and DeBlois (2007) used zero-inflated models to predict tree recruitment. The advantage of zero-inflated models is that they combine the two components of the two-stage approach into a joint distribution of probabilities (Fortin and DeBlois, 2007).

Often, not only is the frequency of recruitment of interest, but also its species composition. Some models stochastically determine amount and identity of species present (e.g. Ferguson *et al.*, 1986). Others have developed equations to predict species proportion from stand competition measures, maximum DBH for a species, and other factors (Hasenauer and Kindermann, 2006). This approach succeeds when the number of species predicted is relatively low (e.g. <5), but becomes increasing more difficult to apply as numbers of species increase. For example, Vanclay (1992) developed a recruitment model for over 100 species in a north Queensland tropical rainforest. To accomplish this, species were grouped into 8 groups and then predictive models were developed, rather than trying to fit over 100 separate equations by species (Vanclay, 1992). Although this approach is effective for short-term projections, Vanclay (1992) suggested that the best approach for long-term projections was to use relative abundance of species in the original stand under the condition that the species persist in the simulated stand.

9.5 Summary

This chapter is concerned primarily with natural regeneration and recruitment in uneven-aged stands. Planting (and to a lesser extent, sowing) after clearfelling is relatively easy to model, and was not considered here. Regeneration within established plantations may be negligible, and many plantation growth models disregard any recruitment. Many models for uneven-aged forests also assume that recruitment is negligible or will not influence short-term estimates of yield. This assumption is unsatisfactory for longer simulations of natural forests, since recruitment may contribute substantially to future stand basal area and thus influence growth and yield forecasts.

As in all model components, there is no single best strategy for all occasions, and the most suitable approach to use depends on the situation. For example, detailed seeding approaches may not be warranted in all yield studies, but may be relevant in models used to analyze silvicultural alternatives for intensively managed stands. However, it is also important to note that seeding is a regular and frequent process, but recruitment only occurs when growing space becomes free. One aspect that has been underrepresented in most regeneration and recruitment models is the influence of animals. The gap model FORGRA is one of the few that actually attempts to mechanistically represent the influence of animal browse (Jorritsma *et al.*, 1999).

Many open questions remain about how to model interactions between regeneration and climate (Price *et al.*, 2001). In modeling recruitment and regeneration, the best approach may be to provide users with an array of options (deterministic, stochastic, etc.) and allow them to assess consequences of these different approaches. Another important consideration when using a regeneration/recruitment model linked to an overstory tree model is to ensure a proper transition between the two models. For example, FVS uses a weighted average of height increment to ensure a proper representation when a tree transitions from a small tree into a large tree (Crookston and Dixon, 2005).

10

Linking growth models of different resolutions

10.1 Introduction

Growth models have both apparent and functional resolution (Ritchie and Hann, 1997a). *Apparent* resolution is the maximum required level of resolution as determined by the model's required inputs. *Functional* resolution is determined by how the model operates and behaves. Both the apparent and functional resolution of a growth and yield model have significant implications for its accuracy. For example, individual-tree growth and yield models can provide unbiased estimates of growth, but can be less accurate in projecting mortality when compared to a stand-level approach. This suggests that there are important advantages and disadvantages of each approach to growth modeling that need to be recognized by both modelers and model users.

Most models have the same apparent and functional resolutions, but there are significant efforts to link growth models of different resolutions in an attempt to avoid the shortcomings of any single approach. The advantages of a linked model are that it can significantly reduce projection errors, constrain predictions, and maintain multiple resolutions. Although there are several different examples of linked growth models, this chapter will focus on the two most common classes: (1) linked stand-level and size-class models and (2) linked stand- and individual-tree-level models.

10.2 Linked stand- and size-class models

Stand-level growth models project changes in total basal area, number of trees, and volume, while size-class models maintain frequencies of individuals by diameter at breast

---

Forest Growth and Yield Modeling, First Edition. Aaron R. Weiskittel, David W. Hann, John A. Kershaw, Jr. and Jerome K. Vanclay.

height (DBH) classes in their predictions. Stand-level models are relatively simple and have low dimensionality, but can provide highly accurate predictions of both growth and mortality. Size-class models have the ability to represent variability of individuals in a stand, which is important for understanding response to management and stand structure. Two primary methods are used for linking stand and size-class models: (1) the use of parameter recovery techniques; and (2) modified stand table projections.

10.2.1 Parameter recovery

Parameter recovery works by providing a parameter estimator of a probability or cumulative distribution that is a function of stand-level variables (Figures 10.1 and 10.2).

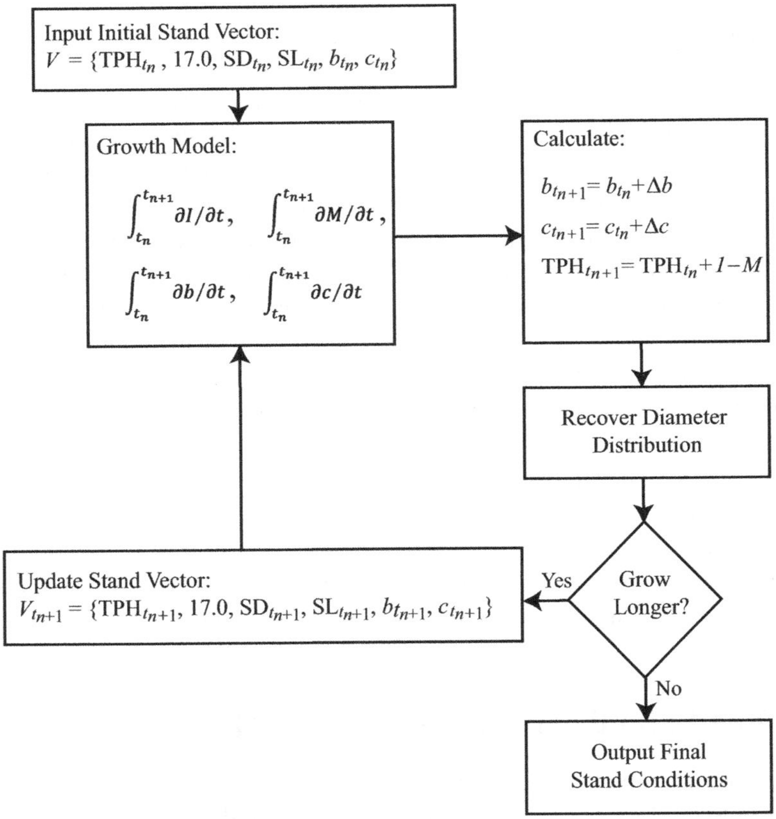

Figure 10.1 Flowchart of parameter prediction model proposed by Hyink and Moser (1983). *V* is total volume (m^3 ha^{-1}), TPH is stem density (no. ha^{-1}), SD is sum of the tree diameters, SL is sum of the tree diameters larger than 18 cm, *b* is the Weibull scale parameter, *c* is the Weibull shape parameter, *I* is ingrowth (no. ha^{-1}), and *M* is mortality (no. ha^{-1}).

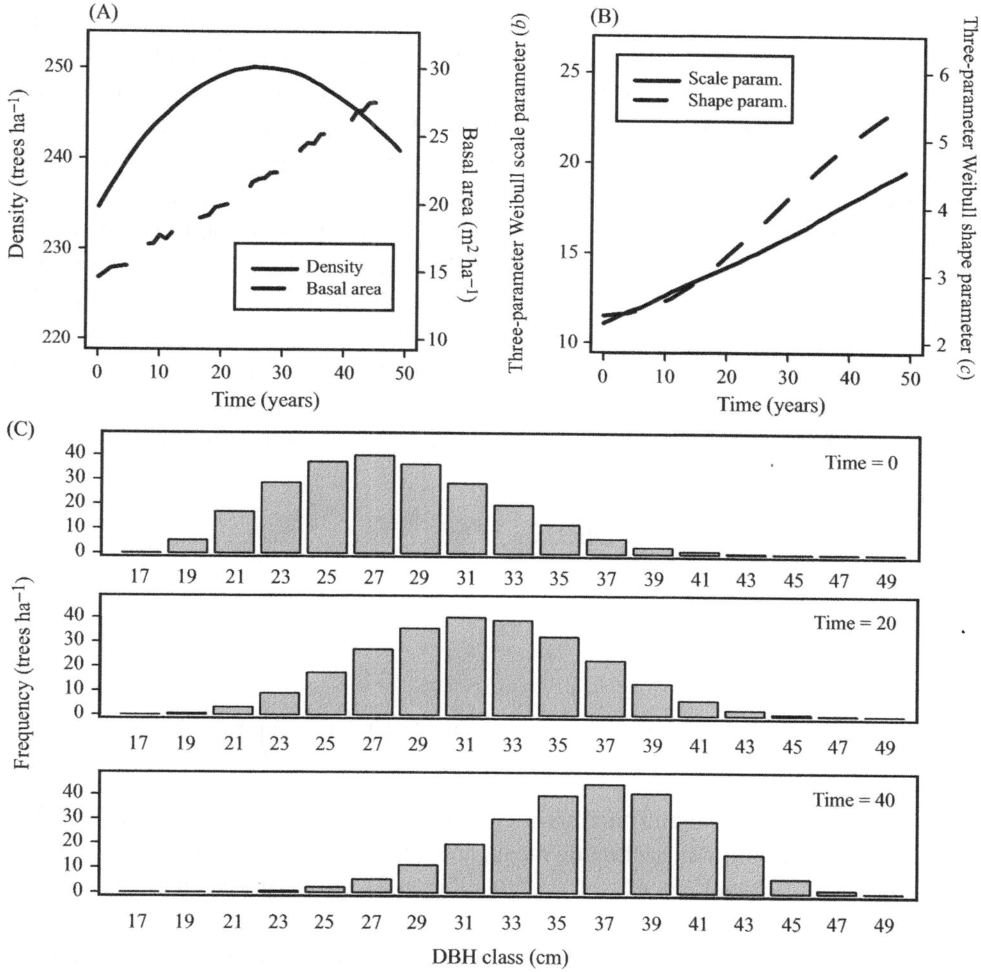

Figure 10.2 Simulation results from the parameter prediction model proposed by Hyink and Moser (1983): (A) Number of trees (ha^{-1}) and basal area (m^2 ha^{-1}) over time; (B) the Weibull scale and shape parameters over time; (C) recovered diameter distributions for three points in time based on projected density, scale, and shape parameters.

The two distributions most commonly used with parameter recovery estimators are the Weibull (Hyink and Moser, 1983) and Johnson's Sb (Knoebel and Burkhart, 1991; Fonseca *et al.*, 2009) distributions. These estimators can be both moment- (Burk and Newberry, 1984) and percentile- (Bailey *et al.*, 1989) based. In the moment-based approach, the parameters are estimated from the first-, second- and third-order moments of the diameter distribution (i.e. mean, variance, and skewness). In the percentile-based approach, the parameters are recovered from various percentiles of the diameter distribution.

For example, Bailey *et al.* (1989) used the quadratic mean diameter, minimum diameter, and the 25th, 50th, and 95th percentiles to estimate parameters of the Weibull distribution.

A hybrid approach that uses both moment- and percentile-based estimators was successfully applied by Baldwin and Feduccia (1987). Although the percentile-based and hybrid approaches perform better than the moment-based approach (e.g. Cao, 2004), moments are the preferred method in growth and yield models because they ensure numeric compatibility and generally require fewer equations. In addition, percentile methods have several other disadvantages such as not providing an unique solution, difficulty in selecting appropriate percentiles, and limited biological interpretability (Fonseca *et al.*, 2009).

Since most stand-level growth models do not predict variance or skewness of diameter distributions through time, alternative estimators are used. For example, Cao *et al.* (1982) iteratively estimated the Weibull parameters based on minimum, arithmetic mean, and quadratic mean diameters:

$$
\begin{aligned}
a &= 0.5 \cdot \mathrm{DBH}_{\mathrm{Min}} \\
b &= \frac{(\overline{\mathrm{DBH}} - a)}{\Gamma\left(1 + \frac{1}{c}\right)} \\
0 &= \mathrm{QMD}^2 + a^2 - 2 \cdot \overline{\mathrm{DBH}} \cdot \left(\Gamma\left(1 + \frac{2}{c}\right)\right)
\end{aligned}
\qquad [10.1]
$$

where a, b, and c are the Weibull parameters; $\Gamma()$ is the gamma function; and $\mathrm{DBH}_{\mathrm{Min}}$, $\overline{\mathrm{DBH}}$, and QMD are the minimum, arithmetic mean, and quadratic mean diameters, respectively. Given that these variables are all well correlated in even-aged stands, equations to predict them from other stand-level variables are easily developed (Fonseca *et al.*, 2009). The utility of this approach is demonstrated several times in long-term growth projections (Cao *et al.*, 1982; Castedo-Dorado *et al.*, 2007b; Newton, 2009). Although the technique is likely more effective in even-aged and single-species stands, Lynch and Moser (1986) demonstrated its usefulness in uneven-aged and mixed-species stands.

When compared to more complex approaches that required a tree list, Qin *et al.* (2007) found that the parameter recovery approach produced similar results in terms of precision, but gave biased predictions of total and merchantable volume. Knowe *et al.* (1997) found that an individual-tree approach gave the best representation of observed diameter distributions because parameter recovery techniques showed a systematic bias with planting density. Similar results were also obtained by Borders and Patterson (1990). However, the technique is effective when a tree list is not available.

An alternative approach is to project a diameter distribution through time by using a theoretical diameter growth equation derived from a diameter probability distribution. This technique was first demonstrated by Bailey (1980), who assumed that trees did not change their relative sizes over time. The implied diameter growth equation for the Weibull distribution is:

$$\mathrm{DBH}_2 = a_2 + b_2 \cdot \left(\frac{\mathrm{DBH}_1 - a_1}{b_1}\right)^{\left(\frac{c_1}{c_2}\right)} \qquad [10.2]$$

where DBH_i is the DBH and a_i, b_i, and c_i are the Weibull parameters at the first ($i = 1$) and second ($i = 2$) measurements. Cao (1997) further extended this technique to account for mortality. Essentially, the approach involves using parameter recovery to describe the initial stand, projecting changes in stand density with a stand-level equation, determining the diameter distribution parameters after mortality using parameter recovery, and then projecting the growth of each diameter class using the generalized equation of Bailey (1980). The approach was found to generally work better than the parameter recovery technique, but the level of improvements were not statistically significant at the 95% level of confidence (Cao, 1997).

In contrast to relying on a predefined probability density function, Tang *et al.* (1997) used a distribution-independent approach to define relationships between current and future stand diameter distributions. A significant limitation of this approach is that the parameters in equation [10.2] may be a function of age, stand height, or other variables that change with age, which ignoring would result in biased predictions (Oscar Gárcia, personal communication).

10.2.2 Modified stand table projection

Stand table projection is a size-class based approach that involves growing trees by moving them from one diameter class to the next. Various approaches are used to implement stand table projections (see Chapter 4), but all require a method for estimating diameter increment and often mortality for longer projections. However, these equations or the data needed to develop them often are not available. To address this limitation, Clutter and Jones (1980) and Pienaar and Harrison (1988) used stand-level growth and mortality equations to update stand tables. Their approach relied on the assumptions that relative tree size can be used to allocate stand-level mortality, relative tree size generally stays constant over time, and trees within each diameter class are uniformly distributed. However, the approach needed an empirical equation that described changes in relative tree size over time.

Instead of relying on an empirical equation, Nepal and Somers (1992) used the Bailey (1980) method to derive a theoretical diameter increment equation. An algorithm was then used to ensure that basal area and survival matched observed values or predictions made by stand-level equations. Nepal and Somers (1992) found that this technique was superior to Pienaar and Harrison's (1988) method and was still effective in stands with an irregular structure. Cao and Baldwin (1999) further improved the algorithm of Nepal and Somers (1992) by using constrained least squares. However, Trincado *et al.* (2003) found that the approach of Nepal and Somers (1992) was superior to the one of Cao and Baldwin (1999) when predicting *Eucalyptus* plantations in Chile, particularly for projections longer than four years.

When individual-tree diameter increment and survival equations are available, Cao (2007) showed that a modified, constrained least squares technique works better than the approach of Nepal and Somers (1992). However, this approach would only be applicable if those individual-tree equations were available and a tree list was not. Otherwise, one of the approaches described below would likely be more effective.

10.3 Linked stand- and tree-level models

Individual-tree models are preferred because they maintain information on each tree in a stand, which provides the highest level of resolution and allows various stand structures to be successfully simulated. However, individual-tree models are an example of passive aggregation (Ritchie and Hann, 1997a), which means that changes in stand-level variables like total basal area are estimated from tree-level projections and not simulated directly. This indicates that individual-tree models are unconstrained and may provide unrealistic projections of stand-level attributes. Three primary approaches are used to link stand- and tree-level models: (1) disaggregation; (2) constrained parameters; and (3) combined.

10.3.1 Disaggregation

Disaggregation is the most common means for linking stand- and tree-level models (see review of Ritchie and Hann, 1997a). In this approach, growth and mortality are projected at the stand-level and then allocated to individual trees. The advantages of this approach are that it provides flexibility and numerical compatibility, and stand-level estimates are not influenced by tree-level projections (Ritchie and Hann, 1997a). A variety of approaches (Table 10.1) are used to disaggregate stand growth to individual trees, including: (1) proportional yield (Clutter and Allison, 1974; Somers and Nepal, 1994); (2) proportional increment (Campbell *et al.*, 1979; Moore *et al.*, 1994); (3) constrained least squares (Matney *et al.*, 1990; Cao and Baldwin, 1999); (4) coefficient adjustment (Qin and Cao, 2006); and (5) additive functions (Harrison and Daniels, 1988; Dhôte, 1994).

In proportional yield allocation, tree-level predictions are adjusted proportionally by their predicted final value and the stand-level estimates (e.g. Somers and Nepal, 1994). This concept was widely used in the past for a range of purposes (Clutter and Allison, 1974; Clutter and Jones, 1980; Pienaar and Harrison, 1988; McTague and Stansfield, 1994; 1995). In proportional increment allocation, stand-level increment is disaggregated to individual trees based on tree relative position in the stand and other variables. The technique is used to disaggregate diameter (Leary *et al.*, 1979), basal area (Campbell *et al.*, 1979; Moore *et al.*, 1994), and volume increment (Dahms, 1983; Zhang *et al.*, 1993). For example, Zhang *et al.* (1993) found that relative size increment was related to a quadratic expression of relative size in the stand:

$$\mathrm{RTG} = b_0 + b_1 \cdot \mathrm{RTS} + b_2 \cdot \mathrm{RTS}^2 \qquad [10.3]$$

where RTG is relative tree increment (tree increment divided by stand total increment), RTS is relative tree size, and the b_is are plot-specific parameters. Zhang *et al.* (1993) indicated that the b_is were dependent on initial number of trees, mean tree size, and the variation of DBH within the stand. However, the limitation of this approach is that it doesn't maintain invariance. Also, Ritchie and Hann (1997b) found that the parameters of equation [10.3] were not significantly related to any stand-level variables, which might explain why it was one of the poorest performing disaggregation approaches they examined.

The constrained least squares method (Matney *et al.*, 1990; Cao and Baldwin, 1999) adjusts tree-level predictions based on constraints of estimated survival, average diameter, and stand basal area from stand-level equations. For example, Qin and Cao (2006) used the following estimator to adjust tree survival probability:

$$\widetilde{\mathrm{PS}}_{i,t+q} = \frac{\left(\widehat{\mathrm{PS}}_{i,t+q} - \left(\sum_j \widehat{\mathrm{PS}}_{j,t+q} - s \cdot \widehat{\mathrm{TPH}}_{t+q}\right)\right)}{n} \qquad [10.4]$$

where $\widetilde{\mathrm{PS}}_{i,t+q}$ is adjusted tree survival probability for the ith tree, t is the beginning of the growth period, q is the length of the growing period, s is the plot size, $\widehat{\mathrm{PS}}_{i,t+q}$ is predicted tree survival probability for the ith tree with a tree-level equation, $\widehat{\mathrm{TPH}}_{i,t+q}$ is predicted number of trees based on a stand-level equation, and n is the number of trees on a plot.

Cao and Baldwin (1999) found that the constrained least squares method significantly improved predictions of future stand development when compared to parameter recovery and stand table projection techniques of Nepal and Somers (1992). The coefficient adjustment method of Qin and Cao (2006) added an additional parameter to each tree-level equation to ensure predictions are identical to those made by stand-level equations.

Additive functions are similar to the coefficient adjustment method of Qin and Cao (2006), but predictions are adjusted rather than model parameters. Examples of additive functions are given in Harrison and Daniels (1988), Dhôte (1994), and Ritchie and Hann (1997b). Additive functions assume that tree increment is linearly related to tree size in any given stand. The relationship has been found to be dynamic and varies from stand to stand, but the changes can be related to stand-level factors. Harrison and Daniels (1988) modeled a stand additivity parameter as a function of site index and dominant height. However, Ritchie and Hann (1997b) found a weak relationship between this parameter and site index and suggested the approach was best suited to young stands (<50 years of age). One limitation of this approach is that it can predict negative increment if not properly specified. The approach of Dhôte (1994) avoided this limitation by employing a truncated distribution of increment such that trees with diameters below an increment threshold are predicted to have increment rates of zero.

Table 10.1 Example of various disaggregation approaches used for predicting tree-level growth, adapted from Ritchie and Hann (1997a).

Source	Dependent variable
Proportional yield	
Clutter and Allison (1974)	Tree basal area
Clutter and Jones (1980)	Tree basal area
Pienaar and Harrison (1988)	Tree basal area
Nepal and Somers (1992)	Tree basal area
Somers and Nepal (1994)	Tree diameter
McTague and Stansfield (1994)	Tree basal area
McTague and Stansfield (1995)	Tree basal area
Qin and Cao (2006)	Tree diameter squared
Proportional increment	
Campbell *et al.* (1979)	Tree basal area increment
Leary *et al.* (1979)	Tree diameter increment

Function[a]
$\mathrm{ba}_{2i} = \overline{\mathrm{ba}}_2 \cdot \left(\frac{\mathrm{ba}_{1i}}{\overline{\mathrm{ba}}_1}\right)^{\left(\frac{A_2}{A_1}\right)^{0.17192}}$
$\mathrm{ba}_{2i} = \mathrm{BA}_2 \cdot \left(\frac{1}{\sum\left(n_{2i} \cdot \frac{\mathrm{ba}_{1i}}{\overline{\mathrm{ba}}_1}\right)^{\left(\frac{A_2}{A_1}\right)^{0.0578109}}}\right) \cdot \left(\frac{\mathrm{ba}_{1i}}{\overline{\mathrm{ba}}_1}\right)^{\left(\frac{A_2}{A_1}\right)^{0.0578109}}$
$\mathrm{ba}_{2i} = \mathrm{BA}_2 \cdot \left(\frac{1}{\sum\left(n_{2i} \cdot \frac{\mathrm{ba}_{1i}}{\overline{\mathrm{ba}}_1}\right)^{\left(\frac{A_2}{A_1}\right)^{0.2333}}}\right) \cdot \left(\frac{\mathrm{ba}_{1i}}{\overline{\mathrm{ba}}_1}\right)^{\left(\frac{A_2}{A_1}\right)^{0.2333}}$
$\mathrm{ba}_{2i} = g \cdot d_{2i}^2 \cdot \left(\beta_0 \cdot \mathrm{e}^{\left(\beta_1 \cdot \breve{d}_{2i}\right)}\right);\ \breve{d}_{2i} = \mathrm{f}(d_{1i}, \theta)$
$d_{2i} = \left(\left(\left(\frac{\delta_f \cdot \left(d_{1f} + \Delta d_f\right) - d_{1f}}{\Delta d_f}\right) \cdot \Delta d_i + d_{1i}\right) \cdot \left(d_{1i} + \Delta d_i\right)^{-1}\right) \cdot \left(d_{1i} + \Delta d_i\right)$
$\mathrm{ba}_{2i} = \mathrm{BA}_2 \cdot \frac{\tilde{d}_{2i}}{g \cdot \sum \tilde{d}_{2i}};\ \tilde{d}_{2i} = \mathrm{f}\left(d_{1i}, \frac{\mathrm{BA}_2}{\mathrm{BA}_1}, \frac{N_2}{N_1}, D_{\mathrm{q}1}, t_2 - t_1\right)$
$\mathrm{ba}_{2i} = \mathrm{BA}_2 \cdot \frac{\tilde{d}_{2i}^2}{g \cdot \sum \tilde{d}_{2i}^2};\ \tilde{d}_{2i} = \mathrm{f}\left(d_{1i}, \frac{\mathrm{BA}_2}{\mathrm{BA}_1}, N_1, \mathrm{BA}_1, \mathrm{sp}, t_2 - t_1\right)$
$d_{2i}^2 = d_{1i}^2 \cdot \left(\frac{40\,000 \cdot s \cdot \mathrm{BA}_2}{\pi \cdot \sum_j^n \left(P_{2j} \cdot d_{2j}^2\right)}\right)$
$\Delta \mathrm{ba}_i = \Delta \mathrm{BA} \cdot r \cdot \mathrm{ba}_{1i}^{0.93}$
$\Delta d_i = \left(N_2 \cdot \overline{\Delta d}\right) \cdot \left(\kappa_1 + 10^{\beta_0 + \beta_1 \cdot \log(\mathrm{GS}_i + \kappa_2) + \beta_2 \cdot \log(\mathrm{GS}_i + \kappa_2)^2 + \beta_3 \cdot \log(\mathrm{GS}_i + \kappa_2)^3}\right)$

(continued)

Table 10.1 (*Continued*)

Source	Dependent variable
Dahms (1983)	Tree volume increment
Zhang *et al.* (1993)	Tree volume increment
Moore *et al.* (1994)	Tree basal area increment
Qin and Cao (2006)	Tree diameter growth squared
Additive	
Harrison and Daniels (1988)	Tree basal area increment
Dhôte (1994)	Tree basal area increment
Ritchie and Hann (1997b)	Tree basal area increment

[a] Where ba_{2i} is the tree basal area for the ith tree at the end of the growth period, $\overline{ba}_2$ is the mean basal area per tree at the end of the growth period, ba_1 is the tree basal area for the ith tree at the start of the growth period, $\overline{ba}_1$ is the mean basal area per tree at the start of the growth period, A_2 is the stand age at the end of the growth period, A_1 is the stand age at the start of growth period, BA_2 is the stand basal area at the end of the growth period, BA_1 is the stand basal area at the start of the growth period, n_{2i} is the predicted number of trees per unit area represented by each tree at the end of the growth period, g is the conversion of diameter squared to basal area in square units, D_q is the quadratic mean diameter, D_{q1} is the quadratic mean diameter at the start of the growth period, N_2 is the number of trees at the end of growth period, N_1 is the number of trees at the start of the growth period, t_2 is the year at the end of the growth period, t_1 is the year at the start of the growth period, sp is species, P_{2j} is the tree probability of survival, Δba_i is the tree basal area increment, $\overline{\Delta ba}$ is mean tree basal area increment, ΔBA is the stand basal area increment, r is an unspecified proportional adjustment in Campbell *et al.* (1979), β_is are parameters to be estimated, $\overline{\Delta d}$ is the mean diameter increment per tree, Δv_i is the tree volume increment, ΔV is the stand volume increment, d_{1i} is the tree diameter for the ith tree at the start of the growth period,

Function[a]
$\Delta v_i = \Delta V \cdot \left(\frac{d_{1i}^{1.818} \cdot h_{1i}^{1.786}}{\sum\left(d_{1i}^{1.818} \cdot h_{1i}^{1.786}\right)}\right)$
$\Delta v_i = \Delta V \cdot \left(\beta_0 + \beta_1 \cdot \frac{v_{1i}}{V_1} + \beta_2\left(\frac{v_{1i}}{V_1}\right)^2\right)$
$\Delta ba_i = \Delta BA \cdot \left(\beta_0 + \beta_1 \frac{ba_{1i}}{BA_1} + \beta_2\left(\frac{ba_{1i}}{BA_1}\right)^2\right)$
$\tilde{d}_{2i}^2 = d_{1i}^2 + \left(\frac{s \cdot \left(\frac{40\,000 \cdot BA_2}{\pi}\right) - \sum\left(p_{2j} \cdot d_{1j}^2\right)}{\sum\left(p_{2j} \cdot \left(d_{2j}^2 - d_{1j}^2\right)\right)}\right) \cdot \left(\hat{d}_{2i}^2 - d_{1i}^2\right)$
$\Delta ba_i = \overline{\Delta ba} + \left(\beta_1 \cdot H^{\beta_3} \cdot e^{(\beta_2 \cdot SI)}\right) \cdot \left(ba_{1i} - \overline{ba}_1\right)$
$\Delta ba_i = \overline{\Delta ba} + \left(f\left(\sum_{\sigma} d_{1i}\right) \cdot \left(d_{1i} - \overline{d}_{\sigma}\right)\right)$
$\Delta ba_i = \overline{\Delta ba} + \beta_0\left(ba_{1i} - \overline{ba}_1\right) + \beta_1\left(cr_{1i} - \overline{cr}_1\right)$

d_{2i} is the tree diameter for the ith tree at the end of the growth period, Δd_i is the diameter increment of the ith tree, d_{1j} is the initial diameter of the jth diameter class, d_{2j} is the diameter of the jth diameter class at the end of the growth period, v_{1i} is the tree volume for the ith tree at the start of the growth period, V_1 is the stand volume at the start of the growth period, H is the dominant height, SI is the site index, $\overline{d}_{\sigma}$ is a specified lower diameter limit for growth (all trees below the limit have no growth, and in the summation, σ is the index value for the tree of the specified lower diameter limit for growth), cr_{1i} is the tree crown ratio for the ith tree at the start of the growth period, $\overline{cr}_1$ is the mean tree crown ratio at the start of the growth period, the κ_is are nonlinear functions of $\overline{ba}$, θ is an array of parameters, h_{1i} is the total tree height for the ith tree at the start of the growth period, GS_i is $\left(\frac{\overline{d}_{1i}}{\sum_i^{10} \overline{d}_{1i}}\right)$ for diameter classes $i = 1, \ldots, 10$, s is size of the plot (ha), δ_f is a growth adjustment for a single tree, d_{1f} is the initial diameter of some selected tree f, Δd_f is the predicted unadjusted diameter growth of some selected tree, and log is the natural logarithm.

Qin and Cao (2006) compared four different methods of disaggregation (additive functions were not examined) and found that they all produced similar results for a range of projection lengths (4–17 years) in loblolly pine plantations. However, when the disaggregation method of coefficient adjustment was compared to an unconstrained individual-tree model, a significant improvement in predicted stand-level attributes was achieved (Qin and Cao, 2006). In Douglas-fir stands of western Oregon, Ritchie and Hann (1997b) compared six different methods of disaggregation (additive and proportional growth) to an unconstrained individual-tree model. In general, they found that the unconstrained individual-tree approach was superior to any of the tested disaggregation approaches for predicting both tree- and stand-level increment (Ritchie and Hann, 1997b). However, the disparity between the disaggregation and the unconstrained individual-tree approach was less evident after crown ratio was eliminated from the individual-tree equations. In addition, they found an additive allocation approach was far superior to simple proportional allocation for predicting basal area increment (Ritchie and Hann, 1997b).

One limitation of Ritchie and Hann (1997b) is that they only looked at short-term (five-year) projections of increment, and the results may not hold for longer projections because of the predominance of mortality in determining net increment. There is also little evidence of how effective disaggregation approaches are in multi-species and uneven-aged stands. McTague and Stansfield (1995) used the approach in uneven-aged and mixed-species stands, but the projections were not compared to an unconstrained individual-tree approach. Ritchie and Hann (1997b) surmised that disaggregation approaches would work better in even-aged, single-species stands because of the limited capability of stand-level equations in multi-species and uneven-aged stands, but this assertion still remains untested.

10.3.2 Constrained

In contrast to the disaggregation approach of allocating stand-level growth and mortality to individual trees, the constrained approach uses a multi-response parameter estimation technique to optimize tree-level predictions at multiple levels. This approach was introduced by Zhang *et al.* (1997b), who used a pair of seemingly unrelated equations, based on the conceptual relationship between basal area of each individual tree and diameter increments of trees in the same diameter class, to constrain predictions:

$$\Delta \mathrm{DBH}_{ji,t} = \mathrm{f}(x_{ji,t} \cdot \beta) + \varepsilon_{ji,t} \qquad [10.5]$$

$$\mathrm{BA}_{j,t+1} = 0.00007854 \cdot \sum_{i=1}^{n_{j,t+1}} \left(\mathrm{DBH}_{ji,t} + \mathrm{f}(x_{ji,t} \cdot \beta)\right)^2 + \varepsilon_{j,t}$$

where $\Delta\mathrm{DBH}_{ji,t}$ is predicted diameter increment (cm) for the ith tree in the jth diameter class at time t, $x_{ji,t}$ is a vector of tree-level covariates, β is a vector of fixed parameters to be estimated from the data, $\mathrm{BA}_{j,t+1}$ is basal area of the jth diameter class at time $t+1$, $n_{j,t+1}$ is the stem density (no. ha^{-1}) and the εs are residual errors assumed to be correlated. Zhang *et al.* (1997b) developed a similar equation for tree survival probability. Results from an unthinned loblolly pine stand indicated that the constrained approach provided predictions of growth nearly identical to an unconstrained model, but improved estimates of total stand basal area after 10 to 12 years of simulations (Zhang *et al.*, 1997b).

Cao (2006) modified the approach of Zhang *et al.* (1997b) by substituting stand-level attributes for diameter-class attributes in both diameter increment and survival equations. This improved predictions of stand-level basal area in loblolly pine plantations compared to the diameter-class constrained approach, but both constrained approaches were found inferior to the disaggregation approach of proportional growth after five years of simulations (Cao, 2006).

10.3.3 Combined

To be effective, disaggregation requires accurate predictions of stand-level growth and mortality, while the constrained approach only influences parameter estimates of the tree-level equations. In both cases, information is not combined from both tree- and stand-levels to improve predictions. The combined approach of Yue *et al.* (2008) uses a composite estimator to join estimates of both tree- and stand-level equations to improve predictions at both levels. The combined approach has three primary steps: (1) preliminary prediction; (2) combination; and (3) feedback modification (Figure 10.3). In the first step, both unconstrained tree- and stand-level growth equations are used to predict basal area growth. These predictions are combined based on a variance and covariance method:

$$\mathrm{BA_C} = \mathrm{BA_S} + \left(\frac{(\sigma_S^2 - \sigma_{TS})}{(\sigma_T^2 + \sigma_S^2 - 2 \cdot \sigma_{TS})} \right) \cdot (\mathrm{BA_T} - \mathrm{BA_S}) \qquad [10.6]$$

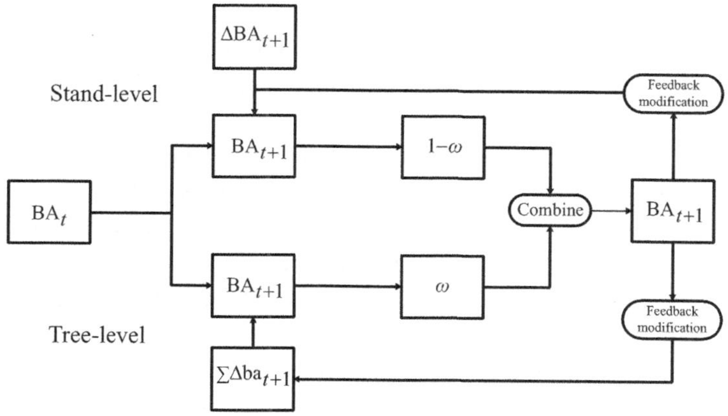

Figure 10.3 Flowchart of the combined stand- and tree-level approach of Yue *et al.* (2008). BA is the stand total basal area, ba is the individual-tree basal area, and ω is a weighting factor.

where BA_C is the combined estimate of total basal area, BA_S is the estimate of total basal area as predicted by the stand-level equation, BA_T is the estimate of total basal area as predicted by the tree-level equation, σ^2_S is the variance of aggregated variables from the stand-level equation, σ^2_T is the variance of aggregated variables from the tree-level equation, and σ_{TS} is the covariance of variables in the tree- and stand-level equations. In the final step, the combined estimate of total basal area is disaggregated to individual trees based on the ratio of the combined total basal area and tree-level predicted total basal area:

$$DBH_{i,t+1} = DBH_{i,t} + \Delta DBH_{i,t+1} \cdot \sqrt{\frac{BA_C}{BA_T}} \qquad [10.7]$$

Yue *et al.* (2008) found the gains in efficiency for predicting total basal area using the combined approach compared to the tree- and stand-level equations only were up to 43 and 16%, respectively, on an independent dataset. The combined approach is a relatively new and novel concept for growth modeling that deserves further attention, particularly in modeling individual-tree survival.

10.4 Summary

All models have imposed limitations based on their resolution. Various methods are used to link models of varying resolutions to take advantage of features of each approach. Although the primary methods of combining models are presented in this chapter, a variety of other proposed model combinations are used. For example, Qin *et al.* (2007) found that combining an individual-tree model with a diameter-distribution model was the most effective approach they tested, while Härkönen *et al.* (2010a) found that an unconstrained individual-tree model predicted stand volume better than a constrained model. In addition, as discussed in Chapter 8, several individual-tree growth models like PROGNAUS (Monserud *et al.*, 1997) and ORGANON (Hann, 2011) impose constraints so that the predictions are consistent with a stand-level maximum size–density line.

The combined stand-level and size-class approach of Daniels and Burkhart (1988) is effective when a tree list in not available, but becomes increasingly more difficult as stand structures become more varied due to management or the presence of multiple species. The combined individual-tree and stand-level approach has received significant attention, particularly disaggregation. However, gains using the disaggregation approach are mixed in the limited number of studies that have rigorously examined them, and their performance is likely dependent on an array of factors. Other approaches like the combined approach deserve further attention and should see increased use in the future.

11

Modeling silvicultural treatments

11.1 Introduction

The practice of forestry is primarily centered on selecting and applying the best set of silvicultural treatments needed to meet the objectives of the landowner and society. Therefore, one of the key uses of forest growth and yield models is projecting the long-term consequences of these various stand-level silvicultural decisions. This is an important use of models because installing and maintaining long-term field trials is expensive and time-consuming. In addition, not all treatment combinations can be replicated across the landscape. Finding an optimal silvicultural system, therefore, is often dependent on projections obtained from a growth model. Forest product companies can make significant financial decisions based on the output of a single growth model. Thus, understanding how growth models represent different treatments is crucial for evaluating the credibility of the simulations.

Stand responses to treatment produce both a direct and an indirect effect (Miller and Tarrant, 1983; Auchmoody, 1985). In the context of fertilization, Opalach and Heath (1988) defined the direct effect as "... that part of the response due to improved nutrition ...," and the indirect effect as "... the remaining portion of the response due to altered stocking brought on by fertilizer in previous growing seasons." Ignoring the difference between the direct and indirect effect of a treatment can lead to unreliable estimation of the size and duration of the treatment's response (Auchmoody, 1985).

Here, the direct effect of any treatment will be defined as the difference between the increment and/or mortality rate of the treated plot's tree- and/or plot-level attributes and an untreated plot with identical tree and/or plot structural attributes. Finding untreated plots or trees with identical structural attributes can be a challenge (Duzan *et al.*, 1982), and differences between untreated and treated plot structural attributes can mask the effects of

---

Forest Growth and Yield Modeling, First Edition. Aaron R. Weiskittel, David W. Hann, John A. Kershaw, Jr. and Jerome K. Vanclay.

a treatment (Wells *et al.*, 1976). Separating the direct and indirect effects can also be a challenge for analyses involving multiple growth periods after application of the treatment. Even if the untreated and treated plots on an installation were exactly the same at application, the direct effect will have produced differences in the plots' stand- and tree-level attributes by the end of the first growth period. Therefore, differences between the untreated and treated plots in the second and subsequent growth periods will be a combination of the direct and indirect effects.

Forest modelers have used considerably different approaches to represent silvicultural treatments in both tree- and stand-level models. Three commonly used approaches are to develop: (1) separate equations for untreated and treated plots or trees (hereafter called the "treated equation" and the "untreated equation," respectively, for brevity); (2) a single equation that incorporates appropriately behaved treatment predictor variables and fitted to the combined untreated and treated datasets; or (3) a treatment modifier equation that is either a multiplier upon or added to a separately developed untreated equation. All three methods have used either linear regression techniques applied to a log–log linearization of the underlying nonlinear model form, or nonlinear regression techniques, to estimate the parameters of the equations.

The development of separate treated and untreated equations obscures the difference between the direct and indirect effects of treatment response and it can lead to inconsistent estimates of treatment responses. The second approach can be executed in a manner that confounds not only the direct and indirect effects, but also the treatment response and the untreated response, such that separating the various elements of response becomes impossible. However, the second approach can also produce a component of the equation that serves as a multiplicative or additive treatment modifier. In that case, the second method simultaneously estimates the parameters for both the untreated tree/stand portion of the equation and the treatment modifier. The third method can estimate the parameters of the treatment modifier equation in one step, or it can sequentially estimate the untreated equation parameters first and the treatment modifier parameters second in two or three steps.

Using the second, simultaneous approach to parameterize the modifier component in a log–log linearization of a nonlinear model could result in the following.

1. Less precise predictions of untreated responses if the addition of the treatment data and the associated predictor variables results in increasing the level of ill-conditioning in the correlation matrix of the predictor variables for the combined dataset (Kmenta, 1997). In such a situation, the parameters of the untreated equation can change substantially when the data for the treated trees/plots are added to the modeling dataset.

2. Biased and inconsistent predictions of untreated responses if the treatment predictor variables introduced into the combined model are underspecified (Kmenta, 1997), unless the two sets of predictor variables are orthogonal to each other (Draper and Smith, 1998).

3. Unbiased and more precise predictors of untreated responses, if neither of the above are problems, because of the increased sample size from combining the untreated and treated data.

Unfortunately, problems with ill-conditioning in the correlation matrix of the predictor variables also carries into nonlinear models (Seber and Wild, 2003), and misspecification of nonlinear model forms can also lead to biased and inconsistent predictions (Kmenta, 1997; Ritz and Streibig, 2008). Consequently, simultaneously estimating the parameters of both an untreated equation and a treatment modifier can be considered successful if it leads to an equation that accurately partitions the overall response of the attribute into the portion both due and not due to treatment. A successful partition will result in a modifier that characterizes the direct effect of treatment.

The third approach fixes the parameter estimates of the untreated equation to those estimated from only the untreated data. It can be useful if application of the second approach increases the ill-conditioning in the correlation matrix of the predictor variables and, as a result, leads to an inaccurate partitioning of the response. The third approach is also useful if the resulting treatment modifier will be applied to untreated equations that were developed in previous studies or in geographic regions of the species that do not have adequate treatment data (e.g. Pienaar and Rheney, 1995; Hynynen *et al.*, 1998; Amateis, 2000; Amateis *et al.*, 2000; Westfall and Burkhart, 2001; Hann *et al.*, 2003; Westfall *et al.*, 2004). The third approach has been used to develop treatment modifier equations that characterize either the combined direct and indirect effects or just the direct effect.

The combined direct and indirect response variable is usually estimated by taking the difference of the measured treated attribute minus the measured untreated attribute. This will be labeled the "one-step method" for calculating the treatment response variable. This one-step method provides an accurate estimate of the combined treatment response variable if the size and structure of the untreated and treated trees/plots are exactly the same at the time of treatment. Any differences between the untreated and treated trees/plots will be reflected in an increase in the amount of unexplained variation about the modifier equation, and less precise parameter estimates.

Differences between the measured untreated and treated trees/plots used to form the modifier response variable also affect the estimation of the direct effect. This difficulty has often been managed by using a predicted instead of the actual attribute for the untreated plot or tree in forming the direct effect treatment response variable (Scanlin and Loewenstein, 1979; Miller *et al.*, 1988; Carson *et al.*, 1999; Hann *et al.*, 2003; Gould *et al.*, 2008). Wang (1990) called this the "multi-step method" for calculating the treatment response variable because it is performed sequentially in two or three steps: (1) develop or select a region-wide untreated equation for predicting the attribute of interest; (2) optionally, calibrate the regional equation by scaling it to the untreated plot(s) on each of the installations in the dataset (e.g. Hann *et al.*, 2003); and (3) calculate the difference between the measured tree/plot treatment attribute and the possibly scaled predicted tree/plot untreated attribute, and use this difference to form the treatment response variable for every post-treatment growth period on an installation.

The second and third steps are performed for each of the installations and then the data are accumulated across installations to form the treatment modifier modeling dataset. At this point, an assessment can be made on the data as to whether or not the untreated equation adequately characterizes the treatment response. This option is not available when using the one-step method. If the untreated equation is inadequate for characterizing treatment response, then the accumulated data are used to develop the treatment response modifier equation. This approach removes the indirect effect of treatment because the treated tree or plot's predictor variables at the start of the second and subsequent growth periods after treatment incorporate both the direct and indirect effects of the previous growth periods. The multi-step method can also help reduce variation in the first growth period's treatment response caused by structural differences between the treated and untreated plots (Scanlin and Loewenstein, 1979; Hann *et al.*, 2003).

Stand- and tree-level response to silviculture varies both in scale and temporally. Snowdon (2002), as well as Nilsson and Allen (2003), identified two general types of response to treatment. Snowdon's (2002) Type 1 response produces a relatively short-term effect of treatment, while the Type 2 response is a result of a long-term change in productivity from the application of the treatment. In other words, a Type 2 response leads to fundamental changes to the carrying capacity of the site (Carlson *et al.*, 2008). Examples of where Type 1 responses from treatment have been observed are the treatment of herbaceous vegetation and the application of nitrogen fertilizer; examples of where Type 2 responses to treatment have been observed are the treatment of competing hardwood tree species, the application of phosphorus to a site with a severe deficiency of phosphorus, and the use of genetically improved growing stock (Snowdon, 2002; Nilsson and Allen, 2003).

Consequently, identifying a Type 1 or 2 response is central to deciding how to model the direct effect of the treatment. Because indirect effects of treatment are a consequence of previous direct effects, the type of response behavior for the combined direct and indirect effects should resemble that of the direct effect but with possibly longer duration. For example, the duration of the direct effects for a Type 1 response will usually be shorter than the duration of the combined direct and indirect effects. For these reasons, different approaches should be used to model these two types of responses to treatment.

A model form characterizing a Type 1 response should incorporate some measure of the duration of the treatment response. This can be a function of elapsed time since treatment or a function of the amount of increment in a plot attribute, such as top height, since treatment. Conversely, a Type 2 response should be incorporated in a manner that maintains a direct effect response throughout the projection. Therefore, a treatment modifier without a measure of treatment duration is assuming, rightfully or not, a Type 2 response.

The choice of whether to develop a treatment modifier equation that predicts the combined direct and indirect effects or just the direct effect of treatment does affect how the resulting treatment modifier is applied to the untreated equation. A growth model that incorporates a treatment modifier which predicts both the direct and indirect effects will, conceptually, need to make two sets of predictions for the length of the projection: (1) a projection of the initial plot or tree development over the desired number of growth cycles without application of treatments; and (2) a projection of the treatment modifier's direct and indirect treatment responses over the desired number of growth cycles. The total treated tree or plot's response for each growth period is then formed by summing the two predictions.

While the need to make two sets of predictions may not be too time-consuming for a whole-stand model, its application to a tree-level model may considerably slow processing time. In contrast, a growth model that incorporates a treatment modifier which predicts only the direct effects will apply the untreated equation to the tree and/or plot predictor variables both before and after treatment; the modifier equation then predicts any necessary changes to the untreated tree/plot predictions in order to estimate the total treated tree or plot's responses to treatment. As a result, application of the untreated equation after treatment will include the indirect effect in its predictions.

Not all model forms are amenable to expressing treatment behavior as a modifier that can be added to, or multiplied with, an equation for untreated trees/plots. For example, it would be difficult to derive a multiplicative or additive modifier for application to the nonlinear logistic equation frequently used in tree-level modeling without affecting the asymptotic behavior for which that model form is frequently selected to characterize. In this type of situation, incorporation of treatment response predictor variables directly into the equation, and the subsequent simultaneous estimation of the parameters for the untreated and treated predictor variables in the equation, is probably the best option.

It is important to mention that many of the growth and yield models used to make silvicultural decisions do not explicitly represent all treatments alternatives in them (Table 11.1). The apparent assumption is that the equations developed from the untreated plots are adequate for representing the treatment. For example, less than half of the commonly used models in the US Pacific Northwest explicitly represent fertilization, despite the wide use of this silvicultural practice in the region (Robinson and Monserud, 2003). The implications of this are illustrated by Johnson (2005), who compared the long-term projections of various treatments alternatives from six commonly used growth models in the US Pacific Northwest. The conclusions of this analysis were that the models showed a wide range of responses to the treatments, and few of them adhered to all of the general research findings on the treatments, which suggests that users need to be careful when making decisions based on model simulations (Johnson, 2005).

This chapter will focus on the common treatment activities represented in growth models including: (1) genetic improvements; (2) early stand treatments; (3) thinning;

Table 11.1 Handling of forest silvicultural treatments in several example statistical growth and yield models.

Growth model	Genetic improvements	Vegetation management	Thinning	Fertilization	Harvesting
Whole stand					
DFSIM (Curtis *et al.*, 1981)	—	—	✓	✓	✓
Scube (García, 2011)	—	—	✓	—	✓
Size class					
FIBER (Solomon *et al.*, 1986)	—	—	✓	—	✓
Vanclay (1989a)	—	—	✓	—	✓
Distance-dependent individual tree					
SILVA (Pretzsch *et al.*, 2002)	—	—	✓	—	✓
TASS (Mitchell, 1975)	✓	—	✓	—	✓
Distance-independent individual tree					
ORGANON (Hann, 2011)	✓	—	✓	✓	✓
PROGNAUS (Monserud *et al.*, 1997)	—	—	✓	—	✓

(4) fertilization; (5) the combination of thinning and fertilization; and (6) harvesting. Approaches taken at both the stand- and tree-levels will be described.

11.2 Genetic improvements

Geneticists have made significant strides in breeding commercially important tree species for certain attributes, particularly growth. Gains in growth of 5 to 25% above unimproved stock have been reported for many species, and even higher gains have been achieved (White *et al.*, 2007; pp. 298–299). However, these estimated gains are often obtained from individual trees on research trials with limited regional replication, which makes it difficult to understand the long-term implications of these gains at the stand-level. Consequently, a variety of approaches have been used to represent genetic improvements in growth models. All of the approaches taken below have assumed a Type 2 direct response due to genetic improvement.

11.2.1 Stand-level

At the stand-level, two primary approaches for representing stand response in plantations of genetically improved trees have been used, namely: (1) modification of the site index, and (2) growth equations specific to a certain family or provenance. Nance and Wells (1981) were some of the earliest to recognize that modifying the site index could account for much of the gain in growth and yield from genetic improvements. Buford and Burkhart (1987) confirmed these findings and indicated that no other major modifications were necessary to make models sensitive to genetic improvements.

However, later studies suggested that simply adjusting the level of site index was not sufficient for representing genetic improvement in growth models (Xie and Yanchuk, 2003; Adams *et al.*, 2006). Instead, Adams *et al.* (2006) found it necessary to use growth and surival equations that had family-specific parameters and included measures of stand density. Adams *et al.* (2006) were also some of the first to develop family-specific survival equations, which is important because some studies indicate that the intercept of the self-thinning line can vary by seed source (e.g. Buford, 1989).

Like Adams *et al.* (2006), several other studies have presented growth equations with parameters specific to individual families (Kurinobu and Shingai, 1987; Danjon, 1995), particularly in loblolly pine (Knowe and Foster, 1989; Sprinz *et al.*, 1989). The limitation of this approach is that the resulting equations are family specific and, therefore, the approach requires extensive datasets to obtain reliable parameter estimates for each of possibly many families.

In contrast, the approach of genetic-gain multipliers requires relatively little modification to existing growth models. Carson *et al.* (1999) illustrated this approach for *Pinus radiata* in New Zealand, as they developed time-scaling multipliers for the mean top height, basal area per hectare, and stocking equations in an existing growth model. These multipliers compress the amount of time required to achieve a predicted value from the existing growth model, with the amount of compression based upon the size of genetic improvement. They found that genetic improvement significantly increased the rate of basal area per hectare growth, but did not significantly influence the rate of height increment (Carson *et al.*, 1999), which further emphasizes the limitation of simply adjusting site index when the selection criterion is based on something other than height.

11.2.2 Tree-level

Individual tree growth models have been used to project stand-level gains using various hypotheses generated from different field trials (Rehfeldt *et al.*, 1991; Hamilton and Rehfeldt, 1994). However, the development of individual-tree growth equations that incorporate the effects of genetic improvements is limited due to the lack of data and knowledge of how to best model the changes (Talbert and Hyink, 1988).

Xie and Yanchuk (2003) used the procedure of Lambeth (1980) to account for the declining nature of genetic gain in dominant height over time. In addition, they quantified the genetic worth of each seedlot in their analysis (Xie and Yanchuk, 2003), which allowed the development of more generalized equations that are still applicable to seedlots not available in the original dataset. Their genetic worth values were incorporated into TASS, and associated TIPSY, growth and yield system.

Gould *et al.* (2008) used a modified two-step method to develop multiplicative growth modifiers for estimating changes in the increment of Douglas-fir individual-tree growth due to genetic improvements. The genetic gain multipliers were found to be positively related to the seedlot's genetic worth for both diameter and height increment (Gould *et al.*, 2008; Figure 11.1). Gould and Marshall (2010) demonstrated how these multipliers can be used in the ORGANON (Hann, 2011) and FVS (Wykoff *et al.*, 1982) regional growth models, and found that projected volume per hectare gains were nearly doubled when the genetic worth was increased from 5 to 10%.

How to project the long-term influences of genetic improvements is still an open question. Continued improvement in modeling genetically improved trees will likely occur due to the increased availability of data. This will be an important issue to resolve due to the continued investment in tree breeding programs and the greater application of clonal forestry. However, the limitation will continue to be in trying to scale the individual tree plots used by most geneticists up to the stand-level.

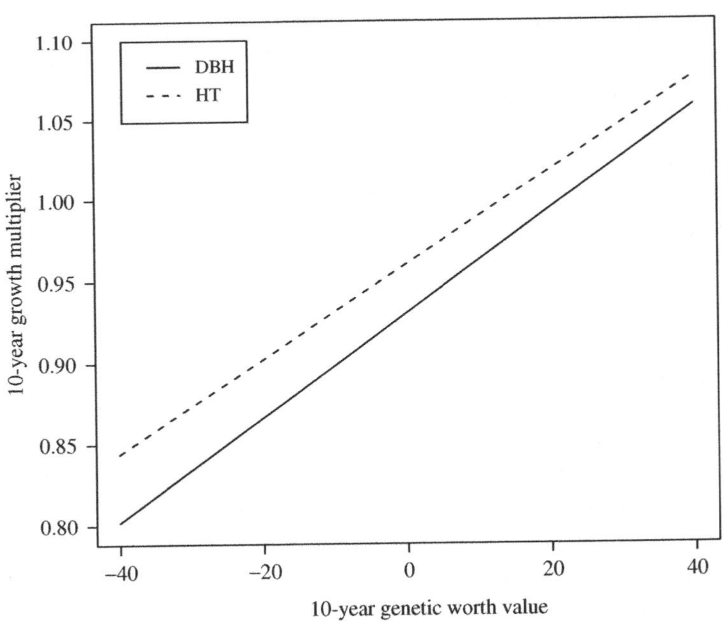

Figure 11.1 Modifiers for 10-year diameter (solid line) and height (dashed line) individual-tree increment for coastal Douglas-fir, over genetic worth value. Based on the equations of Gould *et al.* (2008).

11.3 Early stand treatments

The early stand treatment of weed control and cleaning, particularly with the use of herbicides, is an effective method for significantly increasing long-term yields in a variety of forest types (Wagner *et al.*, 2006). Representing the long-term influence of early stand treatments in growth models is difficult because most models do not simulate non-tree growth, and the treatment primarily influences early stand dynamics, which may not be recognized in most simulators that simulate diameter increment only at breast height. Richardson *et al.* (2006), as well as Mason and Dzierzon (2006), have reviewed the ways that early stand treatment influences on growth have been modeled, and found that it has been represented to varying degrees of success.

11.3.1 Stand-level

Pienaar and Rheney (1995) developed additive modifiers for the growth of slash pine basal area per hectare, and dominant height, that characterized the direct effect of both bedding practices and the application of herbicide to control competing vegetation. They assumed a Type 1 duration of response, and estimated the parameters of the untreated component and the modifier component simultaneously using nonlinear regression. Quicke *et al.* (1999) altered the equations of Pienaar and Rheney (1995) to include initial density, and then used them to characterize the direct effect of vegetation control on the growth of loblolly pine basal area per hectare, and dominant height. Mason and Milne (1999) generated the difference form of the modifier component from the Pienaar and Rheney (1995) equation and used it to capture the direct effects of weed control, fertilization, and site preparation on the basal area per hectare growth of radiata pine.

However, these studies did not model mortality. Mason *et al.* (1997) developed a stand-level model that predicted growth, survival, and size-class distributions for the first five years after planting, and they explored how it might be linked to an existing growth model. Similar initial stand growth models exist for other regions (Belli and Ek, 1988; Payandeh and Haig, 1991), but significant biases can occur if they are improperly linked to existing growth and yield models (Mason and Dzierzon, 2006). In addition, these previous studies generally accounted for vegetation control with a simple indicator variable rather than a continuous measure like percentage control, which limits their ability to represent treatments not covered in their original datasets.

Knowe (1994a) developed a stand-table projection system that projected the effects of vegetation management on early stand development, and the change in quadratic mean diameter was sensitive to the intensity of the treatment. Knowe and Stein (1995) expanded this model by developing equations to predict the influence of competing vegetation on survival, dominant height increment, and development of woody vegetation. Either measured tree-list data or predicted tree-list data using a Weibull diameter distribution model could be used as input to the model developed by Knowe and Stein (1995). While the usage of measured tree-list data proved to be more accurate at younger ages, the difference in accuracy was largely eliminated at age 10 years after planting.

The data used by Knowe and Stein (1995) came from repeated measurements over 10 years, which were taken on experimental plots consisting of six site preparation treatments applied to four locations. The early work of Knowe (1994a) and the subsequent model of Knowe *et al.* (2005) were constructed from single two-year-long growth measurements taken on 84 plots located in operationally planted stands that had previously received site preparation, hardwood release, pre-commercial thinning, or no treatments. Knowe *et al.* (2005) had to use simpler characterizations of the previous treatments than Knowe and Stein (1995) because information concerning the previous treatments was limited. In addition to predicting the impact of treatments and the amount of residual hardwood basal area per hectare upon stand-level conifer basal area per hectare growth, hardwood basal area per hectare growth, and conifer survival, Knowe *et al.* (2005) also developed static equations for predicting the height, crown width, and crown ratio of trees generated from their Weibull diameter distribution model in order to provide tree lists for input into existing single-tree growth and yield models such as ORGANON (Hann, 2011) and FVS (Wykoff *et al.*, 1982).

11.3.2 Tree-level

Individual-tree growth models sensitive to early stand treatments are relatively rare. Knowe (1994b) quantified the effects of vegetation control treatments on individual-tree height–age and height–diameter relationships in young Douglas-fir plantations, which could be used in the stand table projection system of Knowe (1994a). Richardson *et al.* (1999) developed alternative indices of surrounding plant competition that were evaluated by assessing the amount of variation they explained in the height and diameter increment of radiata pine. Kimberley and Richardson (2004) extended the work of Richardson *et al.* (1999) by incorporating the influence of seasonal growth patterns, but the parameters were only determined for one location in New Zealand.

Westfall *et al.* (2004) developed a set of young stand diameter and height increment modifier equations for characterizing the direct and indirect effects upon loblolly pine of site preparation and herbaceous control, using the one-step approach and nonlinear regression to estimate the parameters of the modifier equations. All of their modifiers assumed a Type 2 duration of the treatments. Their results indicated that both tree diameter and height increment were highly sensitive to herbaceous weed control, but they did not evaluate the influence of varying levels of control.

Ritchie and Hamann (2006; 2008) developed the CONIFERS young stand growth model that incorporated individual plant diameter, height, and crown width increment equations for three hardwood tree species, three shrub species, young Douglas-fir, and young ponderosa pine in northern California and southwest Oregon. The equations incorporated the of amount crown closure at the top of the subject tree (CCH; see the discussion of this variable in Chapter 2) in conifers, hardwoods, and/or shrubs. Vegetation control was simulated indirectly by reducing the appropriate type of crown closure. Therefore, the user must possess independent information on the efficacy of the particular treatment they wish to apply to the targeted competitors.

Vaughn (2007) extended the CONIFERS model to young Douglas-fir in northwest Oregon and western Washington, and found that height, but not diameter increment, was sensitive to the percentage cover of competing vegetation. The CONIFERS young stand model was designed to pass a tree list on to regional growth models such as ORGANON (Hann, 2011) and FVS (Wykoff *et al.*, 1982) once the young trees had achieved dominance over the competing vegetation.

Statistical modeling of the influence of early stand treatments is challenging because of the inherent high variability of stand establishment dynamics and the difficulty of quantifying the influence of competing vegetation on available site resources like moisture. In addition, most herbicide trials are commonly established on small plots or without tagged trees, which makes developing individual-tree predictive equations difficult. Both Richardson *et al.* (2006) and Mason and Dzierzon (2006) suggest that process-based and hybrid growth models might be the best approach for predicting the influence of early stand treatments. Approaches for accomplishing this are illustrated by Watt *et al.* (2003) and Mason *et al.* (2007), which deserve additional attention. The advantages and disadvantages of these approaches are further outlined in Chapters 13 and 14.

11.4 Thinning

Thinning is one of the most commonly employed forest silvicultural treatments and has been extensively studied. Thinnings can differ by their timing, intensity, and type, and each can have a substantial impact upon the expected residual stand and tree responses to thinning. Tappeiner II *et al.* (2007) describe five types of thinning: (1) thinning from below; (2) thinning from above; (3) crown thinning; (4) free thinning; and (5) strip or row thinning.

Thinning from below extracts the suppressed, intermediate, and smaller codominant trees in order to provide more growing space to the dominant and larger codominant trees. As a result, the average DBH and crown size of the residual trees increases with thinning from below. Thinning from above removes a portion of the dominant and codominant trees and, therefore, the average DBH and crown size of the residual trees decreases. A crown thinning focuses the cutting around selected high-quality trees in the main canopy in order to release them from competition, which can decrease the average DBH and crown size of the residual trees. A free thinning uses tree species, spacing, and crown class criteria to select trees for release from competition with varying impact upon average DBH and crown size of the residual trees.

Finally, a strip or row thinning ignores tree species, tree size, and crown class criteria by selecting trees for removal based solely upon residual tree spacing. Consequently, the average DBH, crown size, and species composition of the residual trees should remain the same as the stand before thinning. Even if the residual stocking levels are the same, these five types of thinning will produce different tree- and stand-level responses to thinning because the attributes of their residual tree populations will be different.

All stand development models incorporate equations for predicting the development of untreated stands, and all of these models include measures of competition in most if not all of their dynamic equations (see Chapter 2). Therefore, untreated stand/tree models will predict a response to thinnings because of the reduction in competition. Whether or not this predicted behavior is adequate to characterize the effects of thinning depends upon the species; the type, timing, and intensity of thinning; and the particular stand or tree attribute being examined. Assuming the untreated stand or tree equations are adequate may be a necessity if no or inadequate data are available from thinned plots. If adequate thinning data are available, then it is possible to statistically examine whether or not the untreated stand/tree equations are adequate.

The three basic approaches described at the start of this chapter have been used when the untreated equations are found to be inadequate. Several multiplicative and additive thinning modifiers will be examined in detail to compare and contrast their behavior (Tables 11.2 and 11.3). The original expressions of these modifier equations were reformulated using common terms and algebra in order to facilitate comparisons. The type of response effect used to form the thinning response variable, the type of duration assumed by the model form, and the fitting procedures used to estimate the parameters of the model form have also varied (Table 11.4). Predicted behavior of key elements composing the various thinning modifiers is examined in Tables 11.5 and 11.6. Examples of other approaches to characterizing thinning response will also be explored.

11.4.1 Stand-level

At the stand-level, the impact of thinning is to reduce growing stock and, as a result, the growth rate for those attributes of the residual stand that are directly related to level of growing stock (e.g. basal area per hectare, volume per hectare) will most often be smaller immediately after thinning than if the stand had not been thinned. The particular growth rate of the residual stand will depend upon the amount of growing stock before and immediately after thinning, the type of thinning, the age of thinning, the time since thinning, and the productivity of the site. Other stand attributes may or may not exhibit change due to thinning. As examples, the quadratic mean diameter of the stand could increase, decrease, or remain the same after thinning depending upon the type of thinning employed, and the top height (H100) or average dominant height (HD) of the stand can be reduced by thinning shock (i.e. reduced growth following thinning) (e.g. Sharma *et al.*, 2006).

The representation of thinning at the stand-level is challenging because thinning treatments are designed to selectively remove individual trees, which greatly alters the residual stand structure. Initially, it was common to assume that the growth and yield of a thinned stand was similar to an unthinned stand of the same age, site index, and basal area per hectare (Clutter and Jones, 1980; Cao *et al.*, 1982; Matney and Sullivan, 1982; Burkhart and Sprinz, 1984). However, it quickly became apparent that the assumption was incorrect at the stand-level, and subsequent studies introduced thinning-related predictor variables to incorporate the type, intensity, and/or timing of thinning.

In particular, four stand-level multiplicative thinning modifiers and two additive modifiers that illustrate approaches taken to characterize thinning response are presented in Table 11.2. The four multiplicative modifiers predict the direct effect of the thinning response, while the two additive modifiers predict the combined direct and indirect effect (Table 11.4). The parameters of all of the multiplicative modifiers were fit simultaneously using either the log-linearization and linear regression or nonlinear regression. Both of the additive modifiers used the one-step method and nonlinear regression to estimate the parameters. Thinning modifier ST-1 assumed a Type 2 duration of treatment response, and the remaining modifiers assumed a Type 1 duration of treatment response (Table 11.4).

Modifier ST-1 uses the ratio of quadratic mean diameter of thinned trees to the quadratic mean diameter of all trees prior to thinning (QMD_t/QMD_b) to characterize the type of thinning. A ratio greater than one indicates a thinning from above, a ratio less than one indicates a thinning from below, and a ratio of one indicates a proportional thinning. The QMD_t/QMD_b ratio has been used both to reduce top height of a Douglas-fir stand after thinning if the ratio was greater than one (Curtis *et al.*, 1981) and as part of a modifier in a volume per hectare increment equation (Miller *et al.*, 1988). Modifier ST-1 was used to predict both basal area per hectare and surviving number of trees per hectare at the end of the growth period. The modifier predicts higher ending BA and lower ending survival (TPH_2) than the untreated equations when thinning from below, and the opposite behavior when thinning from above. In addition to type of thinning, the size of the response is dependent upon the age of the stand at the start and end of the growth period and the age at time of thinning.

The other five thinning modifiers incorporate intensity of thinning (defined as the proportion of the stand removed in thinning) and time since thinning (defined as the number of years since thinning) but not type of thinning (Table 11.2). Two of the multiplicative modifiers (ST-2 and ST-3) use number of trees per hectare before thinning (TPH_b) and removed in thinning (TPH_t) to express thinning intensity, while ST-4 uses basal area per hectare both before (BA_b) and removed in thinning (BA_t; Table 11.5). The two additive modifiers (ST-5 and ST-6) use the ratio of TPH_t to TPH_b and the difference of BA_b minus BA_t to express thinning intensity. As discussed in Chapter 2, number of trees per hectare is a true measure of density but it is less sufficient at characterizing competition than basal area per hectare (Zeide, 2005), and the level of insufficiency increases with the

Table 11.2 Stand-level multiplicative and additive thinning modifiers.

Modifier designator	Modifier equation[a]
Multiplicative	
ST-1	$e^{\left(\pm b_0 \cdot \left(1.0-\left(\frac{\mathrm{QMD_t}}{\mathrm{QMD_b}}\right)\right) \cdot \frac{A_2^{-1}-A_1^{-1}}{(A_t \cdot A_2)^l}\right)}$
ST-2	$e^{\left(-b_0 \cdot \left(\frac{\left(\frac{\mathrm{TPH_t}}{\mathrm{TPH_b}}\right)}{\left(1.0-\left(\frac{\mathrm{TPH_t}}{\mathrm{TPH_b}}\right)\right)}\right) \cdot \left(\frac{A_t}{(A_t+Y_t)}\right)\right)}$
ST-3	$e^{\left(-b_0 \cdot \left(\frac{\mathrm{TPH_t}}{\mathrm{TPH_b}}\right) \cdot \left(\frac{A_t}{A_t+Y_t}\right)\right)}$
ST-4	$\left(1.0-\left(\frac{\mathrm{BA_t}}{\mathrm{BA_b}}\right)\right)^{\left(+b_0 \cdot \left(\frac{A_t}{(A_t+Y_t)}\right)\right)}$
Additive	
ST-5	$-\mathrm{BA_t}+\log(X_1)\cdot X_2;$ $X_1=\left((A_t+Y_t)^{-b_0}\cdot(\mathrm{BA_b}-\mathrm{BA_t})^{+b_1}\cdot \mathrm{TPH_b}^{-b_2}\cdot \mathrm{TPH_t}^{+b_3}\cdot\left(1-\frac{\mathrm{TPH_t}}{\mathrm{TPH_b}}\right)^{+b_4}+b_5\cdot I_t\right)\cdot(Y_t+1.0)$ $X_2=\mathrm{BA_t}^{+b_6}\cdot(\mathrm{TPH_b}-\mathrm{TPH_t})^{+b_7}\cdot\left(1.0-\frac{\mathrm{TPH_t}}{\mathrm{TPH_b}}\right)^{+b_8}\cdot Y_t^{+b_9}$
ST-6	$-\mathrm{TPH_t}\cdot e^{\left(-\mathrm{SI}^{+b_0}\cdot\left(1.0-\frac{\mathrm{TPH_t}}{\mathrm{TPH_b}}\right)^{+b_1}\cdot(\mathrm{BA_b}-\mathrm{BA_t})^{+b_2}\cdot \mathrm{TPH_b}^{-b_3}\cdot \mathrm{TPH_t}^{+b_4}\cdot Y_t\right)}$

[a] Where A_1 is the stand age at the start of the growth period, A_2 is the stand age at the end of the growth period, A_t is the stand age at time of thinning, $\mathrm{BA_b}$ is the stand basal area before thinning, $\mathrm{BA_t}$ is the stand basal area removed in thinning, $\mathrm{BA_2}$ is the stand basal area at the end of the growth period, $\Delta\mathrm{HCB}$ is the average crown recession, $I_t = 1.0$ if the second thinning or 0.0 if the first, $\mathrm{QMD_b}$ is the quadratic mean diameter of the stand before thinning, $\mathrm{QMD_t}$ is the quadratic mean diameter of the thinned trees,

Dependent variable	Species	Citation
BA_2 ($-b_0$, $I=1$); TPH_2 ($+b_0$, $I=0$)	Slash pine	Bailey and Ware (1983); Bailey *et al.* (1985)
BA_2	Slash pine	Pienaar and Shiver (1986)
BA_2	Slash pine	Pienaar *et al.* (1990); Pienaar and Rheney (1993)
Annual ΔHCB	Loblolly pine	Short and Burkhart (1992)
BA_2	Loblolly pine	Amateis (2000)
TPH_2	Loblolly pine	Amateis (2000)

SI is the stand site index, TPH_2 is the number of trees in the stand at the end of the growth period, TPH_b is the number of trees before thinning, TPH_t is the number of trees removed in thinning, Y_t is the number of years since thinning, b_i are coefficients with the expected sign of the coefficient indicated before the coefficient, and log is the natural logarithm. Both TPH_t/TPH_b and BA_t/BA_b are indicators of thinning intensity.

Table 11.3 Tree-level multiplicative thinning modifiers.

Modifier designator	Modifier equation[a]
TT-1	$e^{\left(-b_0 \cdot A_t - b_1 \cdot \left(1.0 - \frac{BA_t}{BA_b}\right)\right)}$
TT-2	$\left(1.0 - \frac{BA_t}{BA_b}\right)^{\left(+b_0 \cdot \left(\frac{A_t}{(A_t + Y_t)}\right)\right)}$
TT-3	$\left(1.0 - \frac{BA_t}{BA_b}\right)^{\left(+b_0 \cdot \left(\frac{+b_1 \cdot Y_t - Y_t^2}{(A_t + Y_t)^2}\right)\right)}$
TT-4	$1.0 + b_0 \cdot \left(\frac{\left(\frac{BA_t}{BA_b}\right)}{\left(1.0 - \left(\frac{BA_t}{BA_b}\right)\right)}\right) \cdot (Y_t + 4.0)^{+b_1} \cdot e^{\left(-b_2 \cdot (Y_t + 4.0)^{(+b_1 + 1.0)}\right)}$
TT-5	$1.0 \pm b_0 \cdot \left(\frac{BA_t}{BA_b}\right)^{+b_1} \cdot e^{(-b_2 \cdot Y_t)}$

[a] Where A_t is the stand age at time of thinning, BA_b is the stand basal area before thinning, BA_t is the stand basal area removed in thinning, ΔDBH is the diameter increment of a tree, ΔHT is the height increment of a tree, ΔHCB is the crown recession of a tree, Y_t is the number of years since thinning, and b_i are coefficients with the expected sign of the coefficient indicated before the coefficient. BA_t/BA_b is an indicator of thinning intensity.

widening of the stand's diameter distribution. Therefore, the choice of using number of trees per hectare or basal area per hectare to determine thinning intensity is really a choice of whether thinning response is best characterized by the level of reduction in competition or density. For example, Pienaar *et al.* (1985) also used TPH_t and TPH_b to define thinning intensity in their BA_2 equations, but found predictions could be further improved by also incorporating a suppression index.

Excluding modifier ST-1, the remaining five modifiers can exhibit different behavior when extrapolated to the limits of thinning intensity and duration of the response (Table 11.5).

1. The range of predictions for the three multiplicative modifiers is restricted to non-negative values, and the range of predictions for the two additive modifiers is restricted to non-positive values (Table 11.5).

Dependent variable	Species	Citation
Annual ΔDBH	Loblolly pine	Amateis *et al.* (1989)
Annual ΔHCB	Loblolly pine	Short and Burkhart (1992)
Three-year ΔHCB	Loblolly pine	Liu *et al.* (1995)
Five-year ΔDBH	Scots pine	Hynynen (1995b)
Five-year ΔDBH ($+b_0$); Five-year ΔHT ($-b_0$)	Douglas-fir, western hemlock	Hann *et al.* (2003)

2. All modifiers predict a decrease in the basal area per hectare at the end of the growth period (ST-2, ST-3, and ST-5), number of trees per hectare at the end of the growth period (ST-6), or crown recession (ST-4) with an increase in thinning intensity (Table 11.2), which meets the expected stand-level behavior of a response to thinning.

3. All modifiers predict no response (i.e. a value of one for a multiplicative modifier and a value of zero for an additive modifier) when a thinning is not implemented (Table 11.5), which meets the expected behavior of a response to thinning.

4. Multiplicative modifiers ST-2 and ST-4 predict that the response will go to zero as the thinning approaches the removal all of the trees, while multiplicative modifier ST-3 predicts that the response will approach a small, positive value (Table 11.5).

Table 11.4 Type of response effect and duration as well as fitting technique and strategy for various stand- and tree-level multiplicative and additive thinning modifiers.

Modifier designator	Type of response effect	Type of duration	Fitting procedures	
			Regression technique	Fitting strategy
Stand-level multiplicative				
ST-1 (BA_2)	Direct	Type 2	Log-linear	Simultaneous
ST-1 (TPH_2)	Direct	Type 2	Nonlinear	Simultaneous
ST-2	Direct	Type 1	Log-linear	Simultaneous
ST-3	Direct	Type 1	Log-linear	Simultaneous
ST-4	Direct	Type 1	Nonlinear	Simultaneous
Stand-level additive				
ST-5	Direct + indirect	Type 1	Nonlinear	One-step
ST-6	Direct + indirect	Type 1	Nonlinear	One-step
Tree-level multiplicative				
TT-1	Direct	Type 2	Nonlinear	Simultaneous
TT-2	Direct	Type 1	Nonlinear	Simultaneous
TT-3	Direct	Type 1	Nonlinear	Simultaneous
TT-4	Direct	Type 1	Nonlinear	Simultaneous
TT-5	Direct	Type 1	Nonlinear	Three-step

Modifier ST-4 is used to predict crown recession after thinning, so it is reasonable to expect that the modifier will go to zero for high thinning intensities. The other two modifiers are used to predict stand basal area per hectare at the end of the growth period. While it is reasonable to assume that basal area per hectare at the end of the growth period should approach zero as thinning intensity approaches total removal, the untreated equation to which the thinning modifier is being applied should also provide this behavior. Therefore, when choosing between modifiers ST-2 and ST-3, one has to consider whether the difference in model form of ST-2 needed to force an approach to zero has possibly distorted the predictions for other thinning intensities.

5. The two additive modifiers predict that the modifier will subtract the pre-thinning value of the attribute being predicted as thinning intensity approaches one. Whether this will result in the overall prediction of stand development also going to zero will depend upon the behavior of the untreated equation.

Table 11.5 Predictive behavior of stand-level multiplicative and additive thinning modifiers.

Modifier designator	X value	Limiting values for modifier equations[a]				Y_t value for minimum or maximum modifier
		$X_t = 0.0$	$X_t \rightarrow X_b$	$Y_t = 0.0$	$Y_t \rightarrow \infty$	
Multiplicative						
ST-1	QMD	$e^{\left(\pm b_0 \cdot \left(\frac{\left(A_2^{-1} - A_1^{-1}\right)}{(A_t \cdot A_2)^I}\right)\right)}$	1.0	—	—	—
ST-2	TPH	1.0	0.0	For $X_t = 0$, 1.0 For $X_t \rightarrow X_b$, 0.0	1.0	Minimum = 0.0
ST-3	TPH	1.0	For $Y_t = 0$, $e^{(-b0)}$ For $Y_t \rightarrow \infty$, 1.0	For $X_t = 0$, 1.0 For $X_t \rightarrow X_b$, $e^{(-b0)}$	1.0	Minimum = 0.0
ST-4	BA	1.0	0.0	For $X_t = 0$, 1.0 For $X_t \rightarrow X_b$, 0.0	1.0	Minimum = 0.0
Additive						
ST-5	BA and TPH	0.0	$-BA_b$	For $X_t = 0$, 0.0 For $X_t \rightarrow X_b$, $-BA_b$	$-\infty$	Maximum = 0.0
ST-6	BA and TPH	0.0	$-TPH_b$	For $X_t = 0$, 0.0 For $X_t \rightarrow X_b$, $-TPH_b$	$TPH_t = 0.0$	Minimum = 0.0

[a] Where I is 1.0 if the modifier is for BA_2 and it is 0.0 if the modifier is for TPH_2 (see Table 11.2). A_1 is the stand age at the start of the growth period, A_2 is the stand age at the end of the growth period, A_t is the stand age at time of thinning, BA_b is the stand basal area before thinning, QMD is quadratic mean diameter, TPH_b is the number of trees before thinning, TPH_t is the number of trees removed in thinning, X is the two-sided competition-related variable used in the modifier, X_b is the value of X before thinning, X_t is the value of X at the time of thinning, Y_t is the number of years since thinning, and b_i are coefficients with the expected sign of the coefficient indicated before the coefficient.

Table 11.6 Predictive behavior of tree-level multiplicative thinning modifiers.

Modifier designator	X value	Limiting values for modifier equations[a]				Y_t value for minimum or maximum modifier
		$X_t = 0.0$	$X_t \to X_b$	$Y_t = 0.0$	$Y_t \to \infty$	
TT-1	BA	$e^{(-b_0 \cdot A_t - b_1)}$	$e^{(-b_0 \cdot A_t)}$	—	—	—
TT-2	BA	1.0	0.0	For $X_t = 0$, 1.0 For $X_t \to X_b$, 0.0	1.0	Minimum = 0.0
TT-3	BA	1.0	0.0	1.0	For $X_t =$ 0, 1.0 For $X_t \to X_b$, 0.0	$\text{Minimum} = \frac{+b_1}{(+b_1 + 2.0 \cdot A_t)}$
TT-4	BA	1.0	∞	For $X_t = 0$, 1.0 For $X_t \to X_b$, ∞	1.0	$\text{Maximum} = \left(\frac{+b_1}{+b_2 \cdot (+b_1 + 1.0)}\right)^{\left(\frac{1}{(+b_1+1)}\right)} - 4$
TT-5	BA	1.0	For $Y_t = 0$, $1.0 \pm b_0$ For $Y_t \to \infty$, 1.0	For $X_t = 0$, 1.0 For $X_t \to X_b$, $1.0 \pm b_0$	1.0	Maximum = 0.0

[a] Where A_t is the stand age at time of thinning, BA is the stand basal area, X is the two-sided competition-related variable used in the modifier, X_b is the value of X before thinning, X_t is the value of X at the time of thinning, Y_t is the number of years since thinning, and b_i are coefficients with the expected sign of the coefficient indicated before the coefficient.

6. Modifiers ST-2, ST-3, and ST-4, ST-6 predict that the thinning response will disappear as the amount of time since cutting gets large (Table 11.5), and all of them use the number of years since thinning to index duration of response. Additive modifier ST-6 predicts that the number of trees per hectare removed in thinning will go to zero, and modifier ST-5 predicts that the thinning response will go to negative infinity as the amount of time since thinning gets large, both of which are unreasonable extreme behavior. As a result, Amateis (2000) recommended against using modifiers ST-5 and ST-6 at high thinning intensities.

7. All modifiers predict a maximum and immediate response immediately after thinning, which meets expectation (Table 11.5).

8. Modifier ST-6 predicts that the response to thinning monotonically increases with increasing site index.

The rationale of the two additive modifiers needs further explanation. The predicted values of both modifiers will be added to the parallel predictions of how the stand would have developed if it had not been thinned. Therefore, the additive modifier on stand BA (ST-5) starts with a negative value of the BA removed in thinning and then adds BA back to that negative value so that the sum of the two will go to zero as the amount of time since thinning increases. As a result, the BA in the thinned stand will approach the basal area per hectare in the unthinned stand over time. Similarly, the TPH modifier (ST-6) starts with a negative value for the number of trees per hectare removed, and then reduces that negative number over time so that the unthinned and thinned stands' TPH will converge as the time since thinning increases. While the intended behavior may be rational, the concept of taking the known BA and TPH removed in thinning, and reducing them to achieve that behavior, seems a little less rational. The fact that neither modifier behaves well at higher thinning intensities illustrates the difficulty of guaranteeing rational behavior in all situations.

As an alternative example to additive modifier ST-5, Hasenauer *et al.*(1997) developed a generalized basal area per hectare growth function that confounded the untreated and treated responses in a manner that makes it impossible to separate out a multiplicative (or additive) modifier. While most of the equations in Table 11.2 incorporate time since thinning, Hasenauer *et al.*(1997) used the amount of top height added since thinning. Like modifier ST-5 (when added to an untreated equation), the resulting equation of Hasenauer *et al.*(1997) predicts that the growth of thinned stands slowly converges towards unthinned stands over time, using a model form that is substantially simpler than modifier ST-5.

Several studies have explored the usage of alternative modeling techniques, rather than model forms, for characterizing thinning response. McTague and Bailey (1987) suggested that a BA growth model based on diameter percentiles, which themselves were functions of age, TPH, site index, and indicators for thinning, was effective for projecting both unthinned and thinned stands when information on the thinning did not exist. Chikumbo *et al.* (1999) found a dynamical model dependent on TPH, BA, and the number of thinned trees was more effective at predicting BA growth response in thinned stands than other approaches. Both Barrio Anta *et al.*(2006) and Castedo-Dorado *et al.* (2007a) indicated that it was not necessary to directly account for thinning in stand BA growth equations when the parameters were estimated using the generalized algebraic difference approach (see Chapter 14).

11.4.2 Tree-level

A properly applied thinning that does not damage the residual trees will increase the diameter increment of the residual trees because of the increased availability of moisture, nutrients, and light (Oliver and Larson, 1996; Tappeiner II *et al.*, 2007). Thinning may increase, decrease, or maintain the height increment of the residual trees depending upon tree species, crown class, age, and density of the stand before thinning (Oliver and Larson, 1996; Tappeiner II *et al.*, 2007). The increase in growing space from thinning will lead to an increase in crown width, a reduction in crown recession, and an increase in crown length if height increment is still significant (Oliver and Larson, 1996; Tappeiner II *et al.*, 2007).

Moreover, the improvement of the light environment will increase foliage density within the crown (Oliver and Larson, 1996; Tappeiner II *et al.*, 2007). Finally, a properly applied thinning should reduce the probability of mortality for the residual trees. How quickly these responses will be manifested in the residual trees will depend upon stand structure and tree species, crown size, and crown class of the trees before thinning, with the duration of thinning responses extending well beyond crown closure (Oliver and Larson, 1996).

Tree-level models that incorporate one- and two-sided measures of competition and a measure of crown size into their equations for untreated stands will inherently produce a thinning response that is sensitive to both the intensity and the type of thinning. As with stand-level models, intensity of thinning is reflected in the size of reduction in the competition measures. Unlike stand-level models, type of thinning is reflected in the relative reduction in the two measures of competition often expressed in tree-level models. For example, thinning from below will reduce just the two-sided measure of competition, while thinning from above will reduce both, resulting in larger predicted responses from thinning from above than thinning from below. Furthermore, use of crown size variables in the prediction equations will also provide differential response due to type of thinning if the understory trees have smaller crowns than the overstory trees. On the other hand, smaller size trees can be predicted to have larger potential increments, particularly if a potential-multiplied-by-modifier approach is used for the basic untreated equation.

Because of these inherent thinning responses, a number of the tree-level models, such as FVS and PROGNAUS, have been developed without thinning modifiers. However, these predicted thinning responses may not fully characterize the actual thinning response (e.g. Saunders *et al.*, 2007). Thinning modifiers can be important in tree-level models because: (1) crown density often increases after thinning; (2) damaged trees, which can reduce diameter increment (Hann and Hanus, 2002a) and height increment (Hann and Hanus, 2002b), are often removed in thinning and, as a result, the population is modified; (3) thinning shock can occur after treatment; (4) thinning can increase the susceptibility of the residual stand to damaging agents such as wind and snow; and (5) the dynamics of diameter increment, height increment, and crown recession can accelerate after thinning. The last possibility is of particular concern if the measurement intervals used to collect the modeling data are particularly long, resulting in larger changes in those attributes over the growth period than would be predicted by the untreated equations. To date, only a few studies have examined the need for tree-level modifiers in thinned stands.

Multiplicative thinning response multipliers have also seen use at the tree-level. Five tree-level multiplicative thinning modifiers that illustrate approaches taken to characterize thinning response are presented in Table 11.3. All of the modifiers predict the direct effect of the thinning response (Table 11.4). The parameters of four of the modifiers were fit simultaneously using nonlinear regression, while modifier TT-5 was fit using the three-step approach and nonlinear regression (Table 11.4). Thinning modifier TT-1 assumed a Type 2 duration of treatment response, and the remaining modifiers assumed a Type 1 duration of treatment response (Table 11.4).

Unlike the stand-level modifier equation, all tree-level equations use the ratio of BA_t divided by BA_b to indicate thinning intensity, but they can exhibit different behavior when extrapolated to the limits of thinning intensity and duration of the response (Table 11.6).

1. The diameter increment modifiers (TT-1, TT-4, and TT-5) predict an increase in increment with an increase in thinning intensity, while the crown recession modifiers (TT-2 and TT-3) and the height increment modifier (TT-5) predict a decrease in the size of the direct response as thinning intensity increases (Table 11.3). These behaviors meet the expected direct responses to thinning.

2. All modifiers except TT-1 predict no direct response (i.e. a value of one) when a thinning is not implemented (Table 11.6), which meets the expected behavior of a response to thinning.

3. Both modifiers TT-2 and TT-3 approach zero as thinning intensity approaches one (Table 11.6), which is appropriate behavior in their application to modeling tree crown recession. Such behavior is not expected when modeling diameter or height increment

of trees because residual trees in the younger stands where thinning is usually applied should still be growing. Modifiers TT-1, TT-4, and TT-5 predict approaches to non-zero values as thinning intensity increases (Table 11.6). Because of the structure of the thinning intensity transformation, the TT-4 modifier predicts unreasonably large values as the thinning intensity approaches one. The structure of the TT-1 modifier restricts it to only approach values under one, while the structure of the TT-5 modifier allows it to approach values greater than or less than one as thinning intensity approaches one. Modifier values under one reduce the predicted increment from the untreated plot equation, while values above one increase it.

4. All of the modifiers except TT-1 predict that the direct thinning response approaches one (no direct response) as time since thinning gets large (Table 11.6), which meets the expected behavior of a response to thinning.

5. Modifiers TT-2, TT-4, and TT-5 show an immediate direct thinning response (i.e. a multiplicative modifier value $\neq 1.0$) when time since thinning is zero, while modifier TT-3 shows no additional thinning response. (Table 11.3 and Table 11.6). The predicted value of modifier TT-3 starts at one and decreases to a minimum value before returning to one with increasing time since thinning (Table 11.6). Modifier TT-4 starts at a value larger than one and increases to a maximum value before returning to one with increasing time since thinning (Table 11.6). Evaluation of the reasonableness of these behaviors will depend upon the form and type of attribute being modified. An immediate thinning response for a time since thinning value of zero would be reasonable if the equation to which the modifier was applied predicted future increment in diameter or height, particularly for multi-year long growth periods.

The added flexibility of modifier TT-5 has resulted in parameter estimates significantly different from one that increased predicted diameter increment (Figure 11.2) and reduced predicted height increment of the residual trees following thinning for two long-lived conifer species growing on highly productive sites (Hann *et al.*, 2003). A very recent analysis of a short-lived and fast-growing hardwood species found that modifier TT-5 was necessary for accurately estimating the reduced height increment of residual trees, but that it was not necessary for estimating residual tree diameter increment (Hann *et al.*, 2011).

Hann *et al.* (2003) also demonstrated how modifier TT-5 could be altered to account for multiple thinnings by discounting densities removed in more distant thinnings to the time of the most recent thinning, and then adding those discounted measures of density to the density before the most recent thinning. This discounted and accumulated amount of previous removals is then used to make future projections of thinning response. The basic assumption of this approach is that trees in stands for which a new thinning is conducted after the direct effect of the previous thinning has expired will behave as if the stand had

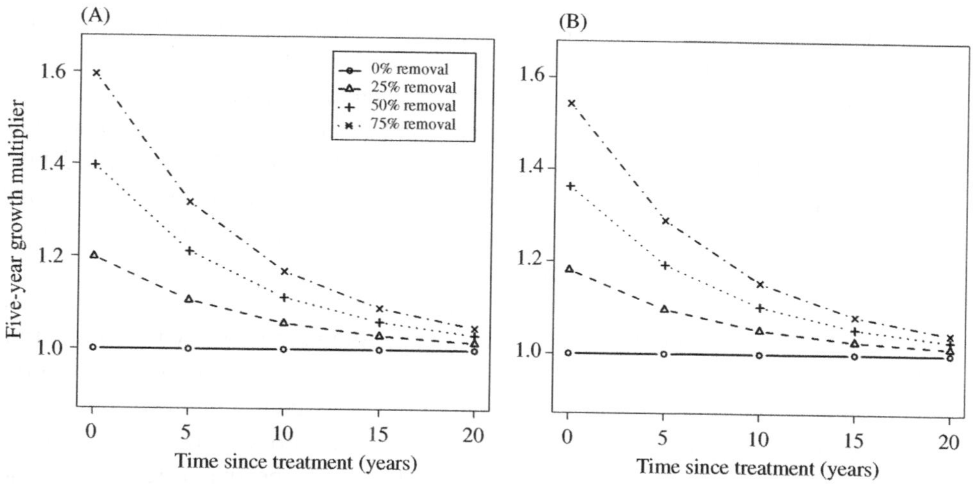

Figure 11.2 Modifiers for five-year diameter increment of various thinning intensities (percentage basal area removed) and time since treatment (years) for Douglas-fir (A) and western hemlock (B). Based on the equations of Hann *et al.* (2003).

not been previously thinned, and that multiple thinnings conducted very close in time will behave like a similar thinning composed of the accumulated removals.

Certain model forms, such as the logistic, exhibit expected asymptotic behavior that can be compromised by an external multiplicative modifier function. In those situations, it has been found best to incorporate thinning-related variables within the equation. Hynynen (1995a) used an exponential model form when developing a crown ratio equation for both thinned and unthinned Scots pine, which predicted larger crowns in thinned stands. Hann *et al.* (2003) used a modifier within a logistic model form to account for the influence of thinning on Douglas-fir and western hemlock height to crown base. The Hann *et al.* (2003) modifier was constructed in a manner that guaranteed height to crown base was predicted to be the same immediately before and after thinning, and that height to crown base would start to increase as tree and stand attributes developed and as the time since thinning increased. Both approaches involved adding back to the residual competition variable(s) the value(s) removed in thinning, but discounted with the amount of time (or dominant height development) since thinning. An alternative approach is to restrict the crown ratio or height to crown base equation in a manner such that predictions of crown recession immediately after thinning are zero (e.g. Ritchie and Hann, 1987).

The logistic equation form has also seen wide usage in modeling tree mortality/survival. Most studies have found that a properly formulated individual-tree mortality equation often does not require a thinning modifier (Hamilton, 1986; Westfall and Burkhart, 2001; Hann *et al.*, 2003). However, Bravo-Oviedo *et al.* (2006) found that their unthinned plot survival equation for Scots pine should not be applied to thinned plots, and Karlsson and Norell (2005) fit separate survival equations to different thinning treatments and discovered that survival rates varied by intensity and type of thinning.

Several studies have found that static total height to diameter equations can characterize the relationship in both thinned and unthinned stands without the need to incorporate additional thinning descriptors in the equation (Zhang *et al.*, 1997a; Dorado *et al.*, 2006). However, this finding is contrary to those studies which found that thinning affected height increment. Perhaps the amount of variability in predicting height from diameter overwhelms the impact of thinning upon height increment.

Predicting the long-term consequences of various thinning regimes in individual-tree growth models is difficult, but trying to implement the thinning properly in a simulation can also pose a significant challenge. Söderbergh and Ledermann (2003) reviewed the different thinning algorithms for five commonly used individual-tree growth models in Europe, and found that the algorithms could be grouped into six different categories based on: (1) the type of data used during development; (2) the spatial resolution; and (3) actual implementation in the model. Most thinning regimes can be applied using both an analytical algorithm and a distance-independent growth model, but representing geometric thinnings like strip thinnings can be difficult with this approach (Söderbergh and Ledermann, 2003), particularly if there is substantial spatial variation within the stand.

Representing the influence of thinning in growth models has ranged from assuming growth is the same as an unthinned stand of the same density, to individual-tree growth modifiers dependent on the intensity, time since treatment, and number of thinnings. Although diameter distribution equations have been modified to account for thinning (e.g. Murray and Von Gadow, 1991; Álvarez González *et al.*, 2002), the approach can predict trees to be present when they never existed or were removed in the thinning. Thus, an individual-tree approach is recommended, particularly in mixed-species forest where thinning can favor the removal of a particular species. However, most of work on modeling the influence of thinning has occurred in single-species plantations, and very different techniques from those presented here might be required in mixed-species and naturally-regenerated stands.

Although the focus in this chapter has been on statistical models, process-based and hybrid models have also been modified to account for thinning (Landsberg *et al.*, 2005; Petritsch *et al.*, 2007; Wang *et al.*, 2011).

11.5 Fertilization

Fertilization increases tree and stand growth rates by supplying nutrients that are limiting growth. Unlike thinning, which often reduces stand increment but increases tree increment, fertilization of responding stands will increase both immediately after application. Likelihood of a response to fertilization does vary between stands (Peterson *et al.*, 1986), and it can be difficult to predict. The size of the direct effect response will depend upon tree species, the type and amount of fertilizer applied, the time since application, and the underlying productivity of the site, with poor-quality sites often having larger relative responses than high-quality sites.

The same three basic methods described at the start of this chapter have also been used to model fertilization response. As with thinning, several multiplicative and additive fertilization modifiers will be examined in detail to compare and contrast their behavior. Therefore, Tables 11.7 and 11.8 describe the model forms used for the fertilization modifiers and how they were applied. Again, the original expressions of these modifier equations were reformulated using common terms and algebra in order to facilitate comparisons. Table 11.9 describes the type of response effect used to form the fertilization response variable, the type of duration assumed by the model form, and the fitting procedures used to estimate the parameters of the model form. Finally, Tables 11.10 and 11.11 examine the predicted behavior of key elements composing the various fertilization modifiers.

Fertilization is an important management tool in several regions, particularly in the western (Tappeiner II *et al.*, 2007; pp. 242–251) and southern United States (Jokela *et al.*, 2004). However, like vegetation management, it can be hard to predict empirically because of high variability and the limiting factors to growth not being reflected by site index. Response to fertilization is not only dependent on the amount and type of fertilizer, but the weather conditions at time of application, understory vegetation characteristics, and soil attributes, which are often not characterized in most statistical growth models.

11.5.1 Stand-level

Bailey *et al.* (1989) accounted for the effects of both nitrogen and phosphorus fertilization on loblolly pine stand BA growth, TPH, and dominant height increment using differential yield equations. However, they used a model form that confounded the untreated component and the treated component such that a treatment modifier could not be identified. The parameters of the resulting combined model form were estimated simultaneously. Hann *et al.* (2003) examined the impact of fertilization upon the maximum size–density trajectory of Douglas-fir stands and found that fertilization accelerated the rate that a stand moved along the trajectory but not the configuration of the trajectory. Their results paralleled previous findings that site quality increased mortality rates but did not affect the configuration of the maximum size–density trajectory (White and Harper, 1970; Smith and Hann, 1984), producing the so-called Suchatschew effect (Harper, 1977).

The three stand-level multiplicative modifiers and three additive modifiers described in Table 11.7 illustrate the wide range in model forms used to characterize response to fertilization for just three tree species. The three multiplicative modifiers predict the direct effect of the fertilization response, while the three additive modifiers predict the combined direct and indirect effect (Table 11.9). The parameters of two of the multiplicative modifiers were fit simultaneously using the log-linearization and linear regression to

Table 11.7 Stand-level multiplicative and additive fertilization modifiers.

Modifier designator	Modifier equation[a]
Multiplicative	
SF-1	$e^{\left(\frac{\left(+b_0\cdot\left(\log\left(\mathrm{NF}\cdot\left(\Delta\mathrm{H100_f}\cdot e^{\left(\frac{-\Delta\mathrm{H100_f}}{+b_1}\right)}\right)^{\left(\frac{3}{2}\right)}+1.0\right)\right)\right)}{\mathrm{H100_f}}\right)}$
SF-2	$e^{\left(\frac{\left(+b_0\cdot\left(\log\left(\mathrm{NF}\cdot\left(Y_f\cdot e\left(\frac{-Y_f}{(+b_1-b_2\cdot\mathrm{SI})}\right)\right)^{3}+1.0\right)\right)^{2}\right)}{\mathrm{SI}}\right)}$
SF-3	$+b_0\cdot(\mathrm{NF}+1)^{+b_1}\cdot\mathrm{ftn}_1(Y_f,\mathrm{SI})\cdot\mathrm{ftn}_2(\mathrm{SI},A_f)\cdot\mathrm{ftn}_3(\mathrm{RD}_f);$ $\mathrm{ftn}_1(Y_f,\mathrm{SI})=e^{\left(\frac{\left(-b_2\cdot(Y_f+0.5)^2+(-b_3+b_4\cdot\log(\mathrm{SI}))\right)}{(Y_f+0.5)}\right)}$ $\mathrm{ftn}_2(\mathrm{SI},A_f)=e^{\left(\left(\frac{-b_5}{\mathrm{SI}^2}\right)+\left(\frac{-b_6+b_7}{(A_f+Y_f+0.5)}\right)\cdot\mathrm{SI}\right)}$ $\mathrm{ftn}_3(\mathrm{RD_f})=\mathrm{RD_f}^{-b_2}\cdot e^{(+b_9\cdot\mathrm{RD_f})}$
Additive	
SF-4	$\left(\left(\frac{+b_0}{A_f}\right)+b_1\cdot\mathrm{BA_f}-b_2\cdot\mathrm{SI}\right)\cdot\mathrm{NF}^{-b_3}\cdot e^{(+b_4\cdot\mathrm{NF}+b_5\cdot\mathrm{PF})\cdot Y_f^{+b_6}}\cdot e^{(-b_7\cdot Y_f)}$
SF-5	$X_1\cdot Y_f^{+b_0}\cdot e^{(X_2\cdot Y_f)};$ $X_1=\left(1.0-e^{(-b_1\cdot\mathrm{NF})}\right)\cdot\mathrm{HD_f}^{+b_2}\cdot A_f^{-b_3}\cdot\mathrm{SI}^{-b_4}\cdot\mathrm{TPH_f}^{+b_5}+b_6\cdot I_{d1}$ $X_2=-b_7+b_8\cdot\log(1.0+\mathrm{PF})$
SF-6	$X_1\cdot Y_f^{+b_0}\cdot e^{(X_2\cdot Y_f)};$ $X_1=\left(1.0-e^{(-b_1\cdot\mathrm{NF})}\right)\cdot\mathrm{HD_f}^{-b_2}\cdot\mathrm{TPH_f}^{-b_3}\cdot\mathrm{BA_f}^{+b_4}$ $X_2=-b_5+(+b_6-b_7\cdot I_{d1}+b_8\cdot I_{d2})\cdot\log(1.0+\mathrm{PF})$

[a]Where A_f is the stand age at time of fertilization, $\mathrm{BA_f}$ is the stand basal area at fertilization, $\mathrm{BA_2}$ is the stand basal area at the end of the growth period, ΔBA is the basal area increment of the stand, $\Delta\mathrm{H100_f}$ is the change in top height of the stand since fertilization, ΔQMD is increment in quadratic mean diameter of the stand, ΔV is the volume increment of the stand, $\mathrm{H100_f}$ is the top height of the stand at time of fertilization, $\mathrm{HD_f}$ is the average dominant height of the stand at time of fertilization, $\mathrm{HD_2}$ is the average dominant height of the stand at the end of the growth period, I_{d1} is 1.0 for very poorly drained soils and 0.0 otherwise, I_{d2} is 1.0 for well-drained soils

Used for predicting	Species	Citations
Juvenile QMD	Douglas-fir	Curtis *et al.* (1981)
Annual gross ΔV, annual gross ΔBA, and annual net ΔQMD	Douglas-fir	Curtis *et al.* (1981)
Annual gross ΔV	Douglas-fir	Miller *et al.* (1988)
Annual ΔBA	Radiata pine	Lowell (1988)
HD_2	Loblolly pine	Amateis *et al.* (2000)
BA_2	Loblolly pine	Amateis *et al.* (2000)

and 0.0 otherwise, NF is the weight of nitrogen per unit area applied at fertilization, PF is the weight of phosphorus per unit area applied at fertilization, QMD is the quadratic mean diameter of the stand, RD_f is Curtis (1982) relative density of the stand at time of fertilization, SI is site index of the stand, TPH_f is the number of trees in the stand at fertilization, Y_f is the number of years since fertilization, b_i are coefficients, with the expected sign of the coefficient indicated before the coefficient, $ftn_i()$ is the specified function evaluated at the values within the parentheses, and log is the natural logarithm.

Table 11.8 Tree-level multiplicative and additive fertilization modifiers.

Modifier designator	Modifier equation[a]	Used for predicting	Species	Citations
Multiplicative				
TF-1	$1.0+(+b_0+b_1\cdot I_{PF})\cdot NF\cdot(Y_f)^{+b_2}\cdot e^{\left(-b_3\cdot Y_f^{(+b_2+1)}\right)}$	Two-year ΔDBH and the ΔHD component of two-year ΔHT	Loblolly pine	Hynynen *et al.* (1998)
TF-2	$1.0+b_0\cdot NF^{+b_1}\cdot e^{\left(-b_2\cdot Y_f-b_3\cdot(SI-1.37)^{+b_4}\right)}$	Five-year ΔDBH and the ΔH100 component of five-year ΔHT	Douglas-fir	Hann *et al.* (2003)
Additive				
TF-3	$(b_0\cdot I_{NF}+b_1\cdot I_{PF}+b_2\cdot(1.0-I_{PF})\cdot I_{KF})\cdot e^{\left(-b_3\cdot(1.0+Y_f)^{+b_4}\cdot SI\right)}$	Annual ΔHT	Loblolly pine	Westfall *et al.* (2004)
TF-4	$(b_0\cdot I_{NF}+b_1\cdot I_{PF}+b_2\cdot I_{KF})\cdot(1.0+Y_f)^{+b_3}\cdot e^{\left(-b_4\cdot(1.0+Y_f)-b_5\cdot SI\right)}$	Annual ΔDBH	Loblolly pine	Westfall *et al.* (2004)

[a] Where ΔDBH is the diameter increment of a tree, ΔHT is the height increment of a tree, ΔH100 is the top height increment of the stand, ΔHD is the height increment of the dominant trees in the stand, I_{KF} is 1.0 for application of potassium fertilizer and 0.0 otherwise, I_{NF} is 1.0 for application of nitrogen fertilizer and 0.0 otherwise, I_{PF} is 1.0 for application of phosphorus fertilizer and 0.0 otherwise, NF is the weight of nitrogen per unit area applied at fertilization, SI is site index of the stand, and Y_f is the number of years since fertilization, and b_i are coefficients with the expected sign of the coefficient indicated before the coefficient.

Table 11.9 Type of response effect and duration as well as fitting procedures for various stand- and tree-level multiplicative and additive fertilization modifiers.

Modifier designator	Type of response effect	Type of duration	Fitting procedures	
			Regression technique	Fitting strategy
Stand-level multiplicative				
SF-1	Direct	Type 1	Log-linear	Simultaneous
SF-2	Direct	Type 1	Log-linear	Simultaneous
SF-3	Direct	Type 1	Log-linear	Two-step
Stand-level additive				
SF-4	Direct + indirect	Type 1	Nonlinear	One-step
SF-5	Direct + indirect	Type 1	Nonlinear	One-step
SF-6	Direct + indirect	Type 1	Nonlinear	One-step
Tree-level multiplicative				
TF-1	Direct	Type 1	Nonlinear	Simultaneous
TF-2	Direct	Type 1	Nonlinear	Three-step
Tree-level additive				
TF-3	Direct + indirect	Type 1	Nonlinear	One-step
TF-4	Direct + indirect	Type 1	Nonlinear	One-step

estimate the parameters, while the other multiplicative modifier used the two-step method with a linearized model form fit using stepwise linear regression. All three of the additive modifiers used the one-step method and nonlinear regression to estimate the parameters, as well as assuming a Type 1 duration of treatment response (Table 11.9).

Despite the varied model forms and parameter estimation techniques used to create the modifier equations, most of them produce similar predictive behavior:

1. All modifiers except SF-4 predict an increase in stand size or increment with an increase in the amount of fertilizer applied (Table 11.7), which meets the expected behavior of a response to fertilization. Modifier SF-4 predicts a decline in increment at low levels of fertilization before reversing and predicting increasing increment at higher levels of fertilization.

2. All modifiers except SF-3 predict no response (i.e. a value of one for a multiplicative modifier and a value of zero for an additive modifier) when no fertilization is applied (Table 11.10), which meets the expected behavior of a response to fertilization.

Table 11.10 Predictive behavior of stand-level multiplicative and additive fertilization modifiers.

Modifier designator	Limiting values for modifier equations[a]				X_f value for maximum modifier
	NF = PF = KF = 0.0	X_f value	$X_f = 0.0$	$X_f \to \infty$	
Multiplicative fertilization modifiers					
SF-1	1.0	$\Delta H100_f$	1.0	1.0	Maximum $= +b_1$
SF-2	1.0	Y_f	1.0	1.0	Maximum $= +b_1 - b_2 \cdot SI$
SF-3	$b_0 \cdot ftn_1(Y_f, SI) \cdot ftn_2(SI, A_f) \cdot ftn_3(RD_f)$	Y_f	$b_0 \cdot (NF + 1.0)^{b_1} ftn_1(0, SI) \cdot ftn_2(SI, A_f) \cdot ftn_3(RD_f)$;	0.0	Maximum $= \left(\frac{(+b_3 - b_4 \cdot SI)}{+b_2}\right)^{\left(\frac{1}{3}\right)} - 0.5$
Additive fertilization modifiers					
SF-4	0.0	Y_f	0.0	0.0	Maximum $= \left(\frac{+b_6}{+b_7}\right)$
SF-5	0.0	Y_f	0.0	0.0	Maximum $= \left(\frac{-b_0}{-X_2}\right)$
SF-6	0.0	Y_f	0.0	0.0	Maximum $= \left(\frac{-b_0}{-X_2}\right)$

[a] Where A_f is the stand age at time of fertilization, RD_f is Curtis (1982) relative density of the stand at time of fertilization, SI is site index of the stand, NF is the weight of nitrogen per unit area applied at fertilization, KF is the weight of potassium per unit area applied at fertilization, PF is the weight of phosphorus per unit area applied at fertilization, X_f is the variable selected to indicate fertilization duration, $\Delta H100_f$ is the change in top height of the stand since fertilization, Y_f is the number of years since fertilization, and b_i are coefficients with the expected sign of the coefficient indicated before the coefficient. See Table 11.7 for definitions of $ftn_1(Y_f,SI)$, $ftn_2(SI,A_f)$, $ftn_3(RD_f)$, and X_2. The function $ftn_1(0,SI)$ is $ftn_1(Y_f,SI)$ evaluated for a Y_f value of 0.

Table 11.11 Predictive behavior of tree-level multiplicative and additive fertilization modifiers.

Modifier designator	Limiting values for modifier equations[a]				X_f value for maximum modifier
	NF = PF = KF = 0.0	X_f Value	$X_f = 0.0$	$X_f \rightarrow \infty$	
Multiplicative					
TF-1	1.0	Y_f	1.0	1.0	$\text{Maximum} = \left(\frac{+b_2}{(+b_3 \cdot (+b_2 + 1))}\right)^{\left(\frac{1}{(+b_2+1)}\right)}$
TF-2	1.0	Y_f	$1.0 + b_0 \cdot \text{NF}^{+b_1} \cdot e^{\left(-b_3 \cdot (\text{SI} - 1.37)^{+b_4}\right)}$	1.0	Maximum = 0.0
Additive					
TF-3	0.0	Y_f	$(+b_0 \cdot I_{NF} + b_1 \cdot I_{PF} + b_2 \cdot (1.0 - I_{PF}) \cdot I_{KF}) \cdot e^{(-b_3 \cdot \text{SI})}$	0.0	Maximum = 0.0
TF-4	0.0	Y_f	$(+b_0 \cdot I_{NF} + b_1 \cdot I_{PF} + b_2 \cdot I_{KF}) \cdot e^{(-b_4 - b_5 \cdot \text{SI})}$	0.0	$\text{Maximum} = \left(\frac{+b_3}{+b_4}\right) - 1.0$

[a] Where I_{KF} is 1.0 if potassium fertilizer has been applied or 0.0 otherwise, I_{NF} is 1.0 if nitrogen fertilizer has been applied or 0.0 otherwise, I_{PF} is 1.0 if phosphorous fertilizer has been applied or 0.0 otherwise, NF is the weight of nitrogen per unit area applied at fertilization, KF is the weight of potassium per unit area applied at fertilization, PF is the weight of phosphorus per unit area applied at fertilization, SI is site index of the stand, X_f is the variable selected to indicate fertilization duration, Y_f is the number of years since fertilization, and b_i are coefficients with the expected sign of the coefficient indicated before the coefficient.

3. All modifiers except SF-3 predict that the fertilization response will disappear as the amount of time since application gets large (Table 11.10), which meets the expected Type 1 behavior of a response to fertilization. All modifiers except SF-1 use the number of years since fertilization to index duration of response, while modifier SF-1 uses the amount of increment in top height since fertilization as a response variable. Modifier SF-3 will drive the untreated equation to zero as the amount of time since application gets large.

4. All modifiers except SF-3 predict no response immediately after fertilizer has been applied (Table 11.10). This behavior meets expectation when predicting a static variable such as juvenile QMD (i.e. modifier SF-1), but it does not meet expectation when predicting future increment values (i.e. SF-2 and SF-4) and values for the end of growth period (SF-5 and SF-6). Therefore, only fertilization modifier SF-3 meets the expected behavioral response to fertilization of a stand-level increment variable.

5. All modifiers predict a peak in response after fertilization (Table 11.10). Two of the modifiers predict that the peak in response will occur more quickly after fertilization as site index increases. Because top height integrates the effect of time and age, the usage of top height increment in modifier SF-1 as a surrogate for time since fertilization results in a peak that occurs at a constant value of top height increment.

6. All modifiers except SF-3 predict that the response to fertilization monotonically decreases with increasing site index (SF-2 and SF-4), top height (SF-1) at time of treatment, or average dominant height (SF-5 and SF-6) at time of treatment (Table 11.7). Only modifier SF-3 predicts that the response increases to a peak and then declines as site index increases.

7. The three additive modifiers also incorporated the option of fertilizing with phosphorus (Table 11.7). The three multiplicative modifiers were developed for Douglas-fir, and the three additive modifiers were developed for radiata pine and loblolly pine. Unlike the pine species, the response of Douglas-fir to phosphorus has, in general, not been large (Gessel *et al.*, 1979) and, therefore, its application has not been widespread in the Douglas-fir region.

8. All modifiers except SF-2 predict that the size of response varies by stand age and/or top height/average dominant-height at fertilization (Table 11.7). Modifiers SF-1, SF-3, SF-4, and SF-6 predict that the fertilization response declines with increasing age/height at fertilization. Modifier SF-5 predicts that the amount of response declines with age but increases with average height of the dominant trees at fertilization (Table 11.7).

9. Multiplicative modifier SF-3 and the three additive modifiers predict that the size of response also varies by the amount of two-sided competition at fertilization

(Table 11.10). Modifier SF-3 predicts that the size of the fertilization response increases with Curtis's (1982) relative density at fertilization, and modifiers SF-4 and SF-6 predict that the amount of fertilization response declines with increasing BA at fertilization and TPH at fertilization, respectively. Modifier SF-5 predicts that the amount of response declines with BA at fertilization but increases with TPH at fertilization (Table 11.7). The sizes of the parameter estimates and their standard errors indicate that this combination in SF-5 is behaving like the average square BA per tree at fertilization, with the fertilization response increasing with an increase in this average.

The questionable behavior of modifier SF-3 probably indicates that the modifier is overparameterized, which can be a potential problem using a method such as stepwise regression to formulate the model form. On the other hand, the model form of modifier SF-4 seems consistent with that used by others, but the specific parameters are inconsistent, which points to problems with the modeling dataset itself. An examination of the dataset used to parameterize modifier SF-4 shows that the dataset came from only four installations and it contained only one level of nitrogen fertilization.

An alternative to developing equations that include the effects of fertilization or simply adjusting the site index is to the use the age-shift method (e.g. Kimberley *et al.*, 2004; Carlson *et al.*, 2008). This method has also been used for modeling the response to vegetation management (Lauer *et al.*, 1993; South and Miller, 2007), and is simply a reduction in the amount of time a stand takes to reach a certain size when compared to an untreated one. This approach assumes that the shape of the growth curve is not changed by the treatment, and the stand development trajectory is only accelerated. Carlson *et al.* (2008) predicted the age-shift for dominant height, BA, and volume per hectare of fertilized loblolly pine as a function of stand attributes as well as the fertilizer rate and type. The estimated age-shifts were then used to adjust stand age in an existing growth equation and found to give superior predictions compared to growth equations that had directly included the effects of fertilizer (Carlson *et al.*, 2008). The limitation of this approach is that it becomes difficult to estimate the age-shift for a stand which is simultaneously fertilized and thinned.

11.5.2 Tree-level

Daniels and Burkhart (1975) modified the stand's site index to characterize fertilization of loblolly pine. The method of changing site index produces a long-term, Type 2 fertilization response. This strategy was then modified by Burkhart *et al.* (1987, revised 2001) to only apply to fertilizations at the time of planting. The multiplicative modifier of Hynynen *et al.* (1998) was incorporated into the Burkhart *et al.* (1987, revised 2001) model for midrotation fertilizations, which produces a Type 1 response. Many of the efforts at incorporating the tree-level influence of fertilization have assumed or found a Type 1 response.

Both Shaffi *et al.* (1990) and Shen *et al.* (2000) used indicator variables to characterize Type 1 fertilization response of squared diameter increment. The use of fertilization dosage indicator variables can be effective, especially if the length of the growth period corresponds with the duration of the direct fertilization response, as long as they are designed in a manner that allows statistical testing of the significance of the response. Unfortunately, the indicator variables of both Shaffi *et al.* (1990) and Shen *et al.* (2000) were designed in a manner that allowed only very limited statistical testing of the significance of the treatments. The intercept indicators in Shaffi *et al.* (1990) did allow testing the significance of whether the no treatment, fertilized only treatment, or thinned only treatment were significantly different from the combined thinned and fertilized treatment response, but not whether other combinations, such as fertilized only versus no treatment, were significantly different from each other. Examination of the parameter values and their confidence intervals for the model of Shaffi *et al.* (1990) suggests that a significant fertilization response might not have been found in that data.

Similarly, the structure of the intercept indicators in Shen *et al.* (2000) does allow testing the significance of a fertilization response over no treatment, but not whether the larger dosage of fertilizer was significantly different from the smaller dosage. Furthermore, the structure of the indicators used to evaluate whether the effect of one-sided competition, as assessed by $BAL/(\ln(DBH)+1)$, significantly differed by fertilizer dosage level and type of bedrock, was designed in a manner that prevented the statistical testing needed to assess the issue. As with Shaffi *et al.* (1990), examination of the parameter values and their confidence intervals suggests that a significant fertilization response might not have been found for a number of the combinations examined. Finally, the fertilizer response variables in both Shaffi *et al.* (1990) and Shen *et al.* (2000) unnecessarily confounded the treatment variables with the untreated variables so that the effect of treatment cannot be expressed as either a multiplicative modifier or an additive modifier upon the untreated equation.

The additive tree-level modifiers in Table 11.8 also have used indicator variables for characterizing the response of fertilization at the time of planting. The use of indicator variables does restrict application of the equation to the dosages characterized by the indicator variables. Multiplicative modifiers TF-1 and TF-2 do not have these restrictions (Table 11.8). The two multiplicative modifiers predict the direct effect of the fertilization response, while the two additive modifiers predict the combined direct and indirect effect (Table 11.9).

The parameters of one of the multiplicative modifiers were fit simultaneously using nonlinear regression to estimate the parameters, while the other multiplicative modifier used the three-step method with nonlinear regression. Both of the additive modifiers used the one-step method and nonlinear regression to estimate the parameters. All of the tree-level modifiers assumed a Type 1 duration of treatment response (Table 11.9).

The model forms of the tree-level fertilization modifiers produced similar predictive behavior and they are simpler than many of the stand-level modifiers:

1. Both of the multiplicative modifiers predict an increase in increment with an increase in the amount of fertilizer applied (Table 11.8), which meets the expected behavior of a response to fertilization. The additive modifiers used indicator variables for single dosage levels of three types of fertilizer.

2. All modifiers predict no response (i.e. a value of one for a multiplicative modifier and a value of zero for an additive modifier) when no fertilization is applied (Table 11.11), which meets the expected behavior of a response to fertilization.

3. All modifiers predict that the fertilization response will disappear as the amount of time since application gets large (Table 11.11), which meets the expected Type 1 behavior of a response to fertilization. All modifiers use the number of years since fertilization to index duration of response.

4. All modifiers except TF-1 predict a response immediately after the application of fertilizer (Table 11.11). This behavior meets expectation when predicting future increment values.

5. Modifiers TF-1 and TF-4 predict a peak in response after fertilization, and modifiers TF-2 and TF-3 do not (Table 11.11). Modifier TF-2 was used with five-year Douglas-fir increment equations, and Curtis *et al.* (1981) found that the peak in fertilization response occurred two to three years after the application of fertilizer to Douglas-fir.

6. All modifiers except TF-3 predict that the response to fertilization monotonically decreases with increasing site index (Table 11.8).

7. The three modifiers for loblolly pine also incorporated the option of fertilizing with phosphorus and/or potassium (Table 11.8). The other multiplicative modifier was developed for Douglas-fir. As noted under the discussion of the stand-level fertilization multipliers, the response of Douglas-fir to other nutrients has not been large and, therefore, their application has not been widespread in the Douglas-fir region.

The predictive behavior of modifier TF-2 (Hann *et al.*, 2003) is illustrated in Figure 11.3. No additional stand-level or tree-level attributes have been incorporated in the tree-level fertilization modifiers (Table 11.8). Hann *et al.* (2003) did explore a wide range of additional stand-level and tree-level predictor variables, but found none of them contributed significantly to the prediction of fertilizer response. Their results confirmed the earlier findings of Moore *et al.* (1994) that neither thinning nor fertilization changed the relative distribution of growth within a stand. Paralleling their thinning modifier, Hann *et al.* (2003)

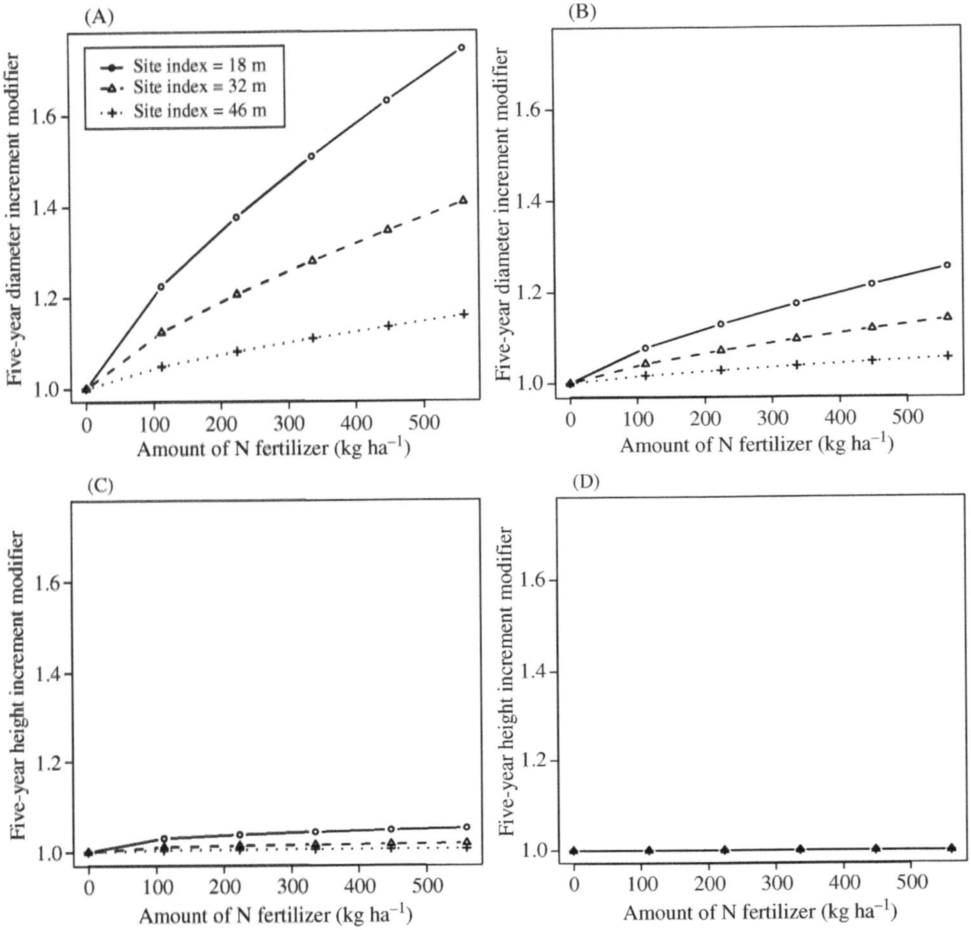

Figure 11.3 Modifiers for coastal Douglas-fir five-year diameter (top) and height increment (bottom) following nitrogen fertilization (kg ha^{-1}) for different site indices and time since treatment. The left panel is immediately after treatment, while the right panel is five years after treatment. Based on the equations of Hann *et al.* (2003).

also demonstrated how modifier TF-2 could be altered to account for multiple fertilizations by discounting the amount of fertilizer applied in more distant fertilizations to the time of the last fertilization, and then adding those discounted measures of dosage to the amount of fertilizer last applied to the stand for future projections of the response to fertilization.

As with characterizing mortality in response to thinning, the expected asymptotic behavior of the logistic equation can be compromised by an external multiplicative or additive modifier function. Both Shen *et al.* (2001) and Hann *et al.* (2003) used a modifier within a logistic model form to account for the direct influence of fertilization on predicted mortality rate of Douglas-fir. Shen *et al.* (2001) used two indicator variables for two levels of nitrogen dosage in the exponential component of the nonlinear logistic equation, and Hann *et al.* (2003) inserted a function of the amount of nitrogen applied and the time since application in the exponential component. Both models predict an increase in mortality rate with fertilization, with larger dosages having higher rates of mortality. The Hann *et al.* (2003) modifier predicts that the increased mortality rate will decline as time since fertilization increases.

Hann *et al.* (2003) also inserted a function of amount of nitrogen applied and the time since application in the exponential component of a static equation for predicting height to crown base. The modifier was structured such that: (1) predicted HCB would be the same immediately before and after fertilization; (2) HCB will be lower for the fertilized tree when compared to a tree not fertilized, until arriving at the time since fertilization where the largest impact of fertilization occurs; and (3) after that time since fertilization, the direct impact of fertilization upon HCB will disappear as time since fertilization continues to increase. The result is an increase in crown length for a period after fertilizing a tree.

Although the direct influences of fertilization are brief (e.g. 2 to 10 years), it often has a significant long-term impact on growth. In the comparison of several growth models in the US Pacific Northwest, Johnson (2005) found that the model predictions differed the most when fertilizer was applied, particularly if combined with a thinning treatment. The models also differed in whether the relative response to fertilization was greater in stands with a lower site index (Johnson, 2005). These results highlight the difficulty in representing fertilization accurately, and suggest that a more mechanistic approach might be more justified.

11.6 Combined thinning and fertilization

While it has been widely accepted that fertilization increases the gross increment of both stands and trees in responding stands, the size and duration of the increase in net increment will depend upon the amount of accelerated mortality that also accompanies fertilization. Therefore, there has been interest in combining thinning with fertilization in order to reduce the loss to mortality and to increase the size and value of the residual stand (e.g. Peterson and Hazard, 1990; Fox *et al.*, 2007). Research on the interaction of thinning and fertilization has been relatively limited, as have the number of growth and yield models that have incorporated both treatments.

11.6.1 Stand-level

Curtis *et al.* (1981) included multiplicative modifiers for the direct effects of both thinning and fertilization on gross and net stand increments. They estimated the parameters of the modifiers simultaneously using the log-linear transformation and linear regression, using a dataset that included untreated, thinning only, fertilized only, and the combination of thinning and fertilization plots. Their assumption, therefore, is that the combination of thinning and fertilization at the stand-level is a multiplicative process producing larger relative increments than the accumulation of the individual relative increments.

Miller *et al.* (1988), on the other hand, characterized the direct effects of thinning only, fertilization only, and the combination of the two on gross stand increment, using three separate multiplicative modifier equations. They used the two-step approach and the log-linear transformation to estimate the parameters with linear regression. They found that fertilizing thinned stands resulted in gross increments of the residuals that fast approached those in unthinned stands. They also found that their combined responses differed markedly from those of Curtis *et al.* (1981).

11.6.2 Tree-level

Most tree-level models that incorporate fertilization have assumed that their indirect estimates of thinning response are adequate, and, therefore, relative responses to fertilization are applicable to both untreated and thinned stands. Shaffi *et al.* (1990) did include the combination of thinning and fertilization as one of their treatment indictor variables in a squared diameter increment equation. However, the structure of their indicator variable allowed only very limited statistical testing of the significance of the treatments (see the tree-level fertilization section for more details).

Hann *et al.* (2003) explicitly examined and characterized the direct effects of combining thinning and fertilization upon both diameter and height increment of Douglas-fir. They estimated the combined response by multiplying their thinning modifier (TT-5 in Table 11.3) by their fertilization modifier (TF-2 in Table 11.8). A comparison of the resulting predictions to the measured combined responses indicated that the product adequately characterized the combined effect. As a result, the combination of thinning and fertilization produced a greater percentage increase in tree increment than the sum of the percentage increases for thinning or fertilization alone.

11.7 Harvesting

Like thinning, harvesting removes trees from a stand, which can alter stand structure and growth dynamics. The distinction between the two is that thinning often focuses on improving the growing conditions for residual trees, while harvesting is generally more opportunistic and concerned with extracting value from the current stand. This can create drastically different stand structure and composition when compared to a traditional thinning. For example, salvage harvesting focuses on the removal of trees damaged by a certain disturbance without consideration of the residual stand stocking, and can thus range from <1 to 100% of the trees in a stand, whereas a thinning is applied to reduce stand density to a desired level and is usually constrained by a target residual density or number of trees per hectare. When compared to the other silvicultural treatments, modeling the influence of harvesting has received less attention.

11.7.1 Stand-level

Simple equations that predict yield as a function of residual basal area per hectare and dominant height and time since logging have been developed and are useful for estimating the time to the next harvest (e.g. Mendoza and Gumpal, 1987). More complex stand-level models have also been used to represent harvesting. Mendoza and Setyarso (1986) developed a transition matrix growth model to evaluate alternative harvesting schemes in Indonesia, and indicated that the current practices were not sustainable. The limitations of a stand-level model or a transition matrix approach to representing harvesting are that harvests tend to be highly variable and selective, which suggests that modeling at the size-class or individual-tree level may be more appropriate.

Hann (1978) examined the transition matrix approach and concluded it was too simplistic to adequately characterize the development of even- and uneven-aged stands of ponderosa pine in Arizona. Therefore, Hann (1980) constructed a size-class growth model that incorporated the amount of time since the last partial cut in the model's diameter-class basal increment equation, diameter-class probability of mortality equation, ingrowth equation, and through-growth equation. Hann (1980) defined through-growth as the proportion of the ingrowth that will grow through the first diameter class and into the second during the growth period.

11.7.2 Tree-level

At the tree-level, several algorithms have been developed to predict the probability of a tree being harvested (Vanclay, 1989c; Ledermann, 2002b). Harvesting can also cause damage to residual trees. Consequently, Vanclay (1989c) also developed a model to predict the probability that a tree would be destroyed during a harvesting operation. The combination of these two equations can be used to reconstruct residual stand structure after harvesting.

Individual-tree growth can also be influenced by harvesting practices. Rondon *et al.* (2009) developed a growth model for strip clear-cuts in the Peruvian Amazon, and found that the system would not be financially profitable at the time of second harvest due to the lack of merchantable trees. Thorpe *et al.* (2010) modeled stand development following different partial harvest scenarios and found that skid trails have significant influence. A spatially explicit modeling structure was needed to represent the treatments due to differences in growth and survival with distance from a skid trail (Thorpe *et al.*, 2010). Accounting for harvesting in growth models will need to continue and expand because of the wider use of alternative harvesting schemes like partial harvesting in certain regions.

11.8 Summary

Silviculture significantly alters stand dynamics, and growth models need to be able to account for these changes. However, methods for representing silviculture in growth models are not straightforward and often limited by the type of data available. Approaches for representing silviculture have included modifying site index or stand age (e.g. age-shift), fitting separate equations by silvicultural treatment, and using modifiers. At both the stand- and individual-tree levels, modifiers are likely the most logical way to represent silvicultural treatments as they don't require modification of the existing equation, can be combined to represent multiple silvicultural treatments, and are relatively easy to estimate. The ability to represent multiple silvicultural treatments is particularly important as the response may be multiplicative. For example, Hann *et al.* (2003) found that the combination of fertilization and thinning significantly increased growth more than either treatment did alone.

Another important consideration in representing silvicultural treatments is the use of continuous measures rather than simple indicators. Rarely are all combinations of a particular treatment available in a dataset. Thus, the use of indicator variables limits predictions to those available in the dataset. For example, an indicator for fertilization like that used by Shaffi *et al.* (1990) wouldn't allow managers to decide between using 200 or 400 kg of nitrogen fertilizer per hectare.

Also, the temporal resolution of the growth model may be an important consideration in representing the effects of silviculture. Although they often have long-lasting indirect effects, the direct effects of silvicultural treatments like thinning and fertilization are generally brief. This indicates that models that have longer time steps may struggle to represent silvicultural treatments accurately. Several silvicultural treatments like vegetation control and fertilization might be best represented by process-based and hybrid models, but these types of models have their own limitations.

Finally, it is important that growth models explicitly represent the silvicultural treatments they simulate. For example, the Northeastern variant of the Forest Vegetation Simulator is capable of simulating various thinning regimes, but assumes growth in thinned stands is similar to unthinned ones. However, Saunders *et al.* (2007) found that the model vastly overpredicted and underpredicted the growth of unthinned and thinned stands, respectively. This is also exemplified by how pruning response is predicted in most growth models by applying the untreated equations to the modified crown length as done in ORGANON (Hann, 2011), which hasn't been well tested. However, one of the few studies to model response to both pruning and thinning found that the stand-level sum of the crown lengths was sufficient, irrespective of the treatment (Fight *et al.*, 1995; Figure 11.4). On the other hand, this may hold at the stand-level, but not the individual-tree level.

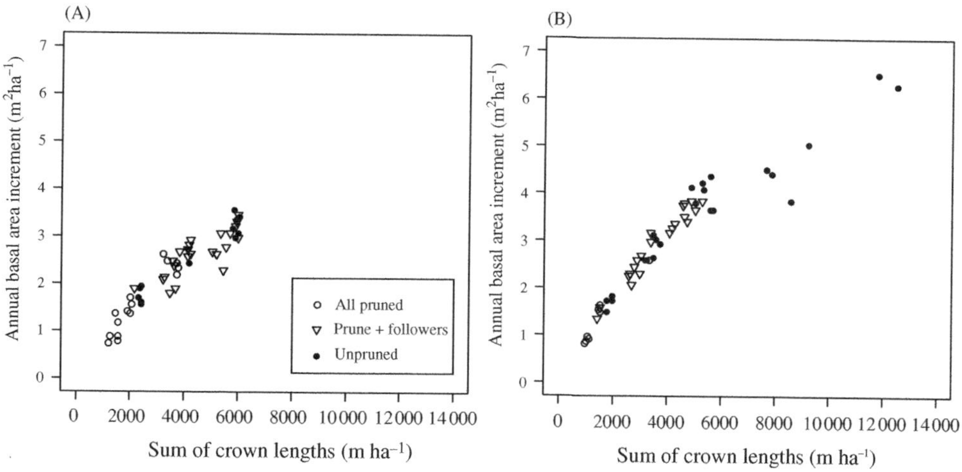

Figure 11.4 Relationship between the sum of crown lengths and annual stand-level basal area increment ($m^2\ ha^{-1}\ yr^{-1}$) for the Kaingaroa (A) and Omarama (B) pruning trials in New Zealand. Adapted from Fight *et al.* (1995).

The representation of silvicultural treatments in growth and yield models is essential for sustainable forest planning. Growth and yield models represent silvicultural treatments in varying ways, which has important implications for their prediction accuracy. Model users should be aware of this and try to understand how their selected growth model represents silvicultural treatments. Future refinements should focus on representing a wider array of treatments and the inclusion of more mechanistic approaches in predicting the response to treatment.

12
Process-based models

12.1 Introduction

Like statistical models, process-based models of forest growth and yield have a long history in forestry. Several conceptual models of forest productivity were developed in the early 1980s, but their scope of application was limited (Landsberg, 2003). Although physiologists have a general understanding of factors that influence forest production, development of process-based models of forests has lagged behind other disciplines like agriculture and horticulture for a variety of reasons including: (1) a lack of knowledge on key tree growth processes, particularly below ground; (2) trees are large, complex, multigenic, and perennial organisms, which makes basic research difficult; and (3) computers with sufficient power to handle the array of processes influencing long-term growth have only recently become widely available (Isebrands *et al.*, 1990). The use and application of process-based models has increased in recent years, and several technical reviews have previously been presented (Ryan *et al.*, 1996b; Battaglia and Sands, 1998; Mäkelä *et al.*, 2000a; Landsberg, 2003). This chapter focuses on some of the key components of process models, describes some commonly used models, and identifies a few important limitations.

Process-based models differ from statistical models in that they attempt to mechanistically represent physiological processes that influence growth, rather than develop models obtained from correlative relationships. Process-based model coefficients are often directly measured from experimental trials, and model forms are based on theoretical plant relations, which allow them to be extended to novel situations. Consequently, process-based models are often used to answer "why" or "how" type of questions rather than "what is" and "what if" questions (Leary, 1985). Key uses of process-based models are researching scientific hypotheses, assessing influences

Forest Growth and Yield Modeling, First Edition. Aaron R. Weiskittel, David W. Hann, John A. Kershaw, Jr. and Jerome K. Vanclay.

of climate and environmental factors, estimating yield for exotic species in new locations, identifying areas of further research, and teaching fundamental physiological principles.

Some important drawbacks to process models are that they are often quite complicated with output that is of little practical interest, which makes them difficult to parameterize or even operate due to high input data requirements. Use of process-based models in an applied forest management context is limited, but has been increasing in recent years because of a new series of challenging questions that forest managers currently face (Mäkelä *et al.*, 2000a).

12.2 Key physiological processes

All process models attempt to mechanistically predict tree growth by representing fundamental physiological processes. The key physiological processes in process-based models are light interception, photosynthesis, stomatal conductance, respiration, carbon allocation, mortality, and soil water and nutrients dynamics. Process-based models differ in their level of spatial and temporal resolution, which can range from individual leaves to whole stands and from hourly to annual time steps. Consequently, process-based models can be classified into two broad categories, *detailed physiological process models* and more simplified *top-down models* (Landsberg, 1986). This chapter will focus primarily on the latter and give an overview of approaches for representing each of the key physiological processes included in most process-based models.

12.2.1 Light interception

Photosynthesis is driven by the amount of light absorbed, and forest growth is often linearly related to absorption. In general, light interception of a plant canopy is influenced by the amount and type of radiation as well as the amount and distribution of leaf area. Nearly all top-down mechanistic models have used the Beer–Lambert law to calculate the amount of light absorption, while other more detailed physiological process models have used two-stream approximation (Sellers, 1985). The Beer–Lambert law relates absorption of light to the properties of the material through which it is traveling, and was originally developed to describe light extinction through a column of water containing suspended particulate (Monsi and Saeki, 1953). It has been preferred in process-based modeling because of its ease of application, as the law simply predicts an exponential decay of light, and rate of decay is dependent on density of the material (Figure 12.1). For use in plant canopies, knowledge of leaf area index (LAI; amount of total projected leaf area per unit of ground) and a light extinction coefficient are required:

$$\mathrm{LI}_N = \mathrm{LI}_0 \cdot \mathrm{e}^{\left(1 - \left(\frac{k \cdot \mathrm{LAI}}{N}\right)\right)} \qquad [12.1]$$

where LI_N is light intensity at the Nth level, LI_0 is light intensity at the top of the canopy, and k is the extinction coefficient.

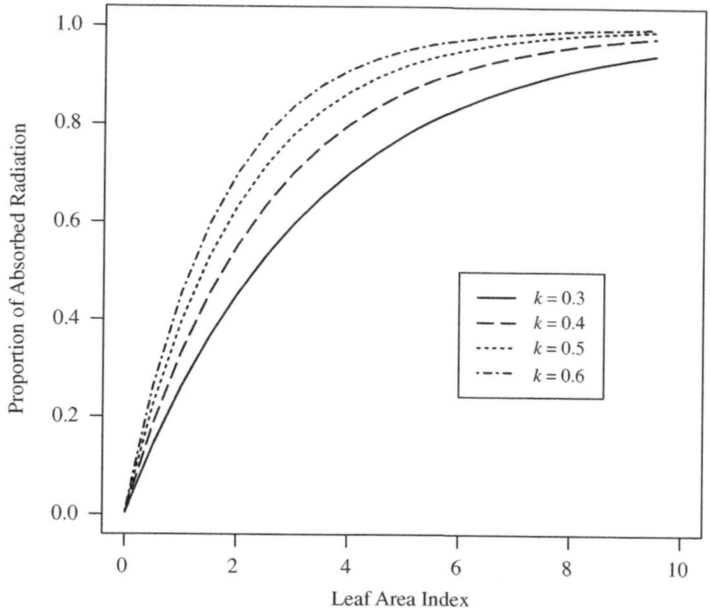

Figure 12.1 Proportion of total incoming radiation intercepted by the canopy for different leaf area indices and light extinction coefficients (k) as predicted by the Beer–Lambert law.

Application of the Beer–Lambert law to plant canopies assumes that leaves are monochromatic and randomly distributed in the canopy, and that absorbance by woody plant elements is negligible. Larsen and Kershaw (1996) found that the Beer–Lambert predictions of average light conditions below the canopy were relatively robust to departures from these assumptions. When compared to other more detailed light interception modeling techniques, Wang (2003) also found that the Beer–Lambert law can be used to accurately estimate amount of absorbed visible radiation, but only if absorption of direct beam and diffuse radiation is considered separately and the extinction coefficients were modified appropriately. However, most top-down mechanistic models treat radiation as homogeneous and use a fixed extinction coefficient of ~0.5 based on the work of Pierce and Running (1988). Diffuse radiation originates from all parts of the sky so the distribution below a given LAI is relatively uniform, and its interception is relatively independent of leaf orientation. In addition, light extinction coefficients are relatively dynamic as they depend on the type of incoming radiation, solar geometry, LAI, and stand structure (e.g. Smith, 1993).There is also some indication that the canopy light extinction coefficient is dependent on stand species composition (Duursma *et al.*, 2003).

Light interception in conifers is particularly influenced by the degree of foliage clumpiness, which can have significant effects on estimates of intercepted radiation

(Kucharik *et al.*, 1999). In general, clumping of foliage allows for a greater penetration of light and thus a greater portion of the canopy is exposed to higher light levels. Correction factors have been widely developed for a variety of forest types (e.g. Chen *et al.*, 2005), while Campbell and Norman (1998) present a simplified equation to estimate a stand-level clumping factor from the ratio of crown length to crown diameter, crown porosity, and a within-shoot clumping factor. Although Kucharik *et al.* (1998) concluded that branches did not intercept a significant amount of radiation in boreal forests, Barclay *et al.* (2000) indicated that contribution of stem boles to light interception can be considerable, particularly in mature, unthinned stands with closed canopies. Hence, the Beer–Lambert law provides a good first approximation of light interception by plant canopies and is adequate for most top-down process-based applications, but its performance in certain situations should always be assessed first.

The integration of light interception is relatively easy for process-based models with sub-daily (e.g. hourly) temporal resolutions. However, estimation of total light absorbance by the forest canopy is more difficult for models with daily or monthly temporal resolutions. This is because both direct and diffuse radiation have both high daily and seasonal variation that must be accounted for. One approximation not widely used in most process-based models that has been shown to work well is to treat total radiation as if it were diffuse over a whole day (Campbell and Norman, 1998).

Several top-down process-based models with a daily resolution estimate diurnal radiation distribution and integrate daily total radiation from the daily mean radiation levels with a sine-curve assumption. In an analysis of this assumption, Wang *et al.* (2002b) found that this approach can cause serious overestimation of radiation, and an unrealistic diurnal pattern. Instead, they recommend differentiating between direct and diffuse radiation and using a cosine function solar angle, which resulted in a much better fit to observed data (Wang *et al.*, 2002b).

In general, light interception in most process-based models has been calculated at the stand scale. A few models have been developed to work for individual trees using ray-tracing techniques, with some of the most notable examples including MAESTRO (Wang and Jarvis, 1990), tRAYci (Brunner, 1998), and CORONA (Groot, 2004). Although these models provide an excellent prediction of individual tree growth (Wang *et al.*, 1991; Brunner and Nigh, 2000), the models are still computationally demanding even for today's technology, which limits their wide applicability.

Other top-down process-based models have modified the Beer–Lambert law to work at the individual-tree-level. For example, Schwalm and Ek (2004) applied the assumption of simplified stand-level light interception models to individual trees, which resulted in a mean absolute prediction error of $0.20\,m^2\,ha^{-1}$ for basal area growth, and 0.12 m for height increment. These results indicate that this might be an acceptable approach, but theoretically this should have significantly overestimated light interception because the relative social position of each tree within the stand was not taken into account by Schwalm and Ek (2004). In contrast, Korol *et al.* (1995) disaggregated stand-level estimates of light absorbance by dividing the canopy into several layers and allocating radiation within each layer based on tree leaf area, which was found to be an effective predictor of individual-tree growth.

Estimation of radiation and its interception is one of the most critical calculations in mechanistic models as it is the primary driver of photosynthesis. Complex models of this process exist and are still computationally demanding for general use, but simplified calculations can be used if: (1) direct and diffuse radiation are separated; (2) light extinction coefficients are allowed to vary by radiation type, solar angle, and stand structure; and (3) the sine-curve assumption is not used for daily integration. Most top-down process-based models use the Beer–Lambert law and are quite sensitive to the light extinction coefficient, so care must be taken when applying these models to novel situations.

12.2.2 Photosynthesis

Estimation of photosynthesis is the single most important calculation in mechanistic models, yet one of the most difficult because of the complexity of factors that influence photosynthesis. In general, photosynthesis is influenced by amount and quality of incoming radiation, leaf temperature, leaf nutrition, leaf water status, and ambient CO_2 concentration. Since photosynthesis shows an optimum response to each of these factors, it has been widely modeled using a rectangular hyperbola approach (Landsberg, 1986). The Farquhar *et al.* (1980) biochemical equation has been parameterized for several species (e.g. Wullschleger, 1993), and has nearly become standard for most process-based models (Landsberg, 2003). The Farquhar *et al.* (1980) equation is primarily driven by two variables: carboxylation (V_{cmax}) and regeneration of ribulose (J_{max}), which themselves are predicted from ambient CO_2, leaf nutrition, and leaf temperature. The general Farquhar *et al.* (1980) equation is:

$$A_n = \min\left(\left(\frac{(V_{cmax}\cdot(C_i-\Gamma^*))}{\left(C_i+K_C\cdot\left(1+\left(\frac{O_i}{K_o}\right)\right)\right)}\right),\left(\left(\frac{J}{4}\right)\cdot\left(\frac{(C_i-\Gamma^*)}{(C_i+2\cdot\Gamma^*)}\right)\right)\right)-R_d \quad [12.2]$$

where A_n is net photosynthesis, C_i is ambient CO_2 concentration, O_i is ambient O_2 concentration, Γ^* is CO_2 compensation point in the absence of mitochondrial respiration, J is rate of electron transport, R_d is rate of mitochondrial respiration, K_C and K_O are the Michaelis–Menten coefficients of Rubisco activity for CO_2 and O_2, respectively. J_{max} and V_{cmax} are influenced by temperature, and this relationship is relatively constant across tree species (Medlyn *et al.*, 2002), but their relationship to other factors is less clear. Gu *et al.* (2010) recently presented a new method for estimating parameters of a Farquhar *et al.* (1980) – like equation, which resulted in improved model performance.

Although not a model of photosynthesis *per se*, the radiation use efficiency (RUE) concept is still widely used in many top-down process-based models. RUE is based on the work of Monteith (1977), who found a linear relationship between absorbed photosynthetically active radiation (APAR) and annual aboveground crop productivity. RUE is essentially a species-specific parameter that relates APAR to annual productivity, and generally ranges from 2 to 4 g C MJ^{-1} for most forests (Landsberg *et al.*, 2003), with intensively managed stands displaying larger values. RUE may be adequate for long temporal periods and large geographic regions (Medlyn, 1998), but it is insufficiently flexible to predict photosynthesis under climate change (Schwalm and Ek, 2001) or other novel situations. For example, Medlyn (1998) indicated that daily RUE was highly variable regardless of canopy structure, and estimates of annual RUE still varied significantly across sites with different LAI or light climates. Accordingly, Duursma (2004) recently indicated that using a mechanistic canopy photosynthesis model increased the precision of predicting stemwood volume growth by 21% when compared to the RUE approach.

While the Farquhar *et al.* (1980) model provides a good estimate of net photosynthesis at the leaf level, it becomes rather difficult to integrate this value over an entire canopy and day. Three general approaches have been used to integrate for an entire canopy: (1) big leaf (the canopy is considered as one leaf); (2) single layer of sunlit (direct + diffuse) and shaded (diffuse only) leaves; and (3) multi-layer. In a recent analysis of several single-layer canopy photosynthesis models, Medlyn *et al.* (2003) found that the single-layer sunlit and shaded approach worked the best at all temporal scales examined (daily, weekly, monthly). In addition, Wang and Leuning (1998) found that the single-layer sunlit and shaded approach produced values within 5–10% of a multi-layered model and was 10-times more computationally efficient. The distinction between sunlit and shaded LAI is usually calculated using an equation outlined in Campbell and Norman (1998), which is based on ray penetration theory and uses the canopy extinction coefficient, LAI, and solar angle. An important refinement to this equation was made by Wang *et al.* (2002b), who suggested that site aspect and slope can significantly alter the fraction of the canopy LAI that is sunlit.

Similar to integration of photosynthesis for a canopy, estimating daily net photosynthesis can be difficult given the highly nonlinear nature of photosynthesis during a normal day. Three primary approaches have been used to address this problem: (1) hourly time steps; (2) Gaussian integration; and (3) analytical solutions to the daily integral of the Farquhar *et al.* (1980) equation. Hourly or even half-hour time steps have been used primarily in research process-based models, while Gaussian integration and analytical solutions are the most efficient for practical applications. Gaussian integration involves breaking the day into a number of relative times (five has generally been used), and assigning a specific weight to each interval. It has been used in several canopy photosynthesis models as it captures diurnal variation in net photosynthesis in a much more computationally efficient manner than hourly time steps (Goudriaan

1986; Leuning *et al.*, 1995; Duursma, 2004). An analytical solution for the Farquhar *et al.* (1980) equation is presented in Chen *et al.* (1999) and used in the model of Schwalm and Ek (2004). This analytical solution involves estimating canopy photosynthesis at solar noon and multiplying this derived value by day length. Gaussian integration and analytical solutions are both superior approaches compared to simply applying a mean value for the entire day.

An additional important aspect in modeling canopy photosynthesis, particularly for conifers, is adjusting net photosynthetic rates for leaf age and water stress. Bernier *et al.* (2001) indicated that canopy net photosynthesis was overestimated by 9% when foliage age effects were not accounted for in the model. A similar conclusion was reached by Ogée *et al.* (2003) for a model developed for Maritime pine in France. In this analysis, root mean square errors increased by 40–70% when no distinction between different foliage age classes was made (Ogée *et al.*, 2003).

Although the importance of water stress on photosynthesis is generally recognized, incorporating it into process-based models has varied because of the complicated plant physiological response to drought and the various other factors to which it is related (e.g. temperature, humidity; Grant *et al.*, 2006). Some models avoid these complications by directly reducing photosynthesis using proportional modifiers based on soil and/or atmospheric water status, but this approach does not mechanistically represent key physiological processes or interactions between soil and atmospheric water deficits.

Photosynthesis is a complex and dynamic physiological process that is essential for accurately predicting forest productivity. Although the use of RUE to estimate photosynthesis is appealing because of its simplicity, the use of a mechanistic equation like Farquhar *et al.* (1980) is optimal because it has been parameterized for a variety of species (e.g. Wullschleger, 1993), is relatively robust, and incorporates most key factors that influence photosynthesis. Also, accurate predictions of photosynthesis require incorporating differences within the canopy and during days and seasons, which is best done with single-layer sunlit and shaded, and Gaussian integration approaches, respectively.

12.2.3 Stomatal conductance

Stomata regulate exchange of water vapor and CO_2 between leaves and the atmosphere. Stomatal conductance is an important parameter in process-based models because it is a major factor controlling both net photosynthesis and degree of canopy evapotranspiration. In general, the same factors that influence photosynthesis also control stomatal conductance since it is related to temperature, vapor pressure deficit, soil water status, and radiation. Damour *et al.* (2010) provided an overview of leaf-level stomatal conductance models. Three approaches have generally been used in top-down process-based models: (1) Ball–Berry (Ball *et al.*, 1986); (2) Leuning (1995); and

(3) Jarvis–Loustau (Jarvis, 1976; Loustau *et al.*, 1997). All of these approaches are semi- to fully empirical models and based primarily on climatic variables. The primary difference is that the Leuning (1995) and Ball–Berry (Ball *et al.*, 1986) models simultaneously solve for both net photosynthesis and stomatal conductance, while the Jarvis–Loustau model has a maximal value that is modified by environmental scalars (0 to 1):

$$g_s = g_{max} \cdot f_1(\text{APAR}) \cdot f_2(T) \cdot f_3(\text{VPD}) \cdot f_4(C_i) \cdot f_5(\theta) \qquad [12.3]$$

where g_s is predicted stomatal conductance, g_{max} is maximum stomatal conductance, and the f_is are nonlinear functions of temperature (T), vapor pressure deficit (VPD), and soil moisture content (θ). The Ball–Berry model uses net assimilation, CO_2 concentration in the leaf boundary layer, and relative humidity in the leaf boundary layer (Ball *et al.*, 1986). Leuning (1995) simply modified this model by substituting vapor pressure deficit at the leaf surface for relative humidity:

$$g_s = g_0 + \left(\frac{a_1 \cdot A_n}{\left((C_s - \Gamma^*) \cdot \left(1 + \left(\frac{D_s}{D_0} \right) \right) \right)} \right) \qquad [12.4]$$

where g_0 is stomatal conductance when A_n approaches 0, C_s is stomatal CO_2 concentration, D_s is stomatal vapor pressure deficit, and a_1 and D_0 are empirical coefficients.

The Ball–Berry and Leuning (1995) stomatal conductance models do not explicitly incorporate the influence of soil water deficits, so they are often reparameterized for these types of situations (Grant *et al.*, 2006). In contrast, the Jarvis–Loustau model often incorporates a specific soil water stress function by including a response function to soil water deficit. Van Wijk *et al.* (2000) compared all three models in a Douglas-fir stand in the Netherlands and found that the Jarvis–Loustau model performed best as root mean square error was reduced by 33 and 14% when compared to the Ball–Berry and Leuning (1995) models, respectively. However, Damour *et al.* (2010) concluded that these models work well under well-watered conditions, but do not adequately address the effects of drought.

Canopy conductance is a function of both total leaf area and stomatal conductance. Quite often in top-down process-based models, canopy conductance is simply calculated as:

$$g_c = g_s \cdot \text{LAI} \qquad [12.5]$$

where g_c is canopy conductance. Similar to the big-leaf approach to photosynthesis, this equation ignores significant within-canopy variation in stomatal conductance rates, and a single-layer two-leaf approach (i.e. sunlit/shaded) is also generally recommended (Wang and Leuning, 1998). Interestingly, Tan and Black (1976) found that mean canopy conductance was generally one-ninth of the mean leaf conductance, except under periods of extended drought.

Estimates of canopy conductance are required in several widely used transpiration equations. The standard transpiration equation used in most process-based models is the Penman–Monteith (Penman, 1948; Monteith, 1965) equation, which is driven by daily mean temperature, wind speed, relative humidity, and canopy conductance:

$$E_t = \frac{k_5 \cdot R_n + k_2 \cdot k_3 \cdot \mathrm{VPD} \cdot g_c}{k_1 \cdot \left(k_5 \cdot k_4 \cdot \left(1 + \left(\frac{g_a}{g_s}\right)\right)\right)} \qquad [12.6]$$

where E_t is canopy transpiration, R_n is net radiation absorbed by the canopy, and g_a is boundary layer conductance for the canopy. The k_is are physical constants that are dependent on air temperature and include the latent heat of vaporization (k_1), the specific heat of air (k_2), the density of air (k_3), the psychrometric constant (k_4), and water vapor content of air held at saturation (k_5). In coniferous forests, the net radiation, boundary layer conductance, and wind speed can largely be ignored, which further simplifies the equation. Other transpiration equations used in top-down process-based model include a simple diffusion equation (e.g. Watt *et al.*, 2003) and the Priestley–Taylor (1972) equation.

Like photosynthesis, leaf and canopy conductance have been extensively studied and are relatively well understood. At the leaf level, the Jarvis–Loustau (Jarvis, 1976; Loustau *et al.*, 1997) approach works well, but requires species-specific parameterization. At the canopy level, leaf conductance is often scaled to the canopy by multiplying the average rate by LAI. This ignores important within-canopy variation and may lead to an overestimation of transpiration, which is often calculated using the Penman–Monteith (Penman, 1948; Monteith, 1965) equation. Future improvements in modeling leaf conductance will involve a better understanding of the influence of drought and the factors that control internal transfer conductance within a leaf (Warren and Adams, 2006).

12.2.4 Respiration

Tree respiration is the release of CO_2 from plant tissue and consists of two components: maintenance and growth. Respiration is a very important calculation in process-based models, but has received significantly less attention than photosynthesis and stomatal conductance. Hence, there is very little consensus on appropriate techniques to accurately model it, particularly for maintenance respiration. Growth or construction respiration used to build woody and foliage tissue can often be estimated from empirical relationships based on heat of combustion, ash, and organic nitrogen content of the tissue (Williams *et al.*, 1987) or on tissue carbon and ash content (Vertregt and Penning de Vries, 1987). Growth respiration has generally been assumed to be between 25 and 35% of total gross production (Ryan, 1990).

Maintenance respiration is the dominant component of total respiration and is used to maintain living biomass. Four primary approaches have been used in top-down process-based models: (1) daily Q_{10} per unit of biomass carbon (Running and Coughlan, 1988); (2) annual Q_{10} per unit of biomass nitrogen (Battaglia *et al.*, 2004); (3) annual nitrogen content and temperature (Ryan, 1991); and (4) constant fraction of gross primary production (Waring *et al.*, 1998).The Q_{10} value describes differences in respiration rates over a 10 °C interval of temperature, and differs by species, season, and biomass type.

Although short-term (i.e. hourly to daily) temperature responses are not well described by a constant Q_{10} of respiration, many process-based models use this approach and ignore the acclimation of respiration to temperature (Wythers *et al.*, 2005), which can lead to a significant overprediction. Kirschbaum (1999) presents a daily respiration model that accounts for temperature acclimation, but few datasets currently exist to parameterize it correctly. Estimating respiration on an annual time step similar to the approaches of Battaglia *et al.* (2004) and Ryan (1991) circumvents the problem of temperature acclimation, but does not capture the inherent seasonal variation of respiration (Vose and Ryan, 2002). Predicting maintenance respiration as a function of nitrogen content rather than carbon has been preferred because typically 90% of the nitrogen in plant cells is in protein, which needs energy replacement and repair, and it is also tightly linked with cellular activity (Ryan *et al.*, 1996b).

Despite the great variability in respiration rates by season and tree tissue type, Waring *et al.* (1998) suggested that net primary production (NPP) is a constant fraction (0.47 ± 0.04) of gross primary production (GPP). This finding has been met with some skepticism as Medlyn and Dewar (1999) suggested it was an artifact of the way NPP and GPP were calculated. Mäkelä and Valentine (2001) indicated that this ratio generally decreased with increasing tree height. Similarly, Lai *et al.* (2002) found that the ratio of NPP to GPP can vary between 0.35 to 0.65 for young loblolly pine stands, and the ratio exponentially decreased with increasing aboveground biomass. However, two extensive reviews by Gifford (2003) and Litton *et al.* (2007) have concluded that partitioning to respiration is conservative across a variety of forest ecosystems and does not vary with changes in resources or forest age within a site. Due to the concept's appealing simplicity, a number of top-down process-based models are now using a constant ratio between NPP and GPP to estimate respiration.

Respiration is difficult to model, and even measure, as a physiological process (Gifford, 2003); a variety of approaches have been used to model it with varying degrees of success. Earlier process-based models avoided its calculation by just focusing on NPP, but most top-down process-based models now prefer to model GPP and assume NPP is a constant fraction. This assumption will need further testing with experimental data, given its relative importance in process-based models.

12.2.5 Carbon allocation

Similar to respiration, carbon allocation is a critical physiological process and is not well understood. Carbon allocation describes the manner in which NPP is distributed to tree biomass components (e.g. foliage, roots, stem). Several comprehensive technical reviews pertinent to this topic and written with modeling in mind are presented in Cannell (1989), Cannell and Dewar (1994), Lacointe (2000), Barton (2001), and Litton *et al.* (2007). Most process-based models assume there is a priority of carbon allocation, with stems having the lowest priority after respiration, foliage, and fine roots. Models differ in relative importance and designations of these biomass components, time scale of allocation, and response to resource availability. Allocation parameters can either be static or dynamic. Three primary approaches have been used in top-down process-based models and include: (1) pipe model; (2) functional balance; and (3) optimal partitioning. Although these approaches are rather similar in nature, their differences and level of refinement are worth noting.

The pipe-model theory was first formulated by Shinozaki *et al.* (1964), and simply views a stem as a series of pipes that link roots with foliage. The theory states that trees maintain a constant amount of leaf dry matter per unit of cross-sectional area of active pipes throughout the crown.

Functional balance theory states that there is a strong relationship between key physiological processes. For example, the commonly assumed root: shoot functional balance concept suggests that total carbon fixation by the foliage is proportional to total nitrogen acquired by the root system. The concept also suggests that trees attempt to maintain mechanical functional balance, as the support of foliage must be related to stem growth to maximize safety from buckling. The pipe model and the functional balance model are quite similar in nature in that they are based on trees having certain structural goals and modifying their behavior accordingly. The theories just differ in what the primary goal is.

These theories have mostly been used independently by Mäkelä (1986; 1997; 2002) and Valentine (1985; 1988; 1997). A hybridization of their approaches was presented in Valentine and Mäkelä (2005). Essentially, their approach predicts that the amount of carbon allocated to foliage is driven by the current amount of foliage, while functional balance maintains that a constant proportion of carbon must be allocated to match foliage growth. Carbon is then allocated to the woody components of the tree (stem, branches, and coarse roots) according to the allometric model of constant proportionality.

Limitations with this type of carbon-based allocation have long been noted (e.g. Basslow *et al.*, 1990). Three primary problems exist with this approach: (1) trees are rarely constant; (2) effects of climate are not represented; and (3) there is no mechanistic connection to aboveground (light) or belowground (nutrients) resource availability (Le Roux *et al.*, 2001).

Trees are very dynamic and sensitive to an array of factors. Although the pipe model assumes a constant ratio between foliage and pipes in the stem, the ratio between sapwood cross-sectional area and foliage biomass is not constant as it generally increases with tree total height (McDowell *et al.*, 2002). Magnani *et al.* (2000) also predicted that as the plant grew taller, allocation was shifted from foliage to transport tissues, most notably to fine roots. Consequently, the pipe-model theory needed to be modified accordingly to account for observed variations in stem profile (e.g. Mäkelä, 2002).

Second, climate can have a large influence on carbon allocation, as a review by Magnani *et al.* (2002) suggests that greater allocation to foliage would occur with increasing temperature, and lower allocation with increasing vapor pressure deficit and soil drought, while allocation to fine roots responded more strongly to drought. Finally, extensive research has indicated that trees generally shift allocation aboveground as belowground nutrition increases (e.g. Keyes and Grier, 1981).

Optimal partitioning offers a more flexible approach to carbon allocation since it allows a connection between the tree, current environmental conditions, and resource availability, while maintaining certain structural constraints on growth. The basic tenet of this approach is that the imbalance between carbon acquisition by foliage, and water and nutrient uptake by fine roots should be minimized, while sufficient transport and structural organs are ensured (Battaglia *et al.*, 2004). Thus, it makes the connection between carbon, water, and nutrients, rather than being solely driven by carbon. Optimal partitioning has been considered to occur over daily (Schwalm and Ek, 2004), monthly (Battaglia *et al.*, 2004), and annual (Running and Gower, 1991) time steps. While different equations have been used, allocation is basically determined as the minimum of carbon, water, or nitrogen availability. The difficulty lies in determining what trees are actually optimally partitioning, since growth, water transport (Magnani *et al.*, 2000), and uniform bending stress along the stem (Dean, 2001) have all been suggested. Co-allocation between carbon and nitrogen has also been considered, achieved using stand-level carbon and nitrogen balances and allocation principles as constraints (Mäkelä *et al.*, 2008).

Carbon allocation is a difficult-to-understand and complex physiological process that is influenced by a range of factors such as tree age and resource availability. The influence of competition has been debated, as some studies found it to have a significant influence (Giardina and Ryan, 2002), while others have found it to be independent of tree density (Litton *et al.*, 2007). Regardless, most top-down process-based models have used "goal-seeking" approaches such as pipe-model and functional balance theory, which have important limitations. More complex approaches to carbon allocation exist (Thornley, 1991), but this complexity is often unwarranted for most practical applications. Continued refinements to carbon allocation modeling will greatly help strengthen the robustness of process-based models.

12.2.6 Soil water and nutrients

Availability of soil water and nutrients is clearly linked to key physiological processes, but few process-based models represent belowground processes very well because of their inherent complexity. Both soil water and nutrient availability are driven by soil texture and depth, climate, and plant uptake. Several comprehensive technical reviews of the incorporation of soil water and nutrient availability into process-based models are given in Homann *et al.* (2000), Kirschbaum (2001), and Saint-André *et al.* (2004). Since measurements of maximum soil water holding capacity are generally unavailable, it is often predicted from soil depth, rock content, and texture (e.g. Saxton and Rawls, 2006).

Most top-down process-based models use a single-layer tipping bucket approach. This method treats the soil as a homogeneous material, and soil water availability is estimated by adding daily rainfall and subtracting transpiration, evaporation, and runoff from an estimated maximum soil water holding capacity:

$$\mathrm{PAW}_i = \min\left(\mathrm{WHC}, \mathrm{PAW}_{i-1} + P_i - E_{\mathrm{t},i} - E_{\mathrm{tw},i} - E_{\mathrm{s},i} - F_i\right) \qquad [12.7]$$

where PAW_i is plant-available water for the ith day, WHC is maximum soil water holding capacity, P_i is daily precipitation, $E_{\mathrm{t},i}$ is canopy transpiration on the ith day, $E_{\mathrm{tw},i}$ is evaporation of intercepted rainfall from the tree canopy, $E_{\mathrm{s},i}$ is evaporation from soil, and F_i is drainage from the root zone. Since LAI of forests is generally above three (indicating total ground cover), most models ignore soil evaporation in this calculation.

Paul *et al.* (2003) presented a multi-layer model of soil water dynamics, which extended the single-layer tipping bucket approach to a cascading bucket approach. Unlike many other top-down process-based models, the model accounted for the influence of litter, understory vegetation, and site mounding. This showed good agreement with observed data across a range of forest types (Paul *et al.*, 2003). Multi-layer models of soil water dynamics have generally been preferred over the single-layer approach, but Weiskittel *et al.* (2010) found that a single-layer model showed excellent agreement with observed soil moisture content across multiple years at a site in western Washington. This suggests that the complexity of the soil water modeling approach should match the desired resolution of the predictions, with a single-layer approach being appropriate for most practical applications.

Nutrient cycling still remains the most difficult soil attribute to model accurately. At the fundamental level, nutrient supply is composed of three pools: (1) mass of an element that is readily available in the soil solution; (2) mass of an element in labile pools that may be accessed through the activity of microbes and plants; and (3) mass of an element actually acquired by plants (Fisher and Binkley, 2000). Representation of these pools in forest growth models is complicated by the tremendous variability of nutrient pool size and dynamics, its inherent sensitivity to environmental conditions, and limited physiological understanding of roots. Hence, several different modeling approaches have been suggested.

For example, the Nutrient Budget Model (NUTREM) is a simplified model of nutrient uptake, retranslocation, and removal for loblolly pine (Ducey and Allen, 2001). From basic stand-level data (e.g. age, dominant height, basal area per hectare), NUTREM estimates annual uptake for each major nutrient and total nutrient content of the stand using a three-stage approach. First, stemwood biomass increment and canopy structural variables are estimated from empirical equations. Second, the construction rate of woody and non-woody tissues is estimated from gross stemwood increment. Finally, uptake from the soil for each nutrient is calculated as total construction cost minus retranslocation from foliage.

In contrast, the Nutrient Cycling Model (NuCM; Verburg *et al.*, 2001) estimates forest nutrient requirements from a specified optimal growth rate. This model simulates the cycling of nitrogen, phosphorous, potassium, calcium, magnesium, sodium, and sulfur based on user-defined stand growth, litterfall, weathering, nitrogen and sulfur mineralization rates, soil minerals composition, initial litter, soil organic matter pools, and carbon to nitrogen ratios. In addition, fertilization and changing atmospheric deposition of nutrients are dealt with in a mechanistic manner. Other important differences between NuCM and other previously described models are: (1) soil is divided into 10 layers; (2) movement of water is simulated using the continuity equation: (3) decomposition is a simplified three-stage process (i.e. litter moves into fine litter and two soil organic fractions); and (4) mineral weathering rates depend on the mass of the mineral present and the dissolved proton concentration raised to a fractional power.

Similar highly detailed mechanistic representation of soil nutrient dynamics can also be found in Kirschbaum and Paul (2002) as well as Corbeels *et al.* (2005a, b). Given that most forests are limited by nitrogen availability, a simplified model, presented by Paul *et al.* (2002), predicts daily nitrogen mineralization from a specific base rate, soil temperature, and relative soil water availability. This type of approach has been incorporated in CABALA (Battaglia *et al.*, 2004) and DF.HGS (Weiskittel *et al.*, 2010). For DF.HGS, predictions of annual soil nitrogen mineralization calculated with the approach of Paul *et al.* (2002) matched observations across a range of soil types in Oregon reasonably well.

12.3 Example models

Like statistical models, a plethora of process-based models currently exist. For example, Le Roux *et al.* (2001) reviewed 27 commonly used individual-tree process-based models. An even greater number of stand-level process-based models exist. This section focuses on three example process-based models that are still widely used today, have been used in practical applications, and represent a range of approaches to modeling physiological processes (Table 12.1).

12.3.1 Forest-BGC

The Forest-Bio-Geo-Chemical (Forest-BGC) Model has a long history of development and use. It was first presented by Running and Coughlan (1988). The original Forest-BGC model was further refined by Running and Gower (1991). The model started off as a stand-level tool to be used for understanding the carbon, water, and nitrogen biogeochemical cycles.

Table 12.1 Representation of key physiological processes in the three example process-based models of Forest-BGC (Running and Coughlan, 1988; Running and Gower, 1991), CenW (Kirschbaum, 1999), and BALANCE (Grote and Pretzsch, 2002).

Physiological process	Model		
	Forest-BGC	CenW	BALANCE
Light interception	Beer–Lambert law; fixed extinction coefficient; separation between direct and diffuse	Beer–Lambert law; fixed extinction coefficient; no separation between direct and diffuse	Beer–Lambert law; fixed extinction coefficient; adjusted for individual-tree competition; no separation between direct and diffuse
Photosynthesis	Farquhar model adjusted by temperature and nutrition	Rectangular hyperbola approach sensitive to radiation, CO_2, temperature, and nutrition	Radiation use efficiency concept
Stomatal conductance	Jarvis model using radiation, VPD, temperature, and soil water potential	Ball–Berry model	Jarvis model using radiation, VPD, temperature, and soil water
Respiration	Constant growth respiration; daily maintenance respiration function of biomass and temperature	Constant growth respiration; daily maintenance respiration function of nitrogen content and temperature; Q_{10} adjusted for seasonality	Constant growth respiration; daily maintenance respiration function of biomass and temperature modified by nitrogen percentage
Mortality	Carbon balance	Daily fractional loss by carbon pool	Carbon balance with adjustment for storage
Carbon allocation	Optimal partitioning based on carbon, water, and nitrogen	Multiple pools; functional balance dependent on plant height and nutrition	Functional balance and pipe-model theory
Soil water	Daily multi-layer tipping bucket	Daily multi-layer tipping bucket	Daily single-layer tipping bucket
Soil nutrients	Decomposition; soil nitrogen and carbon mineralization	Highly detailed; multiple pools; decomposition; soil nitrogen and carbon mineralization	Dynamics for nitrogen assuming a constant mineralization rate

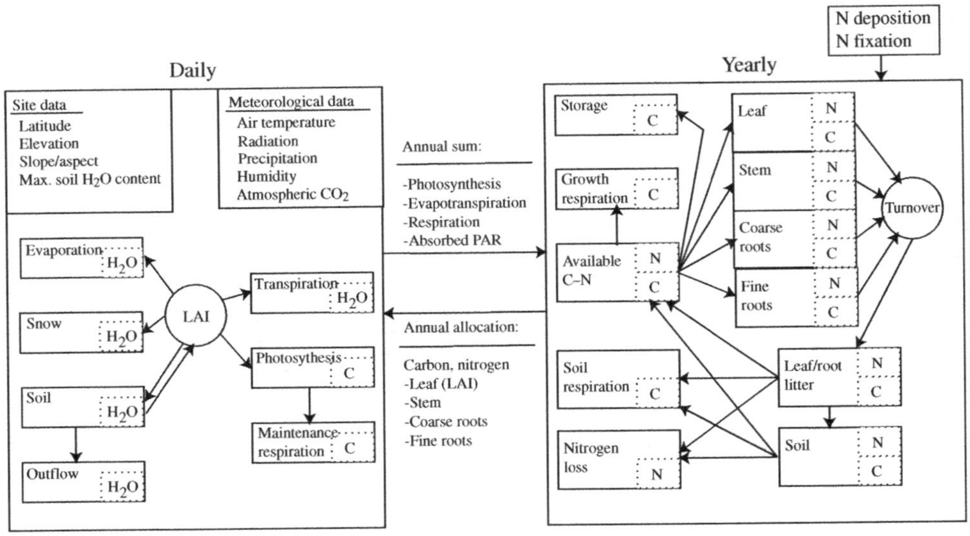

Figure 12.2 Flowchart of key model processes for the Forest-BGC model. Adapted from Running and Coughlan (1988) and Running and Gower (1991).

The model has been extensively tested, and performs well across a range of forest types (Running, 1994). Several variants of Forest-BGC exist and include an extension to individual trees (TREE-BGC; Korol *et al.*, 1991) and large geographic regions (BIOME-BGC; White *et al.*, 2000). Most recently, Forest-BGC has been used to examine forest growth response to thinning (Petritsch *et al.*, 2007) and long-term impact of management on yield (Merganičová *et al.*, 2005). The model has been widely used because it made some key simplifying assumptions. The model has dual time steps (daily and annual), is driven primarily by LAI, only requires basic site and meteorological information, and focuses on the nitrogen budget rather than full nitrogen cycle (Figure 12.2). The model also does not define species explicitly, but species-specific physiological characteristics can be represented.

The model estimates transpiration, photosynthesis, and total respiration on a daily basis, while carbon allocation, nitrogen mineralization, and soil decomposition are simulated on a yearly basis. This provides the flexibility to represent all the key physiological processes, but minimizes the complexity of modeling soil processes. Although the original model only operated at a single-canopy layer, the current version now takes the sunlit and shaded leaf area approach in estimating both photosynthesis and stomatal conductance. Based on soil texture, the amount of available soil water determines predawn leaf water potential, which is combined with vapor pressure deficit, temperature, and radiation to estimate stomatal conductance. From leaf nitrogen concentration and the estimated stomatal conductance, the photosynthetic assimilation rate is determined using the model of Farquhar *et al.* (1980). This leaf-level estimate of photosynthesis is combined with the phenological equations for light and temperature to scale the estimate to the canopy.

The model uses a single-layer tipping bucket approach for estimating soil water availability and the maximum water holding capacity of the soil is determined from equations presented in Cosby *et al.* (1984) and Rawls *et al.* (1992). Potential soil evaporation and canopy transpiration are estimated from the Penman–Monteith equation. The days since rainfall and an effective day length determine final rates. Daily maintenance respiration is calculated with a constant Q_{10} value that varies by biomass component and mean daily temperature. For foliage, both a day and nighttime maintenance respiration are estimated.

On the annual scale, total litterfall, decomposition, available soil carbon and nitrogen, plant carbon allocation, and growth respiration are calculated. Litterfall is determined by constant turnover fractions that vary by biomass component. Decomposition rates of the soil and litter are calculated as annual integrated fractions of daily weather conditions divided by optimum air temperature and soil water availability values for decomposition. Mineralization of soil nitrogen is set as a constant proportion of litter carbon decomposition, which, combined with retranslocation of foliage nitrogen, determines amount of available nitrogen. A carbon allocation ratio for foliage and fine roots is calculated from the amount of available nitrogen and a water stress integral (Myers, 1988). The remainder of the carbon available for growth is allocated to stems and coarse roots. Growth respiration is estimated as a constant fraction of the biomass component.

For predictions at the individual-tree-level, Korol *et al.* (1995) determined stand-level GPP using the approach of Forest-BGC, and then allocated it to individual trees based on an inter-tree competition algorithm. The algorithm allocates based on tree height, relative RUE, and APAR, and was a significant improvement to proportionally allocating based on tree leaf area. Tree diameter and height increment were predicted using an allometric relationship between biomass increment and tree dimensions. The maintenance respiration rate of the stem was determined as a function of the sapwood and phloem volume rather than total biomass. Allocation to carbon storage was estimated as a constant fraction of GPP. Tree mortality occurred when maintenance respiration demands exceeded the sum of carbon allocated to the tree in the current year and the carbon storage from the previous year. Predictions of individual tree diameter and height increment made by the model showed a strong correlation with observed increment in two stands of varying density (Korol *et al.*, 1995).

Overall, Forest-BGC is one of the first-generation process-based models developed, and has been shown to perform well across a range of forest types (Running, 1994). The model continues to be refined, and the basic model structure has been used across a range of scales from predicting individual-tree growth to global biomes. The model has proven useful because it balances model complexity with solid representation of the key physiological processes.

12.3.2 CenW

The Carbon, Energy, Nutrients, and Water (CenW) Model was developed and described by Kirschbaum (1999). This model is highly detailed with respect to several physiological processes, particularly those belowground. Like Forest-BGC, CenW works at the stand-level (based on the mean tree), but nearly all processes are simulated on a daily basis. The model has been used to investigate influences of irrigation and fertilization on growth (Kirschbaum, 2000a), relationships between growth, climate, and soil characteristics (Simioni *et al.*, 2008), mechanisms controlling growth decline with age (Kirschbaum, 2005), and long-term implications of climate change (Simioni *et al.*, 2009). The model has been primarily parameterized for radiata pine (Kirschbaum, 1999), but the model architecture was intended to be generic and easily applied to other species.

Similar to Forest-BGC, CenW requires the user to input biomass and daily weather information. Photosynthesis is determined based on equations of Sands (1995), which assume that all photosynthetic parameters are constant during the day, radiation varies sinusoidally, maximum photosynthesis rate (A_{max}) is dependent on absorbed radiation, temperature, and CO_2 concentration, and radiation is absorbed according to the Beer–Lambert law. A_{max} is also dependent on previous temperature damage caused by frost or heat scorch and possible reductions due to insects, age, and nitrogen or water limitations. The effective light extinction coefficient is calculated based on the maximum extinction coefficient with a uniform foliage distribution, the relative range of variation in the extinction coefficient, and the percentage of canopy cover. The Ball–Berry (Ball *et al.*, 1986) approach is used for stomatal conductance, but is modified by a water limitation factor. Maintenance respiration is based on a daily respiration rate per unit of nitrogen at 25 °C, a base rate of respiration, and the sum of nitrogen in the plant (except foliage). The base rate is modified by a short-term temperature response function. Foliage respiration is calculated only at nighttime because respiration during the day is included in the photosynthesis calculations. Growth respiration is a constant fraction of growth.

Daily carbon allocation is based on several considerations, with a constant fraction allocated to reproductive organs after a certain age, as well as stemwood and roots. Allocation to fine roots and foliage depends on tree height and nutritional status. Allocation to stemwood is divided between branches, stem, and bark, based on constant ratios. The model also predicts nitrogen allocation in a similar fashion to carbon allocation, but accounts for current concentration of nitrogen in each of the biomass components. Diameter and height increment for the mean tree in the stand are predicted using a modified allometric approach of Korol *et al.* (1995), except for small trees. Daily senescence rates are constants by biomass component.

Although its representation of aboveground processes is relatively similar to most process-based models, CenW is unique in its treatment of belowground processes since it is based on a modification of the highly detailed CENTURY model (Kirschbaum and Paul,

2002). The important modifications made to the CENTURY model are: (1) inclusion of additional pools for litter obtained from dead stems, branches, and roots; (2) the ratio of carbon to nitrogen is allowed to vary depending on the supplying soil pool; (3) nitrogen exchange is assumed to occur only between the active organic matter and the mineral nitrogen pool; and (4) mycorrhizal uptake of soil nitrogen is accounted for. For soil water balance calculations, the model uses a cascading bucket approach and divides the soil into the litter layer and a user-defined number of soil layers, which can vary in depth and water holding capacity. Evaporation is calculated for both the litter and soil layers based on the Penman–Monteith equation.

For soil nutrient dynamics, pools are simulated for the litter layer and three soil organic matter pools (active, slow, and resistant). There are a total of seven litter layers with one for fine wood, coarse aboveground wood, and coarse belowground wood, while foliage and fine roots are each divided into metabolic and structural litter layers. Finally, daily soil temperature and moisture modifiers on decomposition are based on a generalized equation of Kirschbaum (2000b) and relative soil moisture content (rather than a simple ratio of actual to potential evapotranspiration used in the original CENTURY model), respectively. These modifications significantly improved model predictions under a range forest soil conditions (Kirschbaum and Paul, 2002).

CenW is a forest management-oriented process-based model that has been used primarily in Australia. The model's detailed treatment of soil nutrient dynamics makes it well suited to understand potential influences of fertilization and climate change. Compared to other existing process-based models, the computer interface is user friendly, freely available, and well documented. CenW has been used in a variety of simulation situations ranging from forecasting growth response to fertilization (Kirschbaum, 2000a) to understanding soil nutrient dynamics after land use change (Kirschbaum *et al.*, 2008). The range of applications illustrates the model's robustness, flexibility, and generality. Although the model is still quite sensitive to difficult to obtain parameters, like carbon allocation ratios and the relationship of foliage nitrogen to photosynthetic parameters (e.g. Kirschbaum, 1999), it represents a good modeling framework for how these type of models can be useful for a variety of purposes.

12.3.3 BALANCE

Unlike Forest-BGC and CenW, the BALANCE model simulates all physiological processes at the individual-tree-level and is distance dependent (Grote and Pretzsch, 2002). The model has been used to examine influences of environmental change on forest stand health (Rötzer *et al.*, 2005) and to estimate increases in carbon sequestration potential under climate change (Pretzsch *et al.*, 2008). Although BALANCE is intended to be generic in nature, the model was originally parameterized for Norway spruce (*Picea abies* L.) and European beech (*Fagus sylvatica*) in southern Germany.

The structure of BALANCE is quite unique not only because it makes all calculations at the individual-tree-level, but it is comprised of three primary model components that operate on different time steps. This allows it to be flexible in its representation of key physiological processes and still maintain a high spatial and temporal resolution. The model attempts to "balance" the calculation of tree development on environmental conditions, and vice versa (Pretzsch *et al.*, 2008).

The model reconstructs each tree in three dimensions based on its initial dimensions and the competitive position of the tree in the stand. Both the tree crown and roots are divided into eight separate layers, which drive calculations of microclimate factors on a daily basis. All physiological and biogeochemical processes are then calculated on a monthly basis from values aggregated from daily calculations. Light availability is calculated three times yearly (beginning of the year, start and end of vegetation period). Light absorption in the crown is estimated for each layer with a modified Beer–Lambert equation that is driven by foliated crown volume, amount of leaf area, and a parameter describing the degree of clustering. Photosynthesis is determined from average radiation, temperature, CO_2 concentration, a species reduction factor, and various other stresses (i.e. drought, nutrient limitations, pollution). Growth respiration is considered a constant fraction of GPP, and maintenance respiration is calculated based on total biomass, temperature, and a respiration rate that is dependent on nitrogen concentration.

Allocation of carbon is based primarily on the approach of Grote (1998), which is dependent both on nutrient availability and tree structure. The model allocates to stem-wood, foliage, buds, fine roots, branches, and coarse roots. Senescence of foliage and fine roots is determined by constant fractions of current biomass. Twigs are shed in proportion to the loss of foliated volume and foliage density. Annually, individual tree diameter and height dimensions are updated based on a constant height-to-diameter relationship. Basically, the total increase in stem biomass is distributed along the stem so that the sapwood area increase below the crown is the same at any particular height. Total tree height is then predicted from the new diameter at breast height. The horizontal extension of each crown and root segment is based on the net amount of carbon (foliage) and nitrogen (roots) assimilated in each layer. The extension of new segments is based on efficiency of the old segments and distance from the new to old segments. Crown extension is limited by mechanical competition effects determined by the distance to the two nearest neighbor trees.

Soil water and nitrogen availability are estimated on a daily basis and determined by amount of occupied soil volume and fine root density. Like CenW, available soil water is calculated using a cascading bucket approach and determined by precipitation, canopy interception, and soil water movement. Soil nitrogen availability is determined from mineralization of two litter layers (coarse and fine), weathering, and deposition. For each soil layer, water and nutrient uptake by individual trees are driven by their availability, amount, and specific uptake rate of the roots in that layer.

Although BALANCE is a highly conceptual model, it can represent reality quite well, and differences between measured and simulated individual-tree diameters after seven years of simulations were relatively small (Rötzer *et al.*, 2005). The model also gave good predictions of daily soil water balance and senescence of foliage when compared to observed values (Rötzer *et al.*, 2005). The degree of BALANCE's spatial and temporal resolution makes it useful for a variety of situations, particularly for assessing long-term influences of species mixtures on stand development or influences of multi-layer stand structures on local microclimate conditions (Pretzsch *et al.*, 2008).

12.4 Limitations

Process-based models offer an appealing mechanistic approach to predicting forest growth, but have multiple limitations and a limited number of uses just like statistical models. Theoretically, process-based models can be applied to novel situations and provide better predictions when compared to purely statistical models because of their mechanistic basis. The performance of process-based and statistical models has been compared, and similar results have been found (e.g. Miehle *et al.*, 2009). For example, Matala *et al.* (2003) found that a process-based and a statistical model gave similar predictions of relative growth rate, but the statistical model was more stable in its predictions and not as sensitive to initial stand conditions and silvicultural treatments.

Schmid *et al.* (2006b) suggested that process-based models be preferred for simulations up to 100 years because they account for changing climatic conditions. On the other hand, they found process-based models to be quite sensitive to uncertainties in model structure, parameter values, and input data, which can be difficult to handle in large-scale applications (Schmid *et al.*, 2006b). As identified by these examples, the key limitations of process-based models are initialization, parameterization, scale, and sensitivity.

12.4.1 Initialization

Process-based models are driven primarily by LAI, climate, and several site factors, which are generally not captured in routine forest measurements. Both direct and indirect measurements of LAI are difficult, time-consuming, and expensive (Jonckheere *et al.*, 2004). Most process-based models estimate LAI from a species-specific leaf area parameter and a stand foliage biomass estimate. This approach avoids difficulties in measuring LAI, but ignores the tremendous variation that can occur within a canopy due to differences in location and foliage age (e.g. Weiskittel *et al.*, 2008). Measuring biomass is also likely more difficult than LAI, as direct measurements are logistically complex and indirect estimates are relatively inaccurate.

Regional allometric biomass equations exist for several species (Jenkins *et al.*, 2004), but they are generally based solely on DBH and may not be accurate when applied to an individual stand. For example, Grier *et al.* (1984) found that a regional equation differed from a local equation by −33 to +93% for fertilized plantations of Douglas-fir. As noted by both Matala *et al.* (2003) and Schmid *et al.* (2006b), process-based models are often quite sensitive to the input data, particularly LAI and foliage biomass. Given the difficulty and uncertainty in measuring both LAI and stand biomass, this makes it difficult to apply process-based models on a routine basis or across a range of sites.

One alternative to measuring LAI directly or using tree-level allometric equations is to reconstruct the crown for each tree on a tree list and estimate biomass at this resolution, similar to the approach taken in BALANCE. Weiskittel *et al.* (2010) used this approach in Douglas-fir because it reduced the root mean square error by 30 to 42% when compared to other approaches and was unbiased across a range of silvicultural treatments. In addition, significant advances have been made in using remote imagery such as LandSat and LiDAR to estimate biomass (Lim *et al.*, 2003). The limitations with this approach are that it often needs local calibration and can be expensive to collect.

Process-based models often require climatic and soils information. Obtaining local estimates of climate has become much easier. In the past, climate data were often obtained from the nearest weather station, but this approach ignored differences in elevation and aspect that might influence local weather patterns. To address these limitations, Running *et al.* (1987) developed an approach for adjusting nearby observations of weather for local conditions (MT-CLIM), and used it estimate local climate patterns in mountainous terrain. These relationships were further generalized by Thornton *et al.* (1997) and implemented in DAYMET (www.daymet.org/), which provides point estimates of daily temperature, radiation, and vapor pressure deficit in the United States for the years 1980 to 2003 at a 1 km^2 spatial resolution. This spatial resolution is adequate for most applications, and can be refined with a digital elevation model (DEM) that is of a higher resolution.

Soils information can still be difficult to obtain, particularly in areas of complex terrain. In the United States, county soil surveys conducted by the USDA Natural Resources Conservation Services are now available electronically (http://websoilsurvey.nrcs.usda.gov). The limitations of this data are that it can be of a coarse spatial resolution and the methodology is primarily based on techniques used in agricultural soils. Recent advances have been made in using remote sensing data (Ichii *et al.*, 2009) and existing geographic information systems (GIS) layers such as DEMs (Frazier *et al.*, 2009) to estimate key soil properties.

Process-based models differ in the amount of soils information required. Some require depth, texture, and percentage coarse fragments, while others just need a water holding capacity. Often, water holding capacity can be readily estimated from generalized equations based on texture (e.g. Saxton and Rawls, 2006). Some models require estimates of percentage organic matter and nutrient content, which significantly limits the scope of these models.

Process-based models provide a more localized estimate of growth because they require more detailed site information such as climate and soils, which can be difficult to obtain at a regional scale. Technology has made this information more available and will continue to do so. However, the accuracy and resolution of this information will always be the primary factor influencing process-based model predictions.

12.4.2 Parameterization

The advantage of process-based models is that they are built on mechanistic equations and can be extrapolated to new situations, whereas statistical models are often limited to the data conditions used to develop the model. Although many physiological processes have standard equations that are widely used, like the Farquhar *et al.* (1980) photosynthesis model, they often require species-specific parameters. Even for widely studied species like Douglas-fir, specific parameters for all physiologically important factors can be difficult to locate just from the general scientific literature (Weiskittel *et al.*, 2010).

In addition, like permanent measurement plots, physiological measurements are expensive and time-consuming to make because they are often highly variable and require specialized equipment. Physiological measurements are often made on seedlings in a greenhouse setting, which can be drastically different from individuals observed in a field setting. Thus, parameterizing a process-based model can be a difficult and time-consuming process, particularly for species that have not received much attention in the scientific literature. In addition, unlike a parameter estimate obtained using regression, a mean physiological parameter is often reported, and it can be hard to ascertain the inherent variability of this parameter.

An interesting alternative to parameterizing each individual equation used in a process-based model from the literature or with new data is the use of a Bayesian optimization technique. This technique has been demonstrated several times with promising results (Van Oijen *et al.*, 2005; Svensson *et al.*, 2008; Deckmyn *et al.*, 2009). In this approach, Markov chain Monte Carlo simulation is used to vary model parameters and calibrate model predictions to observed data. The procedure begins by quantifying uncertainty in the initial parameter values, which are then specified in the form of a prior probability distribution. Based on the calibration dataset and Bayes' theorem, the parameter values are updated and the model simulation rerun.

This type of calibration is much more informative than simple parameter 'tweaking' because it can quantify parameter uncertainty, the degree of correlation between parameters, and output uncertainty (Van Oijen *et al.*, 2005). Optimizing the model with this approach using different calibration datasets can produce different parameter value distributions, but this can also highlight model sensitivity, and argues for calibrating based on several output variables rather than just one. For example, Van Oijen *et al.* (2005) used both measured individual tree heights and annual NPP to parameterize a process-based model, and found that the estimated model predictive uncertainty was significantly reduced when only one output variable was used in the calibration. Given today's computing power, this approach should receive more attention in future efforts to parameterize process-based models, and is a good way to link the strong empirical datasets of traditional growth models with the more theoretical approach of process-based models.

12.4.3 Scale

Since physiological data are often limited in scope, process-based models are commonly parameterized using data from sites where both meteorological and physiological measurement are intensively made, such as eddy-flux towers. This can limit the scale at which process-based models can be applied because eddy-flux towers often do not cover the range of variability that a particular ecosystem encompasses. This is in contrast to statistical models, which are often parameterized with extensive regional datasets and can be applied at a large spatial scale. At the other extreme, some process-based models like Forest-BGC have been simplified to work across extensive biomes, which can limit the models' capacity to simulate dynamics at high spatial scale, like an individual stand.

Process-based models often vary in the temporal scales at which they operate, ranging from hours to months. This too has important implications for the model's scalability. For example, when estimating NPP for an extensive region, a daily or even monthly resolution would be appropriate because this would likely capture the climate variability that drives production. However, when trying to compare alternative silvicultural regimes on the same site, daily or even an hourly temporal resolution is necessary since very short-term events often determine a stand's response to management. This is also an important factor in simulating influences of climate change, because small changes in daily average temperature can have significant implications if the temperature is near a critical threshold for a key physiological process (Schwalm and Ek, 2001). Consequently, given the number of parameters and their high degree of uncertainty, finding a process-based modeling framework that works across a range of spatial and temporal scales can be difficult.

12.4.4 Sensitivity

Regardless of a process-based model's intended scale, models are almost always sensitive to a relatively small set of key input variables and parameter values (Figure 12.3). For example, most process-based models are highly sensitive to LAI, since it drives both within- and below-canopy microclimate and determines and controls canopy water interception, radiation extinction, transpiration, and carbon gas exchange (Bréda, 2003). Although LAI is a value supplied by the user, sensitivity assessment of several process-based models also reveals model parameters that are highly influential, particularly those that control light interception, photosynthesis, and respiration. For example, White *et al.* (2000) found that varying the percentage of leaf nitrogen in Rubisco by 20% increased the mean annual NPP by 142% for the evergreen needle biome using the BIOME-BGC model.

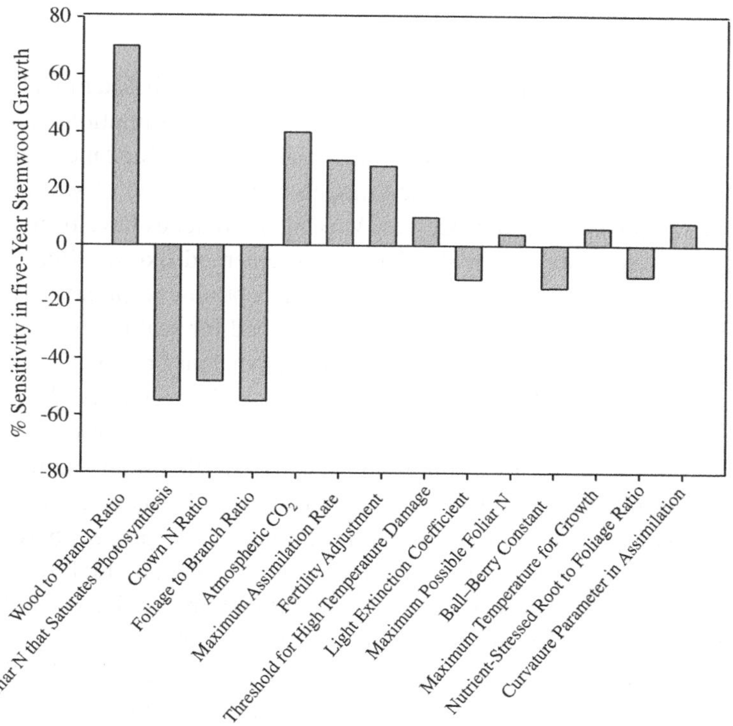

Figure 12.3 Percentage change in predicted five-year total stand stemwood growth for different parameters using the CenW mechanistic model. The parameters were altered by ±25%. Adapted from Kirschbaum (1999).

The degree of sensitivity can vary not only between process-based models, but between different species in the same model. Using BIOME-BGC, Tatarinov and Cienciala (2006) found that spruce steady-state NPP had an absolute relative sensitivity of 0.14, while beech had a relative sensitivity twice as high at 0.30. Kirschbaum (1999) also indicated that sensitivity of CenW depended on silvicultural treatments, because the Ball–Berry stomatal conductance parameter was much more important in situations where irrigation was not applied, but its influence was relatively small in areas where water was not a limiting factor.

These results suggest that even for a given model framework, effects of certain model inputs and parameters can vary significantly and are dependent on values of other inputs and parameters, which emphasizes the significant challenge of proper parameterization and reliability assessment of process-based models. Regardless, understanding a model's sensitivity is helpful in identifying parameters that need the most attention and gives a context for understanding the output of any simulation obtained from a process-based model.

12.5 Summary

From both a scientific and a management perspective, process-based models have a lot of appeal. They avoid limitations associated with statistical assumptions and representativeness of empirical datasets, while providing both a theoretical and mechanistic basis for understanding tree growth. Process-based models are comprised of several sub-models that represent key physiological processes. Varying approaches have been used for each physiological process, and certain standards have emerged like the Beer–Lambert law and Farquhar *et al.* (1980) equations for light interception and photosynthesis, respectively. A range of approaches also have been developed for scaling these processes both spatially and temporally, with most being highly dependent on LAI.

Models such as Forest-BGC, CenW, and BALANCE exemplify the structure and usefulness of the process-based approach. These models have been used to forecast long-term influences of climate change, compare alternative silvilcultural regimes, and explore implications of depleted soil nutrient pools, which are all situations where the use of statistical models would be limited. Important limitations of process-based models are related to their initialization, parameterization, limited scale, and high sensitivity to key variables and parameters. This is particularly important to recognize when these types of models are used to predict forest response to expected climate change (e.g. Schwalm and Ek, 2001). For example, Keenan *et al.* (2011) recently found upwards of a three-fold difference between projections of future forest productivity using a process-based model by just modifying how the influence of CO_2 was represented in the model. However, the great availability of climate and physiological data, as well as newer parameterization techniques like Bayesian optimization, suggests the application of process-based models will continue to increase and expand in the future.

13

Hybrid models of forest growth and yield

13.1 Introduction

Both process-based and statistical growth models have multiple advantages and disadvantages. Hybrid models are a relatively new class of model that attempt to combine the strengths of both process-based and statistical approaches. Although almost all models are hybridized to some degree, hybrid models are distinguished by two or more sub-models that are often running in parallel, with one of the components having a mechanistic focus. Hybrid models offer several advantages that make them appealing to a wide audience (Taylor *et al.*, 2009). The primary advantages of hybrid models are that they: (1) increase flexibility of statistical models to address variability of climate and regional differences; (2) simplify required inputs associated with process-based models; and (3) mechanistically represent influences of silvicultural treatments such as thinning and fertilization.

For these reasons, hybrid models are often described as the future of forest growth modeling (Landsberg, 2003). However, some disadvantages of hybrid models are that they can compound error, still require data not typically available, and the relative gain in accuracy when compared to statistical models is generally small to modest. Although the number of hybrid models is still rather small in the literature relative to strictly statistical or process-based models, they have significantly increased in abundance, particularly following the advent of the Physiological Principles for Predicting Growth (3-PG) framework offered by Landsberg and Waring (1997).

Hybrid models are simply the combination of statistical and mechanistic approaches, often with an applied goal (Johnsen *et al.*, 2001). The main objective with this type of

---

Forest Growth and Yield Modeling, First Edition. Aaron R. Weiskittel, David W. Hann, John A. Kershaw, Jr. and Jerome K. Vanclay.

model is to utilize physiological principles as much as possible, while relying on allometrics and other accepted statistical modeling conventions for the remainder of the attributes of interest. Their scale has mostly been at the stand-level and a monthly resolution, while a few operate at the individual-tree and daily scale. Although the use of these models varies from predicting effects of climate change on tree growth (Baldwin *et al.*, 2001) to the role of silviculture on external log characteristics (Mäkelä and Makinen, 2003), three primary classes of hybrid model frameworks currently exist: (1) statistical growth equations with a physiologically derived covariate; (2) statistical equations with a physiologically derived external modifier; and (3) allometric models.

The degree of hybridization within each of these classes varies greatly, so exact classification of hybrid models is difficult. For example, Milner *et al.* (2003) linked the Forest Vegetation Simulator (FVS) and STAND-BGC in that both models ran simultaneously in parallel and a user selected the degree of coupling. Thus, tree growth could be predicted by FVS, while mortality was predicted by STAND-BGC with this type of approach. Likewise, Peng *et al.* (2002) combined gap, allometric, and detailed mechanistic soil models to predict tree growth; the linked system has been modified to represent key forest silvicultural treatments (Wang *et al.*, 2011). Both of these frameworks are interesting and illustrate a central question driving hybrid modeling philosophy: how the models should interact. The examples given in this chapter attempt to demonstrate the various ways that statistical and process-based models are linked (Table 13.1).

Although some authors have contended that gap models should be considered hybrid models (e.g. Taylor *et al.*, 2009), they are not covered in this chapter because the majority of gap models do not have a statistical component and often struggle to accurately predict growth of overstory trees (Lindner *et al.*, 1997; Yaussy, 2000).

13.2 Types of hybrid models

13.2.1 Statistical growth equations with physiologically derived covariate

A properly specified statistical model is stable, constrained, and often highly accurate for the dataset used for parameterization. Consequently, predicting average regional growth trends should be very reliable. However, site index is often the only factor expressing potential productivity. Hybrid models that rely on statistical growth equations containing a physiologically derived covariate attempt to retain the robustness of statistical models and increase site specificity by including mechanistic output from a process-based model.

This approach is well exemplified by Baldwin *et al.* (2001), who linked an annualized, individual-tree statistical growth model (PTAEDA2) with a highly detailed canopy radiation transfer and photosynthesis model (MAESTRO). Using this modeling approach, MAESTRO was utilized to estimate net photosynthesis for a given climate and this value

Table 13.1 Example of hybrid models and their comparison to statistical models. (MSE is mean square error)

Study	Spatial resolution	Type of hybrid model	Species	Projection length	Degree of improvement
Dzierzon and Mason (2006)	Stand	Statistical growth equation with physiologically derived covariate	Radiata pine	1–20 years	14% (stand basal area) and 8% (dominant height) reduction in MSE
Pinjuv *et al.* (2006)	Stand	Statistical growth equation with physiologically derived covariate	Radiata pine	4–7 years	3% (number of trees), 4% (basal area), and 4% (dominant height) reduction in MSE
Snowdon *et al.* (1998)	Stand	Statistical growth equation with physiologically derived covariate	Radiata pine	16 years	21% (dominant height), 39% (basal area), and 47% (stand volume) reduction in MSE
Baldwin *et al.* (2001)	Individual tree	Statistical growth equation with physiologically derived covariate	Loblolly pine	15 years	97% reduction in mean bias (stand volume)
Henning and Burk (2004)	Individual tree	Statistical growth equation with physiologically derived external modifier	Mixed species	31 years	33% (stand basal area) and 50% (tree diameter) reduction in mean bias
Weiskittel *et al.* (2010)	Individual tree	Statistical growth equation with physiologically derived external modifier	Douglas-fir	4 years	Reduced MSE by 6% (diameter increment) and 12% (height increment), respectively

was related back to site index, which was then used in the statistical diameter and height increment equations (Baldwin *et al.*, 1997). The model predicted that site indices across the region would increase by 0.0847 m year^{-1} due primarily to increased ambient CO_2 concentrations (Baldwin *et al.*, 2001). The limitation with this approach is that site index is influenced by both environmental and silvicultural factors, which can make its prediction difficult. Consequently, Baldwin *et al.* (2001) were only able to explain approximately 40% of the original variation in site index with simulated net primary production (NPP), even after accounting for stand age. This approach is conceptually logical and doesn't require the parameters of the statistical equations to be estimated, but suffers from the difficulty of predicting site index. However, Battaglia *et al.* (1999) found a strong relationship between predicted peak mean annual stand wood volume increment from a process-based model and the site index used in a suite of statistical equations, which significantly improved predictions (Figure 13.1).

In contrast to relating site index to output from a process-based model, refitting a statistical growth equation with physiologically derived covariates was demonstrated by Snowdon (2001), Mason *et al.* (2007), and Pinjuv *et al.* (2006). Mason *et al.* (2007) estimated amount of potentially useable incoming radiation to predict growth for a series of plots in Oregon receiving various levels of vegetation management. The cumulative

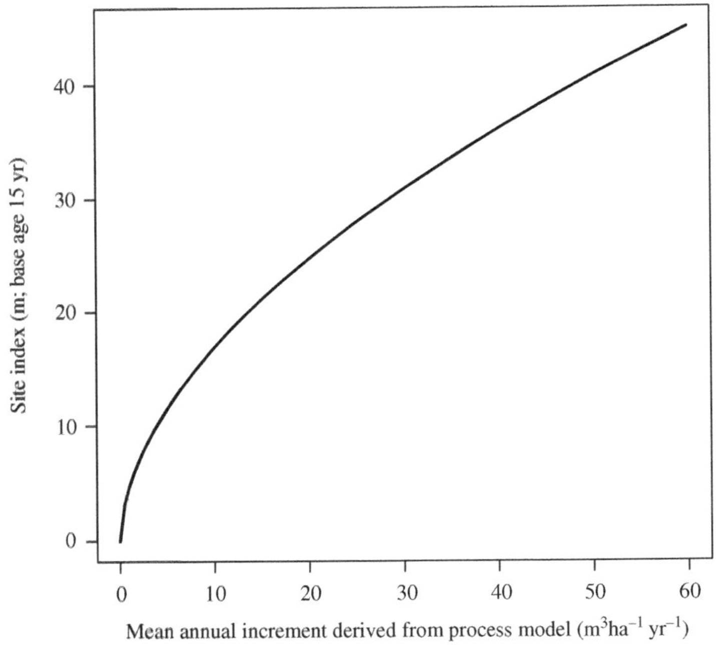

Figure 13.1 Relationship between radiata pine site index (m; base age 15 years) and the mean annual increment derived from a process model (m^3 ha^{-1} yr^{-1}). Based on equations from Battaglia *et al.* (1999).

sums of useable radiation, rather than the traditionally used time variable, were then related to average ground diameter, and model coefficients did not vary significantly between experimental treatments, suggesting that the model effectively captured inherent variation created by the treatments (Mason *et al.*, 2007):

$$\mathrm{GLD}_{t+1} = \mathrm{GLD}_t + b_0 \cdot \left(\frac{R_\mathrm{T}}{1000}\right)^{b_1} \qquad [13.1]$$

where GLD is the mean seedling ground line diameter at time t, R_T is the total useable radiation, and the b_is are parameters. Modeling the influences of these vegetation treatments with a strictly statistical equation would be difficult because an equation for each individual treatment needs to be developed, or additional parameters need to be added to the equation. The approach of cumulative sum of useable radiation was recently extended to radiata pine, and significantly improved predictions (Mason *et al.*, 2011).

Pinjuv *et al.* (2006) determined a mean annual index of soil root zone water over the length of growth projection, which was used as a covariate in a series of nonlinear systems of regression equations. Both Mason *et al.* (2007) and Pinjuv *et al.* (2006) used output from a process-based model, but avoided the difficulties in estimating NPP accurately. In contrast, Snowdon (2001) found the use of NPP from a process-based model significantly improved predictions. However, Snowdon (2001) used a scaled version of NPP that was composed of long-term average NPP and annual deviations, so the index was more a reflection of climatic variation than potential productivity based on NPP. Regardless of whether NPP or other process-based output is used, the physiologically derived covariates are all dynamic and sensitive to fluctuations in annual weather patterns, which can be important for short-term predictions of growth (Snowdon, 2001).

Although detailed light interception models like tRAYci (Brunner, 1998) are insensitive to soil factors and fluctuations in temperature or precipitation, they can be effective predictors of growth. For example, Brunner and Nigh (2000) used tRAYci to estimate amount of annual radiation absorbed by individual tree crowns, then multiplied by tree leaf area, which showed a strong relationship with volume increment. A similar approach has also been used in the IVY growth model for complex stand in eastern Canada (Groot and Saucier, 2007).

Rather than use a process-based or light-interception model, climatic variables can be used in place of site index. For example, Pokharel and Froese (2009) found that mean annual temperature was an as effective or better predictor of individual tree basal area increment as an error-free estimate of site index for three out of the four species they examined in Ontario, Canada. The limitations of these two approaches are that they are insensitive to important site factors and cannot represent the influence of variable weather patterns. In addition, Crookston *et al.* (2007) found that substituting climate variables for the site descriptors used in an individual tree diameter increment equation resulted in strange model behavior, which was attributed to the limited distribution of data points in climate space.

13.2.2 Statistical growth equations with physiologically derived external modifier

An alternative to trying to determine relationships between site index and physiologically derived variables or refitting statistical equations is to develop external growth modifiers. Essentially, this approach involves adjusting growth predicted by a statistical equation with an additive or multiplicative growth index. An example of this type of approach is given in Henning and Burk (2004) who combined an individual-tree, distance-independent statistical model (STEMS) with a generalized process-based model with a monthly time step (PnET-II) for predicting growth in the mixed-species forests of the northern Lake States region in the United States. Similar to Snowdon (2001), Henning and Burk (2004) used a scaled version of net ecosystem production (NEP) that was composed of the long-term average NEP and the estimated NEP for a given growth period, which was then related to tree- and stand-level biases achieved by the statistical model through a simple linear regression model. Henning and Burk (2004) used a five-year time step and found that the NEP index was more effective at the stand-level than the tree-level.

Due to the difficulty of obtaining a long-term regional average of NPP, Weiskittel *et al.* (2010) developed a modifier for intensively managed Douglas-fir plantations based on estimated annual NPP from a process-based model (Figure 13.2). The modifier then was

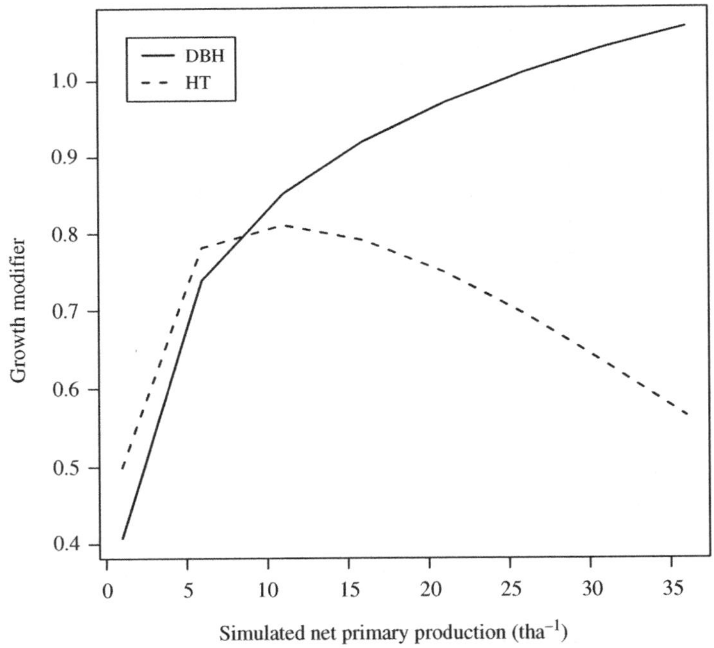

Figure 13.2 Relationship between individual-tree coastal Douglas-fir growth modifiers and stand net primary production in the DF.HGS hybrid model of Weiskittel *et al.* (2010).

applied to individual trees using a multiplicative approach, since that approach performed better than an additive or a stand-level approach (Weiskittel *et al.*, 2010). Henning and Burk (2004) hypothesized that this approach was better suited to fast-growing, single-species, even-aged stands than the mature, mixed-species forest used in their analysis, but they achieved a similar level of improvement as Weiskittel *et al.* (2010) did in Douglas-fir plantations. The apparent limitation in both analyses was that the statistical models had a relatively high accuracy, which makes them difficult to improve upon.

13.2.3 Allometric models

Although allometric models are more closely tied to process-based models than statistical ones, they still rely heavily on statistical equations to relate stem dimensions to biomass, or crown area to photosynthesis. Since the structure of allometric models can vary greatly in their approach, three example models of this type are described in detail. The specific models are CROBAS (Mäkelä, 1997), 3-PG (Landsberg and Waring, 1997), and CABALA (Battaglia *et al.*, 2004).

13.2.3.1 CROBAS

CROBAS is a carbon balance model that was originally developed for Scots pine (Mäkelä, 1997), but has been applied to several other species like Norway spruce (Kantola and Mäkelä, 2004) and jack pine (Schneider *et al.*, 2008). The model is based on three important relations: (1) crown and foliage surface area are allometrically related; (2) a functional balance between foliage and fine roots is maintained; and (3) foliage and stem sapwood area are linearly related (pipe model). The model resolution is the mean tree in each stand and maintains five carbon pools: foliage, fine roots, stem, branches, and transport roots. Initial sapwood area and fine root biomass are calculated as a constant fraction of foliage biomass, which itself is estimated by a statistical equation based on crown surface area. Crown surface area is proportional to crown length squared; assuming that crown shape is constant over time. Consequently, a tree's biomass and sapwood area are driven by total height and height to crown base, as well as several statistical constants. Annual gross photosynthesis is dependent on tree foliage biomass and specific photosynthesis rate that is determined by average branch length in a crown.

Maintenance respiration and senescence are proportional to the biomass pools. Foliage growth is determined by amount of net photosynthesis, height to crown base, and degree of crown self-pruning. Once foliage biomass and crown length are estimated, the remaining biomass pools and sapwood areas are updated based on the allometric relations previously described. Competition in the model is represented by reducing net photosynthesis as leaf area index increases, and invoking higher rates of crown self-pruning based on a statistical relationship to canopy coverage. Stand mortality is also dependent on canopy coverage and empirical constants.

The CROBAS model was extended to run for multiple size classes in a stand by Mäkelä *et al.* (2000b). Valentine and Mäkelä (2005) further generalized the CROBAS model so that the model could be parameterized with traditional forest inventory data and a statistical fitting procedure. Overall, the model has been effective at predicting stem form (Mäkelä, 2002), generating 3D sawlogs (Mäkelä and Makinen, 2003), and optimizing silvicultural systems (Hyytiäinen *et al.*, 2004). Although the model estimates photosynthesis and respiration like a process-based model, the predictions are driven primarily by empirical constants and are insensitive to climate and other site factors. As mentioned in Valentine and Mäkelä (2005), CROBAS could be further hybridized with a more detailed process-based model, but the implications for model behavior and accuracy remain unclear.

13.2.3.2 3-PG

The 3-PG model of Landsberg and Waring (1997) is an allometric model based on key physiological principles, and has been used in a wide range of forest types (Landsberg *et al.*, 2003) (Figure 13.3). Like the original version of CROBAS, the model predicts stand growth based on mean tree. However, unlike CROBAS, 3-PG is sensitive to both climate and site factors such as soil texture and fertility. In essence, 3-PG estimates mean diameter in the stand from stem biomass, and allocates predicted monthly NPP to the various biomass pools based on allometric equations that are a function of mean diameter. NPP is estimated based on the radiation use efficiency (RUE) principle (Monteith, 1972, 1977) and a series of empirically derived 0 to 1 modifiers. The basic equation to estimate NPP is:

$$\mathrm{NPP} = 0.47 \cdot \varphi \cdot \sum_{m=1}^{12} \mathrm{APAR}_m \cdot \min(f_\theta \cdot f_\mathrm{D}) \cdot f_\mathrm{T} \cdot f_\mathrm{F} \cdot f_\mathrm{A} \qquad [13.2]$$

where m is the time interval (months), 0.47 is to convert gross primary production to NPP based on the results of Waring *et al.* (1998), APAR is the absorbed photosynthetically active radiation, φ is the maximum RUE that varies by species, and f_i is a 0 to 1 modifier for soil water availability (θ), vapor pressure deficit (D), temperature (T), frost (F), and age (A).

Soil–water balance is estimated based on soil depth and texture, rainfall, temperature, LAI, and the Penman–Monteith equation for evapotranspiration, while APAR is based on incoming radiation and the Beer–Lambert law (see Chapter 12). Senescence of biomass is a constant proportion of current biomass and varies by biomass type. Stand mortality is based on an empirical $-\frac{3}{2}$ self-thinning equation. Allocation to foliage is driven by an allometric relationship based on DBH, and allocation to roots is based on site fertility, which is based on a user-supplied 0 to 1 value. The remainder of the NPP is allocated to stem growth and converted to DBH based on an additional allometric relationship.

Although 3-PG represents important physiological processes, key simplifications are used to reduce model complexity and input requirements. For example, the model

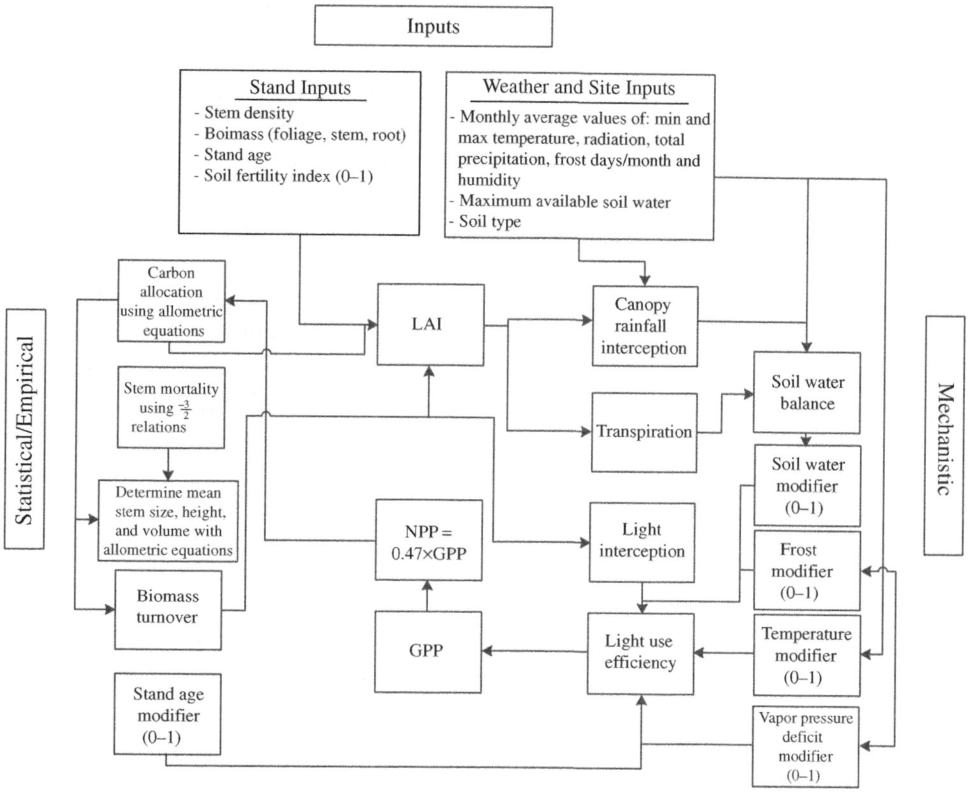

Figure 13.3 Flowchart of key model processes used in the stand-level hybrid model, 3-PG (Landsberg and Waring, 1997).

is insensitive to atmospheric CO_2 concentration, which is an important factor influencing both photosynthesis and stomatal conductance, and limits the model's ability to effectively simulate impacts of climate change (Schwalm and Ek, 2001). Regardless, 3-PG has been used for a wide variety of purposes ranging from simulating long-term influences of a range of silvicultural treatments (Landsberg *et al.*, 2001) to predicting regional variation in productivity (Swenson *et al.*, 2005). The model has been parameterized for several species including loblolly pine (Landsberg *et al.*, 2001), *Eucalyptus spp.* (Almeida *et al.*, 2004), Douglas-fir (Waring and McDowell, 2002), hybrid poplar (Amichev *et al.*, 2010), Scots pine (Landsberg *et al.*, 2005), Sitka spruce (Waring, 2000), tropical forests (Nightingale *et al.*, 2008), and short-rotation willow (Amichev *et al.*, 2011). The model can be run from data obtained from satellites (Coops *et al.*, 2001) and further hybridized with additional models (Peng *et al.*, 2002; Xenakis *et al.*, 2008).

Despite the wide use of 3-PG, the model does make important assumptions that may limit its generality and accuracy. First, the model assumes that the mean of a function is equal to the function of a mean, which, due to Jensen's (1906) inequality, can result in biases as high as 50% when a stand is non-homogeneous in structure (Duursma and Robinson, 2003). Also, as stated by Mason *et al.* (2007), 3-PG is not path invariant and is highly recursive, which means that the model can be fitted to data in a variety of ways and errors may propagate when dependent variables from one month are used as independent variables during the next month.

A sensitivity analysis showed that 3-PG is quite sensitive to the ratio of NPP to GPP, maximum RUE, and maximum canopy conductance (Esprey *et al.*, 2004), which are all parameters used to simplify model complexity. In other words, the model avoids the use of highly detailed mechanistic equations, but prediction accuracy is highly dependent on the single parameters used to represent these processes. Finally, 3-PG is moderately sensitive to user-supplied estimates of site fertility (Esprey *et al.*, 2004), which can allow the model to be "fine-tuned" so predictions better match observations (Miehle *et al.*, 2009). To overcome this important limitation, 3-PG has been combined with soil dynamic models (Peng *et al.*, 2002; Xenakis *et al.*, 2008).

13.2.3.3 CABALA

The CABALA model of Battaglia *et al.* (2004) is more akin to a process-based model like CenW (Kirschbaum, 1999), since it uses mechanistic equations to represent many key physiological processes. However, unlike most process-based models that require stand biomass information for startup, CABALA links stand structure to a function with an array of empirical allometric equations. For example, the model assumes that stand total stem weight can be derived from stand basal area and mean dominant height, while the average ratio of tree height to diameter is a function of the ratio of crown weight to stem weight and total aboveground biomass. In addition, stem sapwood cross-sectional area is empirically related to total tree height and leaf area so that taller trees require more sapwood area to sustain the same leaf area due to hydraulic constraints.

After this initialization, CABALA runs much like a process-based model, as daily net photosynthesis, canopy conductance, and evapotranspiration are all calculated with mechanistic equations. Unlike several hybrid and process-based models that rely simply on the Beer–Lambert law for light interception, CABALA accounts for differences in light interception that might arise due to variations in the inter- and intra-row spacing from various planting arrangements. In addition, CABALA represents biomass allocation based on coordination theory of Chen and Reynolds (1997) that is also bounded by mechanical limitations. In other words, carbon is allocated to minimize differences between actual foliage and root biomass and their respective target biomasses determined by water, nutrient, and light availability. The model has proved useful for simulating long-term outcomes of a range of silvicultural regimes in *Eucalyptus* plantations of Australia (Miehle *et al.*, 2009) and estimating the potential influence of insect outbreaks (Pinkard *et al.*, 2010).

13.3 Comparison to statistical models

Relative to purely statistical models, the degree of improvement in predictions achieved with hybrid models has varied (Table 13.1). At the stand-level, hybrid models are quite effective at improving predictions (Battaglia *et al.*, 1999; Snowdon, 2001; Dzierzon and Mason, 2006), while small to modest gains are achieved at the individual-tree-level (Henning and Burk, 2004; Weiskittel *et al.*, 2010). The level of improvement varies for the same species. Snowdon (2001) found a 23% improvement for predicting radiata pine stand-level basal area growth, while Pinjuv *et al.* (2006) reported only a 4% improvement for the same species.

Improvements are generally greater for basal area than height predictions. For example, Snowdon *et al.* (1998) found that a hybrid approach improved model fits by 13, 33, and 35% for mean tree height, stand basal area, and stand volume, respectively. Also, Dzierzon and Mason (2006) found improvements of 14 and 8% for *Pinus radiata* basal area and dominant height, respectively. Across a range of silvicultural treatments, Weiskittel *et al.* (2010) found that a hybrid model improved predictions of stand volume growth in the control and thinned plots in comparison to a pure statistical model, but the opposite was found when plots were fertilized.

At the tree-level, Baldwin *et al.* (2001) found that an individual-tree hybrid model was significantly superior to a pure statistical model for predicting growth of loblolly pine trees in the southeastern USA, with a 97% reduction in mean bias achieved over a 15-year growth period. In contrast, Henning and Burk (2004) achieved only a moderate improvement in short-term predictions for mixed conifer and deciduous stands in north central USA, with nearly equal mean absolute deviations. However, mean biases of diameter increment predictions were reduced to 0.013 from −0.026 using the hybrid approach, particularly in the aspen–birch forest type (Henning and Burk, 2004). Weiskittel *et al.* (2010) found that a hybridized form of a statistical growth equation reduced MSE by 6 and 12% for four-year individual-tree diameter and height increment, respectively, when compared to a pure statistical approach in intensively managed Douglas-fir plantations.

The range of reported improvements varies widely at both the stand and tree-levels because of the breadth of conditions covered by evaluation data, length of simulations, and differences in the adequacy of the statistical models. For example, Snowdon (2001) found that the degree of improvement achieved by a hybrid model declined with increasing length of prediction period. Interestingly, purely climatic variables may be just as effective covariates as those obtained from a process-based model in certain situations. Henning and Burk (2004) found climate-dependent growth indices almost as effective as process-based ones. However, Snowdon *et al.* (1998) found just the opposite, as the index obtained from the process-based model was better than one derived from annual precipitation. Regardless of the amount of improvement to a purely statistical approach, hybrid models do offer additional output that can be useful for a wide range of purposes.

13.4 Summary

A hybrid modeling framework offers opportunities to address limitations of either a pure statistical or pure process-based approach. However, the history of developing and using hybrid models is much shorter than that of either statistical or process-based models, indicating that there is a range of questions still left unanswered. For example, what are the advantages of hybridizing a statistical and process-based model that both operate at the individual tree? Milner *et al.* (2003) suggested various ways that this type of approach could be implemented, and illustrated that the approach may be particularly beneficial when predicting early stand dynamics.

Despite their relatively short history, a variety of approaches are used to hybridize models. The approaches outlined in this chapter include statistical models with a physiologically derived covariate, statistical models with a physiologically derived external modifier, and hybrid models based on allometric principles, which were illustrated by the models of CROBAS, 3-PG, and CABALA. All of these methods have a variety of advantages and disadvantages, ranging from having to refit statistical equations to significant model sensitivity to a few physiological parameters.

Although the degree of growth prediction improvements achieved by hybrid models varies, hybrid models offer additional output that is useful for a range of other purposes, and allows statistical equations to be extended to novel situations. Philosophically, hybrid models blur the lines between strictly statistical and process-based models, opening the door to a vast array of potential modeling frameworks. Hybrid models will likely continue to increase in the future as our understanding of physiological processes improves and the complexity of questions facing forest managers broadens.

14

Model construction

14.1 Introduction

Although a vast number of growth models currently exist, new ones are continually being developed to address novel situations or in regions where a model did not previously exist. Growth models take a significant amount of resources and time to properly develop. For example, the ORGANON growth model (Hann, 2011) for the US Pacific Northwest was originally developed in the early 1980s and continues to be refined today. Designing and constructing a growth model requires the modeler, in collaboration with potential users, to make multiple decisions in order to be successful.

The first and most important decision is defining objectives and intended uses of the model, which will ultimately drive its structure and resolution. Models are constructed to test hypotheses, compare alternative silvicultural systems, update forest inventories, and determine regional timber harvest schedules. A number of these uses are compatible, but need to be individually assessed to determine the appropriate scale and level of complexity required. Burkhart (2003) provided suggestions for selecting the appropriate level of modeling.

The second decision is determining and evaluating the data needed to meet the objectives and intended uses for the model, which is often the ultimate factor in defining a model's scale and scope. Data come in multiple forms and varying levels of quality that often take a significant investment to maintain, compile, clean, and analyze. Finally, a modeler needs to decide which of an array of statistical techniques will be used to make sense of the data and develop robust equations. These statistical techniques have changed greatly since the formulation of early growth models, and continue to evolve and expand.

Forest Growth and Yield Modeling, First Edition. Aaron R. Weiskittel, David W. Hann, John A. Kershaw, Jr. and Jerome K. Vanclay.

Understanding the advantages and disadvantages of the various statistical techniques is important for deciding the appropriate method to select.

Models generally have three stages in their development: (1) initial construction; (2) evaluation and refinement; and (3) implementation and use. This chapter primarily focuses on the first stage, while Chapters 15 and 16 deal with the second and third stages, respectively. However, it is important to note that models are rarely "finished" and they are often simply put to use when they have proved themselves worthy of overcoming the shortcomings of the previous version or model. Generally, the model will stay in use until sufficient experience, data, or analysis renders improvement, upgrade, or replacement. Hence, there is an important operational feedback loop that is critical to the evolution of model versions or new models.

14.2 Data requirements

Models are often only as good as the data used to construct them. Ideally, modeling and the collection of data should form an iterative process, commencing with model formulation. Model design should dictate data requirements, so that field work provides necessary and sufficient data efficiently. However, it may take several years to obtain all the necessary data. Consequently, most modeling efforts are quite opportunistic and often commence with any data available. The type, breadth, and quality of data have important implications for the modeling effort. Fortunately, data requirements of many modeling approaches are similar and allow a set of minimum data requirements to be defined and standard procedures to be established.

Growth data come in many different forms and include: (1) individual stem analysis; (2) temporary plots; and (3) permanent plots. Raulier *et al.* (2003) illustrates the consequences of fitting models using data from different sources.

14.2.1 Stem analysis

Stem analysis is widely used in the construction of site index curves (e.g. Curtis, 1964b), but is of limited value for many tree species in tropical forests. Even in temperate forests, stem analysis has several limitations, particularly for developing site index curves. First, stem analysis only gives the growth development for an individual tree, and it may not reflect the development patterns of all trees in a stand. Second, a tree must be assumed to be dominant over its entire life, which is often not the case (Raulier *et al.*, 2003). Finally, it is difficult to select an undamaged stem since many trees have damage that is often not visible from first glance (e.g. Nigh and Love, 1999). Consequently, growth curves developed using stem analysis data often overpredict long-term trends in stand-level dominant height growth (e.g. Raulier *et al.*, 2003).

14.2.2 Temporary plots

Temporary plots have been used to develop several growth models. For example, several versions of the ORGANON growth model and many of the western US variants of the Forest Vegetation Simulator (FVS) growth model were developed using temporary plots. The employment of temporary plots is particularly useful when permanent plot data is either inadequate or lacking for modeling the intended population. The usage of temporary plots requires backdating to the start of the previous growth period those tree and stand attributes needed to form the explanatory variables. Both fixed- and variable-radius sampling techniques can be used, but tree cores need to be obtained from all or the majority of trees on a plot to allow for accurate reconstruction of past stand structure. Although randomly placed tree cores are an excellent means to acquire long-term records of growth relatively quickly, they have at least two important limitations with respect to growth modeling.

First, tree cores are only applicable to temperate tree species and can be very difficult to measure accurately in species that are ring-porous, form false rings, or have slow growth. Second, radial increment around the stem can be quite variable, with the percentage of total variation attributable to within-tree variation in radial increment ranging from 8 to 15% depending upon species (Table 14.1). This "measurement error" variation increases the variance of the residuals (Kmenta, 1997) and reduces the achievable maximum value for indices of fit such as the unadjusted and adjusted coefficients of determination.

The coefficient of variation (CV) for within-tree radial increments can range from 0.0 to over 1.0 from tree to tree (Figure 14.1), though the range in the CV values is remarkably similar between species, tree size, and length of growth period. The within-tree variation can be reduced if multiple cores from an individual are collected. For example, the usage of two cores measured 180° apart can reduce the difference between diameter tape and increment core measurements of diameter increment or basal area increment by about 50% (Table 14.2). While the increment core method tended to produce negative biases that were statistically significant in this study, the size of the bias was within the level of accuracy (0.254 cm) to which diameter at breast height (DBH) was measured at the start and end of the growth period (Table 14.2).

There are other difficulties associated with using temporary plots to model tree and stand dynamics. First, the development of height increment models, if desired, will necessitate incurring the expense of felling and measuring past height increment on a sample of trees on each plot. Second, the accurate estimation of the total height (HT), crown size, and timing of past tree mortality can be very difficult. Measurements of HT and height to crown base (HCB) on mortality trees can be corrected for possible bias using measurements taken on living trees exhibiting severe levels of the same damaging agent that killed the trees (Hann and Hanus, 2001), and guidelines on the characteristics and dynamics of snags, such as those of Cline *et al.* (1980), can assist in dating the time of

Table 14.1 Percentage of total variation attributable to within-tree variation in radial increment for five species measured in two studies at different locations.

Species	Location	Number of trees	Number of increment cores	Range in DBH (cm)	Length of growth period (years)	Percentage of total variation attributable to within-tree variation in radial increment
Douglas-fir	Southwest Oregon	1398	4	2.3 to 183.1	5	8.19
Incense-cedar	Southwest Oregon	185	4	2.0 to 92.2	5	7.80
Ponderosa pine	Southwest Oregon	171	4	3.3 to 92.2	5	9.28
Ponderosa pine	Black Hills of South Dakota	192	5	10.2 to 68.6	10	9.58
Sugar pine	Southwest Oregon	103	4	4.8 to 107.7	5	14.82
White and grand firs	Southwest Oregon	342	4	3.3 to 109.7	5	12.28

Based on unpublished data from David Hann.

Table 14.2 Differences in six-year diameter increment and six-year basal area increment when measured with repeated applications of a diameter tape or with one or two randomly placed increment cores on 420 Douglas-fir trees in southwestern Oregon.

Attribute	Measurement technique	Observed		Difference (increment core method minus diameter tape method)	
		Mean	Standard deviation	Mean	Standard deviation
DBH (cm)	Diameter tape	38.68	13.71	—	—
Diameter increment (cm)	Diameter tape	2.84	1.91	—	—
	One increment core towards plot center	2.80	1.96	−0.04	0.92
	One increment core away from plot center	2.74	1.98	−0.11	1.03
	Both increment cores	2.77	1.80	−0.07	0.55
Basal area increment (cm^2)	Diameter tape	300.74	243.00	—	—
	One increment core towards plot center	267.50	250.31	−7.43	99.27
	One increment core away from plot center	261.51	251.80	−13.42	114.34
	Both increment cores	263.87	233.62	−11.07	57.43

Based on unpublished data from David Hann.

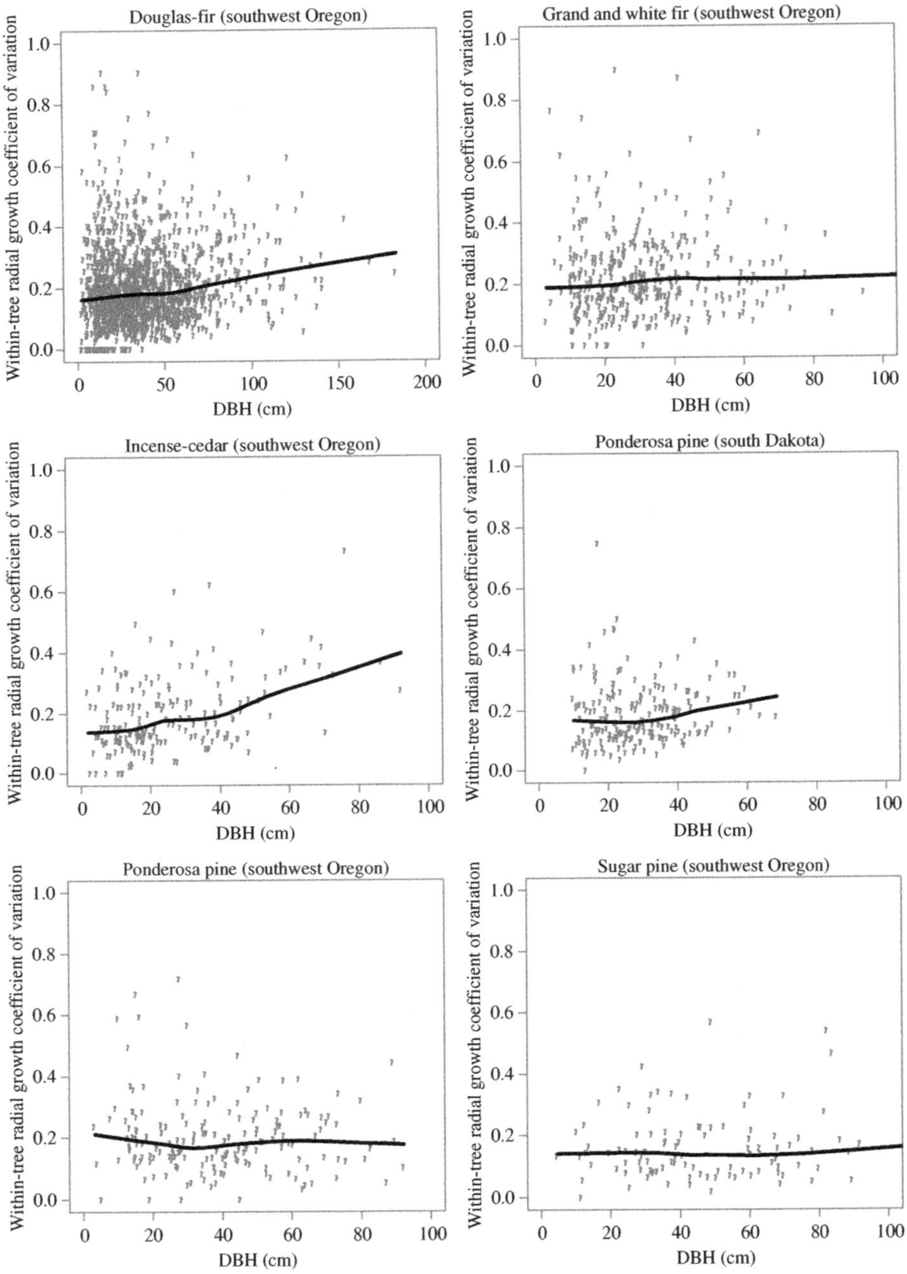

Figure 14.1 Relationship between within-tree radial growth coefficient of variation and diameter at breast height (DBH; cm) for several conifer species in southwest Oregon and South Dakota, USA. The solid line is a lowess regression fit to the data. Based on unpublished data from David Hann.

mortality. Mortality predictions from temporary plots can also be improved by a maximum size–density trajectory, which can be developed from a relatively small sample of repeatedly remeasured permanent plots (Hann and Wang, 1990).

Third, patterns of individual-tree crown recession can be reconstructed on temporary plots, but this requires destructive sampling and a significant investment of time (Maguire and Hann, 1990c). A common alternative to modeling crown recession directly is to use a static HCB equation, which can be developed from temporary plots, to estimate crown recession indirectly. Hann and Hanus (2004) found the indirect method to be unbiased but less precise than a direct estimator of crown recession. Finally, temporary plots are usually taken over a short span of time, which may cause the measurements to be more strongly influenced by short-term weather fluctuations. This is particularly true for dry-climate species such as ponderosa pine, where growth is strongly influenced by variations in rainfall (e.g. Keen, 1937).

14.2.3 Permanent plots

Permanent plots are the best source of data for growth models. Permanent plots can never be completely replaced by temporary plots even for species amenable to stem analysis, because only permanent plots (1) allow satisfactory statistical comparisons within- and between-plots to check adequacy of models; and (2) provide reliable and consistent data on mortality, crown dynamics and stand-level variables. However, permanent plots are expensive to maintain, require multiple measurements before becoming relevant to modeling efforts, and there can be an increased amount of measurement error.

Although the majority of permanent plots are fixed area, variable-radius techniques can be used (see Iles, 2003; pp. 676–685). Curtis and Marshall (2005) provide an excellent description of how to design, install, and maintain permanent plots for research purposes. This chapter will only briefly cover some key issues with permanent plots as they pertain to growth and yield modeling; particularly: (1) plot network design; (2) plot size; (3) variables to collect; and (4) measurement error.

14.2.3.1 Plot network design

A well-designed network of plots is very useful for a range of purposes. To be effective for growth modeling efforts, the network of plots should cover the extremes of site and stand conditions. As a result, such a network differs substantially from a network designed for inventory purposes in that it does not have to characterize the frequency of stand conditions in the existing population, as an inventory would, but rather the full range of stand conditions currently or potentially existing in the target population.

This indicates that locating plots requires multiple levels of stratification rather than just random or systematic locations. Key factors for stratification are: (1) stand age; (2) species composition; (3) site productivity; and (4) geographic location. Often the most useful plots from a modeling standpoint are the ones deemed "too obvious" or "simply worthless" such as old-growth stands or on the top of mountain ridges.

If the goal of the modeling efforts is to model influences of forest management on growth, having a strong experimental design is very important in developing the plot network. Thus, at each plot installation, a control and an array of treatments should be present and replicated when possible. This allows separation of growth caused by site factors and that driven directly by management. The key to installing an effective experimental design at a single location is to keep plots as homogeneous as possible. Many modeling efforts require information on a species' potential productivity across a certain region. Stape *et al.* (2006) describe a twin-plot approach for obtaining this information in a relatively simple and inexpensive manner. The design of any plot network needs to balance cost with being representative of the conditions intended to be considered in the model.

14.2.3.2 Plot configuration and size

Components of both individual-tree and stand growth models rely on unbiased estimates of certain attributes obtained from plots. These attributes are sensitive to both plot configuration and size. For plot configuration, a square plot is generally the most useful and accurate, particularly for larger plot sizes (Curtis and Marshall, 2005). Rectangular plots also are very useful, but should be oriented to reduce variation within plots. Circular plots are effective, but plot boundaries are more difficult to monument. Regardless of plot configuration, all plots require an adequate buffer surrounding the plot's measurement area. This is especially important if multiple forest management treatments are being compared at a given site. The recommended width of the buffer varies, but at least 10 m or the expected crown width of dominant trees at the end of observation period seem adequate (Curtis and Marshall, 2005).

The ideal plot size is often not clear and highly dependent on modeling objectives. Qualitatively, a plot should be large enough to be representative of a stand and its inherent variability, but small enough to ensure high-quality measurements and the capability to be replicated across the landscape. As summarized by Curtis and Marshall (2005), the recommended plot size in hectares has ranged from 0.01 times the dominant height in meters, to 1.0 ha. Larger plot sizes are required in naturally-regenerated, mixed-species, and uneven-aged stands, due to their higher inherent variability compared to planted, pure, and even-aged stands. Some important considerations in selecting a plot size for developing a growth model are: (1) degree of plot permanence; (2) if modeling response to forest management is of interest; and (3) consistency. A larger plot should be used if multiple remeasurements are intended rather than a single remeasurement.

Plot size should be based on the size of trees at the final and not the initial measurement. Curtis and Marshall (2005) showed that the plot size needed for having at least 50 Douglas-fir trees in a plot ranged from 0.05 to nearly 0.45 ha as the quadratic mean diameter increased from 15 to 60 cm (Figure 14.2A), similar relationships are easily derived for other species (Figure 14.2B–D). Forest silvicultural treatments often increase within-stand heterogeneity. Consequently, a larger plot size is recommended, and typically plot sizes of 0.1 to 0.2 ha are used (Curtis and Marshall, 2005).

Although plot size should ideally be adjusted according to stand conditions, a consistent plot size may be preferred to minimize potential confounding effects of variable plot

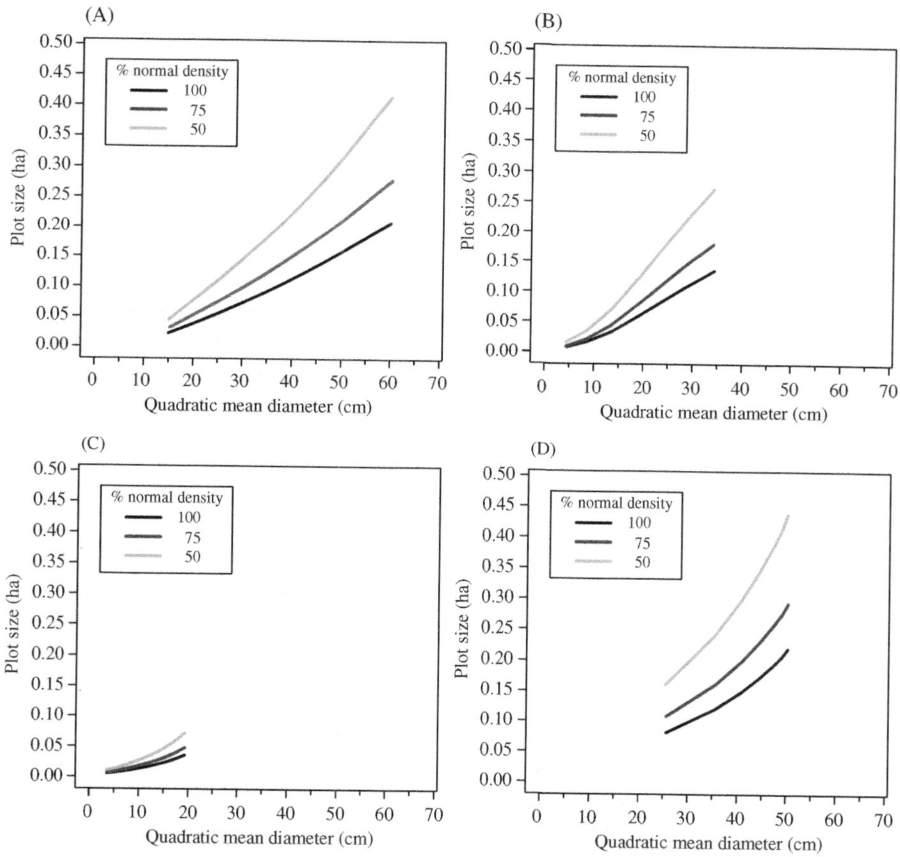

Figure 14.2 Plot sizes required for various species to include 50 trees in stands of different percentage normal density: (A) Douglas-fir (normal density based on McArdle *et al.* (1961)); (B) upland oak (normal density based on Table 2, site index 80, Schnur (1937)); (C) black spruce (normal density based on black spruce, site class 1, Plonski (1974)); and (D) loblolly pine (normal density based on site class 120, Schumacher and Coile (1960)). Adapted from Curtis and Marshall (2005).

sizes. Hann and Zumrawi (1991) demonstrated that growth model predictions are sensitive to plot size, and differences in projections become more severe as plot size diverges from the one used to develop the model. This suggests that a balance between research plot sizes and operational plot sizes is needed.

Plot configuration and size influences the quality of developed growth models in several ways: (1) edge effects; (2) estimation of tree-level measures of competition; and (3) influences on mortality. Distance-dependent models are most sensitive to edge effects because of their need to assess distance and size of competitors (see Chapter 2). Martin *et al.* (1977) demonstrated the influence of plot size on the degree of edge bias and proposed a correction method. Pommerening and Stoyan (2006) reviewed various methods proposed for edge-correction of spatially dependent indices and found that the best method was dependent on the arrangement of trees within a stand. Most proposed correction methods apply better to rectangular plot shapes, but Williams *et al.* (2001) provided an effective one for circular fixed-area plots. A method of avoiding edge bias is to measure trees immediately outside a plot. Most importantly, a plot buffer ensures that edge conditions are similar to those found in the interior of the plot.

Estimates of local tree-level competition are sensitive to both plot size (Lappi, 2005) and plot type (fixed- vs. variable-radius) (Schreuder and Williams, 1995; Stage and Ledermann, 2008). Corrections were proposed to account for different plot sizes (Stage and Wykoff, 1998; Lappi, 2005), but an optimal plot size was not suggested. Lappi (2005) concluded it is safest to measure density from a plot larger than the plot size used for sample tree selection. Stage and Ledermann (2008) suggested that optimum plot sizes for estimating measures of competition correctly are smaller than plot sizes commonly used for estimating stand volume. However, they suggest that using a variable-radius sampling scheme may provide the most effective estimates of distance-independent competition (Stage and Ledermann, 2008), which is similar to the conclusions of Schreuder and Williams (1995) for a distance-dependent measure of competition. A strategy for improving estimates of stand-level competition is to have several randomly placed plots in a stand (García, 2006).

One reason that mortality is hard to predict is because it is a relatively rare event. Estimates of stand-level mortality are often strongly correlated with plot size. On extremely small plots, deaths of a few trees result in large negative increments (Curtis and Marshall, 2005). Consequently, smaller plot sizes should be avoided in estimating and modeling mortality. Curtis and Marshall (2005) recommend that plots be large enough so that catastrophic mortality can be distinguished from regular mortality. Defining a plot size that can accomplish this is difficult and likely dependent on an array of factors. Relatively little work has been done on comparing effectiveness of different plot sizes for modeling mortality. Regardless, a robust mortality model can only be determined from an extensive network of permanent plots that were continually remeasured for a sufficient period of time.

14.2.3.3 Variables to collect

Like plot size, the variables to collect on each plot should ultimately be determined by the objectives of the modeling effort. A stand-level model requires significantly fewer variables than a tree-level model. For both stand- and tree-level models, stand history (e.g. origin, known silvicultural treatments), age, precise location, and physiographic variables (e.g. elevation, slope, and aspect) should be recorded for each plot. If harvests occurred in the past, date of harvest, type of equipment used, and the number, size, and species of trees removed should be determined. Obtaining an initial measurement prior to a treatment is also crucial, even if the plot is going to be measured immediately after treatment. Information on soil depth, parent material, and texture can also be helpful.

To be most useful, each tree in a plot needs to be uniquely identified through time. Without this, effective data editing and quality control become impossible. Re-using tree numbers within a plot may seem like a practical idea, but often leads to problems during analyses. Curtis and Marshall (2005) recommend that plastic barlocks be used to identify each tree and attached to trees with staples. A decision needs to be made about what trees will be uniquely identified. Ideally, all trees taller than 1.3 m are identified on a plot. However, threshold DBH values are often used and can cause problems if set too high (i.e. >2.5 cm). At a minimum, a subsample of trees smaller than 1.3 m should also be identified and followed through time to help understand small tree growth and recruitment dynamics. Once trees are uniquely identified, it is important to permanently mark the location of breast height to ensure that the same location on the tree is measured repeatedly.

Measurements of HT and HCB are crucial for many modeling efforts, but costly to obtain, and are the most prone to measurement errors. Since they are time-consuming measurements, a subsample of trees is often measured for HT and HCB. If development of a height increment model is desired, the same subsampled trees should be measured repeatedly for HT and HCB. However, the relatively recent introduction of handheld laser hypsometers has greatly increased both the speed and accuracy of HT measurements. Consequently, it is now feasible to use much larger HT samples than was common practice in the past.

Selection of trees for HT measurements should ensure that each plot has at least a few measurements and that the observations are obtained across a range of species and DBH classes. Subsampling only dominant and codominant trees can cause serious problems when trying to estimate HT or HCB for smaller trees. The primary limitation in measuring HCB accurately is defining its location so that it is repeatable and easy to determine in the field. The best method is likely to define HCB as the height to the lowest live branch continuous with the main crown. The occasional epicormic branch should generally be ignored, unless they could potentially influence wood quality. Ocularly estimating crown ratio or trying to find the whorl with at least three live branches in some conifer species is difficult, highly inaccurate, and rarely repeatable (e.g. McRoberts *et al.*, 1994).

At a minimum, species and DBH on all trees, and HT and HCB on a subsample of trees should be the measurements made on each plot. Additional measurements depend on model objectives. Some of the more important variables to consider collecting include: (1) crown width; (2) an upper stem diameter; (3) bark thickness; (4) tree age; (5) damage codes; (6) stem location; and (7) understory conditions. In young stands, crown width measurements are important for characterizing site occupancy and the degree of competition (e.g. Ritchie and Hamann, 2008), but become less useful as stands close. However, a subsample of trees should be measured for crown width if maximum and largest-crown width equations are required. Measures of upper stem diameter and bark thickness are important for evaluating and calibrating regional stem taper and bark thickness equations.

As discussed in Chapter 6, many individual-tree equations are dependent on tree age. If this type of approach is desired, sufficient observations of tree age are needed. This is relatively easy to achieve in single-species, even-aged stands, but becomes more difficult and expensive as the number of species and/or cohorts increases. Observations of tree age are also needed to obtain estimates of site index. To get good estimates of site index, at least four dominant, undamaged trees should be measured for height and age, but more measurements are suggested due to the difficulty in selecting undamaged trees and high within-stand variability of site index (Curtis and Marshall, 2005). If these additional measures are obtained on a subsample of trees, it is important that all measurements be obtained on the same subset of trees through time.

The number and quality of damage codes varies, but their collection is highly recommended. Damage codes are helpful in removing trees that might bias static or increment equations and in determining potential causes of mortality. At minimum, damage codes should provide the cause and severity of damage. Curtis and Marshall (2005) provide a comprehensive list of damage codes that attribute damage to logging, wildlife, weather, and insects. Hanus *et al.* (1999a, 2000) also provide a list of damage codes used in the development of the ORGANON growth and yield model, and which were found particularly helpful in estimating unbiased HT and HCB static equations. For example, Hanus *et al.* (1999b) found that undamaged trees were relatively common in their extensive dataset (33 to 64% depending on species), and were able to develop multipliers ranging from 0.0052 to 1.7602 to correct predictions of HT for varying levels of damage. These equations were then used to evaluate and correct for errors that can occur when estimating HT and HCB at time of death for mortality trees found on temporary plots (Hann and Hanus, 2001).

Knowledge of stem locations is required if distance-dependent measures of competition are being used or to understand within-stand patterns of mortality. Although newer technology makes obtaining stem location information easier, it is still a time-consuming and cumbersome task, particularly on larger plots and in steep terrain. However, once stem location is determined for a plot, it doesn't need to be remeasured, unless ingrowth occurs. Advances in remote sensing technologies may make it relatively easy and cost effective to obtain stem locations in the future.

Understory conditions are important for understanding regeneration patterns, site quality, and wildlife habitat suitability. This information is relatively easy to obtain from small, randomly placed subplots, and is particularly useful if modeling development of young stands or in regions where site index information is unavailable. For regeneration, tallies of woody stems by species and height class are the most useful. In young stands and certain site types, knowledge of the amount of competing herbaceous vegetation and shrubs is needed to predict growth accurately (Ritchie and Hamann, 2006). For modeling seedling mortality, Lavine *et al.* (2002) suggested that tagging individual seedlings was not necessary if seedling counts by year were available.

14.2.3.4 Measurement error

Understanding the different types and potential sources of measurement error is important in developing a robust inventory program and growth model. The most important types of measurement error in forest growth modeling are mensuration, grouping, prediction, and sampling (Canavan and Hann, 2004). Mensuration error arises when recorded values are not the same as the true value due to observer, equipment, or recorder biases or mistakes. Grouping error is from rounding data or by defining categories. Prediction error arises if tree attributes are subsampled and the missing values predicted. Sampling error is defined by incomplete knowledge of a population and depends on the number of samples, plot size and shape, the type of sampling, and how the population is defined (Canavan and Hann, 2004).

Canavan (2002) summarizes the influence of these various sources of error in forestry, particularly with respect to growth modeling. Although various correction methods are available (Canavan and Hann, 2004), the simplest way to manage error is to minimize its occurrence. This is accomplished by having standardized plot measurement procedures, checking measurements, ensuring an adequate and representative sample, and using a well-designed database management system.

Mensurational error is unavoidable, but can be managed. Effective ways of accomplishing this are having well-trained crews, using electronic data recorders that check entered data in the field, having previous measurements available at the time of remeasurement, marking the location of measurements, remeasuring at an appropriate temporal scale, and periodically checking measurement quality. A detailed and standardized field protocol guide should be developed and field crews briefed on its importance. This should allow them the ability to quickly answer questions that might arise in the field. Although electronic data records have many limitations, the benefits may outweigh the disadvantages. A well-designed electronic forest-inventory data collection program can ensure that all appropriate measurements are taken on a tree, reject unrealistic measurements, and be used to compare current measurements to previous measurements. However, collecting data on paper still offers some advantages, such as the ability to go back and see recording changes made in the field that may have been made incorrectly.

Cruise checking is an important step in ensuring that measurements are being taken and recorded properly. Simply marking the location where DBH measurements should occur can significantly reduce measurement errors. However, some tree species, particularly in tropical forests, can still remain problematic for DBH measurements due to their tendency to form buttresses. Metcalf *et al.* (2009b) found that this can bias estimates of DBH by up to 10% per meter of height, and presented a statistical method for handling this type of situation.

Finally, remeasurements should be made on a periodic basis that minimizes error and where growth is actually observable. Measurements on a 5- to 10-year period are used in many regions, but annual measurements might be appropriate for some fast-growing plantation species. For example, Snowdon (1987) found that the measurement error in remeasuring radiata pine DBH with a diameter tape in periods of less than four years significantly confounded the relationship between "observed" increment and other variables, which was much stronger when trees were remeasured on intervals greater than four years.

Management of data is often an overlooked and underappreciated activity in developing a model. Grouping error easily arises with improper data management such as not retaining the appropriate number of significant digits or only retaining summarized rather than individual data records. Proper data management is often best done with a relational and standardized database, which allows all data storage in a single location, properly defined column headers, and uniquely identified records. A database should always be designed as if it would be handed to another person at a moment's notice. This requires keeping the database updated, all variables and codes defined, and adequate documentation on how the data were collected. A critical part of data management is a standardized procedure for data editing, which checks for omissions, unreasonable, or impossible values, particularly for changes between successive measurements. As noted earlier, this requires that individual trees be identified by number. Some additional data management practices are covered in Borer *et al.* (2009).

The common usage of subsampling for HT and HCB confounds the resulting statistical analyses by introducing prediction error to the modeling process (e.g. Hasenauer and Monserud, 1997). This is particularly problematic for modeling tree mortality, which utilizes all trees and their attributes on the plot in developing the equation. Problems with prediction error also occur for equations that use stand-level explanatory variables such as crown closure at either the top of the tree or at 66% of the tree height (see Chapter 2), which require the HT and HCB for all trees on the plot in order to be calculated. Plot-level scaling of fixed-effects regional equations for predicting missing HT or HCB (e.g. Hanus *et al.*, 1999a; 2000) or the use of mixed-effects regional equations (Temesgen *et al.*, 2008) can help in reducing the size of bias that results from using estimated rather than measured explanatory variables.

Sampling error has received a significant amount of attention in the forestry literature (Canavan, 2002). Sampling error is generally reduced by selecting an adequate sample size, sufficiently defining the population of interest, and using an appropriate sampling technique (e.g. stratified, systematic, double). These topics are sufficiently covered in Husch *et al.* (2003) and only briefly discussed here. An adequate sample size is determined by population variability and the desired level of precision. In developing a growth model, much larger sample sizes are often required than in experimental studies or operational inventories, because the full range of conditions needs to be covered, and variability is desired rather than controlled for.

14.3 Model form

A model's general behavior and ability to extrapolate is determined primarily by the form of the model. Most biological processes are inherently nonlinear, which indicates that linear model forms should be used with care. There are many considerations required in selecting an appropriate model form. The primary consideration arises from the realization that the datasets used to develop most growth models represent a very small fraction of the population to which the model will be applied. This reality should lead to model forms that are complex enough to accurately and adequately characterize the expected major behaviors of the population, but simple enough to avoid being overly influenced by the sample's peculiarities.

Some of the other important considerations are: (1) whether to use empirical or theoretical models forms; (2) the degree of flexibility desired in the model form; (3) the ease to which the model form can be constrained to exhibit expected behavior; and (4) the criteria used for selecting between alternative model forms. Multiple theoretical model forms exist (Zeide, 1993; Sit and Poulin-Costello, 1994) and are a good starting point for many modelers. However, a well-formulated empirical equation is often just as effective as a theoretical model form (e.g. Martin and Ek, 1984). On the other hand, theoretically based equations may be more reliable for predictions which involve extrapolations beyond the range of observed data (Figure 14.3).

Model form flexibility refers to the ability of the model to take a range of shapes based on the data. Some of the most flexible model forms are sigmoidal curves, which have an inflection point and an upper asymptote. García (2005c) developed a highly flexible unified sigmoidal equation that includes many common theoretical model forms. Although polynomial model forms can be flexible, they do not extrapolate well and should be avoided when possible. Flexibility is important because it allows equations to better represent the data rather than forcing a fixed equation form through the data. Having a constrained model form is often very important in growth models because extrapolations are generally the norm rather than the exception. Asymptotic model forms are logical for most growth equations. The logistic model form is particularly effective for constraining predictions.

Finally, selecting the best model form for the data is often a difficult task. Traditionally, model form selection was based on lack-of-fit statistics such as mean bias and root mean

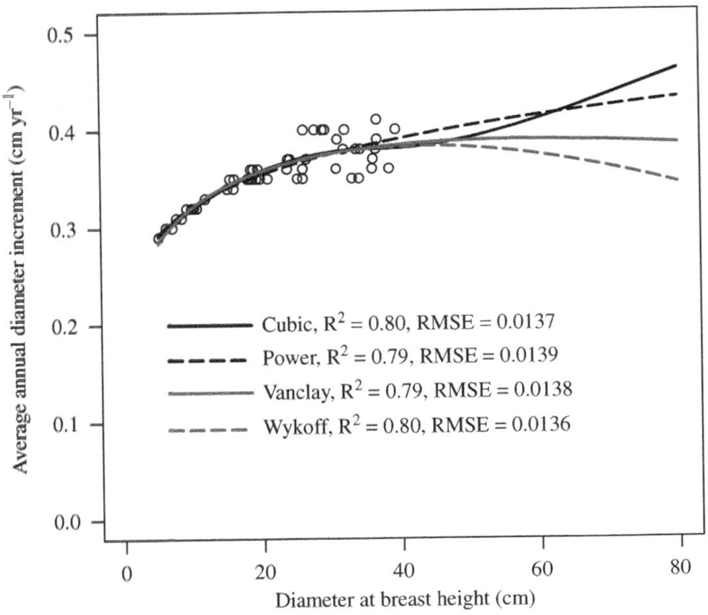

Figure 14.3 Fit statistics and ability to extrapolate for several equation forms for predicting average annual diameter increment (cm yr^{-1}) from diameter at breast height (DBH; cm). Despite fitting the data well, the cubic and power forms do not provide logical extrapolations. Data was generated from observed diameter increment on permanent sample plot data from the New Brunswick Department of Natural Resources.

square error (RMSE). Mean bias is the average difference between the predicted and observed values, while RMSE is the square root of the variance of residuals:

$$\mathrm{MB} = \sum_{i=1}^{n} (y_i - \hat{y}_i)/n \qquad [14.1]$$

$$\mathrm{RMSE} = \sqrt{\frac{\sum_{i=1}^{n} (y_i - \hat{y}_i)^2}{(n-k-1)}} \qquad [14.2]$$

where MB is mean bias, n is the number of observations, k is the number of parameters in the model, y_i is observed value for the ith observation, and $\hat{y}_i$ is the predicted value for the ith observation. Mean bias and RMSE are both absolute measures of model fit (unlike R^2), are in the same units as the original dependent variable, and smaller values indicate a better model fit. RMSE is a good measure of model precision. However, mean bias does not penalize for the number of parameters in a model, while RMSE can heavily weight outliers in the data. Consequently, the use of information criteria like those of Akaike (1973) or Schwarz (1978) is the most common means of comparison today. Akaike's (1973) information criterion (AIC) ranks models based on their precision and complexity:

$$\text{AIC} = 2 \cdot k - 2 \cdot \log(L) \qquad [14.3]$$

where k is the number of parameters and L is the maximized value of the likelihood function for the estimated model. Schwarz's (1978) criterion, which is also known as the Bayesian information criterion (BIC), is similar to AIC, but penalizes more for using additional parameters:

$$\text{BIC} = -2 \cdot \log(L) + k \cdot \log(n) \qquad [14.4]$$

where n is the number of observations in the original dataset. For both AIC and BIC, models with lowest values are preferred.

The limitation of AIC and BIC is that their values are based on the dependent variable, and deciding what constitutes an important difference is difficult. Consequently, AIC and BIC cannot be used to compare various transformations of a dependent variable. A much more effective method for comparing transformations is Furnival's (1961) index (see Table 14.4). In general, an AIC reduction of 10 units is considered significant, while models that differ by less than 10 units are relatively similar (Burnham and Anderson, 2002). However, AIC and BIC only help to determine the best model form based on the available data and do not ensure proper extrapolation. This is well demonstrated in Weiskittel *et al.* (2009a), as the best model according to AIC did not extrapolate well and was consequently not chosen as the final model.

Equations can be empirical or otherwise, but should be carefully formulated to provide sensible predictions across the full range possible for all explanatory variables. Explanatory variables should not be chosen haphazardly; they should be selected deliberately to provide realism and robustness in the model. However, this is often easier said than done. Algorithms for selecting the best set of explanatory variables are often limited to linear models and do not account for interactions between variables.

Data mining techniques are effective at determining key explanatory variables (De'ath, 2007), but are only as good as the data used. Consequently, one of the best approaches for selecting the explanatory variables is comparing variables that were used in the past and are biologically logical. The scientific literature is a good place for finding important explanatory variables used in the past. Biological considerations rely on the modeler to understand the processes being modeled and what variables might best represent them.

14.4 Parameter estimation

Once data are available and appropriate model forms selected, parameters that describe the relationships between the explanatory and dependent variables need to be determined. Various statistical techniques are used to estimate model parameters. This section reviews the most common means used in growth and yield modeling. Advantages and limitations of each approach are highlighted, rather than focusing on technical details. A standard statistical textbook should be consulted for specific details on any parameter estimation technique.

14.4.1 Regression

Simple and multiple linear and nonlinear regression have a long history as the fundamental tools used in developing statistical models. There are three primary techniques used for estimating parameters in regression: ordinary least squares (OLS), generalized least squares (GLS), and maximum likelihood estimation (MLE). OLS parameters are those that minimize the sum of squared residuals, which are the differences between observed values and predictions made by the model. GLS parameters have been estimated in a manner that addresses problems with autocorrelation of residuals or heterogeneous variance in the modeling data. MLE parameters are those that maximize the probability of the sample data. OLS parameter estimates correspond to MLE estimates if certain assumptions are met (Table 14.3).

The most important assumptions are: (1) correct specification of the relationship between independent and dependent variables (i.e. expected value of the error term is zero); (2) explanatory variables are without measurement error; (3) no perfect correlation exists between any linear combination of explanatory variables; and (4) model residuals are independent, have a constant variance (homoscedasticity), and are normally distributed. Multiple methods are used to ensure these assumptions are met and/or minimize the consequence of assumption violations. Many of these are thoroughly covered in Kmenta (1997) and Draper and Smith (1998). This section will provide only a brief overview, particularly with respect to common forestry data.

There are no hard and fast rules or tests for determining whether a relationship is linear or nonlinear or whether the error term is additive or multiplicative. Information useful for making a correct specification of the relationship includes: (1) plots of the data over all possible explanatory variables; (2) summary of model forms used in prior similar analyses; (3) understanding the knowledge of the basic behavior of the system being modeled; (4) fitting candidate model forms and plotting residuals to check for trends; and (5) use of Furnival's (1961) index of fit to determine the form of the best alternative dependent variable.

A common method for improving relationships between two variables is the use of transformations. The natural logarithmic and square root are two common transformations used in forestry. However, transformations often cause problems when predictions are back-transformed. Flewelling and Pienaar (1981) give an excellent summary of correction procedures for log bias, while Gregoire *et al.* (2008) present an unbiased estimator when the square root transformation is used.

Table 14.3 Assumptions of linear least squares regression and properties of the parameter estimates.

Assumption	Properties of parameter estimates		
	Unbiased estimator	Best linear unbiased estimator (BLUE)	Uniformly minimum variance unbiased estimator (UMVUE)
Model is linear and the error term is additive	✓	✓	✓
Number of sample observations is greater than the number of parameters to be estimated	✓	✓	✓
All independent variables are nonstochastic variables measured without error	✓	✓	✓
No perfect correlation exists between any linear combination of the independent variables	✓	✓	✓
The expected value of the error term is zero	✓	✓	✓
Variance about the model is homogeneous		✓	✓
Random errors are uncorrelated		✓	✓
Random errors are normally distributed			✓
Form and number of independent variables in the model are known before parameter estimation			✓

The various forms of measurement error are hard to avoid in forestry, but are often ignored when models are developed. For example, Hasenauer and Monserud (1997) found that measurement errors in remeasured HT were so large that the underlying height increment signal was nearly hidden. Kangas (1998a) demonstrated the significant effects that errors in explanatory variables cause. Correction methods are numerous and covered for linear and nonlinear models in Fuller (1987) and Carroll *et al.* (2006), respectively. Some of the correction methods are the use of: (1) regression calibration; (2) SIMEX algorithm (e.g. Kangas, 1998a); (3) instrumental variables; (4) distribution correction methods (Lappi, 1991b); (5) geometric mean regression (Nigh, 1995a); and (6) all possible differences (Goelz and Burk, 1996).

Multi-collinearity occurs when explanatory variables in a linear model are highly correlated, resulting in parameter estimates that are highly imprecise. Kozak (1997) explored the effects of multi-collinearity on taper functions, and found that very small changes in the data can produce substantial changes in parameter estimates. Multi-collinearity can be suspected when the sign on a parameter is incorrect or the magnitude of the estimate is high relative to other parameters. The degree of multi-collinearity can be examined using variance inflation factors (VIFs) and is generally judged to be problematic if the VIF is greater than 10. Multi-collinearity can be addressed by removing variables with a high VIF or using ridge regression (Bare and Hann, 1981) and/or principal components regression (Morzuch and Ruark, 1991).

A properly specified regression model must also ensure that the residuals are independent, have a constant variance (homoscedasticity), and are normally distributed. Independence is commonly violated in forestry since multiple measurements are often taken on the same individuals or plots, which results in significant autocorrelation. Also, variance often increases with tree size, so model residuals are often heteroscedastic and not normally distributed. Mixed models are an effective method for addressing hierarchical datasets (see below).

The assumptions of independence, constant variance, and being normally distributed are generally evaluated using residual plots, which give the model residuals over the predicted value. Often, model residuals are evaluated using cross-validation or data-splitting techniques, which use a portion of the data for model fitting and the rest for model evaluation (see Chapter 15). However, Kozak and Kozak (2003) found that cross-validation by data-splitting, or double cross-validation, provided little, if any, additional information for evaluating regression models. Instead, they recommended just using the full dataset for model construction and using various traditional model fit or lack-of-fit statistics (Kozak and Kozak, 2003).

Even with the use of mixed effects, significant autocorrelation can still be observed and needs to be addressed (e.g. Garber and Maguire, 2003). Consequently, autoregressive error structures are often employed in forestry, using GLS procedures (Gregoire, 1987; Williams and Reich, 1997). The continuous first-order (CAR1) (Gregoire, 1987; Garber and Maguire, 2003) and the moving average (ARMA) (Monserud, 1986) autoregressive error structures are commonly used in forestry. The CAR1 is preferred because it is computationally efficient and accounts for the irregular nature of most forest measurements.

Heteroscedasticity is often addressed through the application of weighted regression using GLS procedures, which alters the contribution each observation makes to the final parameter estimates. However, selecting an optimal weighting function is difficult, particularly for smaller datasets (<100 observations; Williams, 1994). Although likelihood ratio tests are helpful in determining if weighting is needed, they do have some important limitations, and indices like Furnival's (1961) are more effective (Table 14.4).

Table 14.4 Comparison of root mean square error (RMSE), R^2, Akaike's information criterion (AIC), Bayesian information criterion (BIC), and Furnival's (1961) index (FI) for simple allometric models predicting western hemlock foliage weight (FWT; kg) from diameter at breast height (DBH) and crown length (CL).

Model	RMSE	R^2	AIC	BIC	FI
$FWT = \beta_0 + \beta_1 \cdot DBH + \beta_2 \cdot CL$; unweighted	19.4400	0.94	469.87	477.75	19.44
$\ln(FWT) = \beta_0 + \beta_1 \cdot \ln(DBH) + \beta_2 \cdot \ln(CL)$	0.0639	0.94	−136.61	−128.73	20.18
$FWT = \beta_0 \cdot DBH^{\beta_1} \cdot CL^{\beta_2}$; unweighted	20.1203	0.94	473.50	481.38	20.12
$FWT = \beta_0 \cdot DBH^{\beta_1} \cdot CL^{\beta_2}$; weight = $CL^{-0.5}$	0.8724	0.85	140.76	148.64	15.52
$FWT = \beta_0 \cdot DBH^{\beta_1} \cdot CL^{\beta_2}$; weight = $CL^{-1.0}$	0.0120	0.63	−313.13	−305.25	24.69
$FWT = \beta_0 \cdot DBH^{\beta_1} \cdot CL^{\beta_2}$; weight = $CL^{-1.5}$	0.0057	0.35	−392.40	−384.52	32.17

Based on data from Kershaw and Maguire (1995).

14.4.2 Quantile regression

Quantile regression is useful when predictions other than the conditional mean are desired. In forestry, estimates of maximums are often of interest. Consequently, quantile regression is used to estimate factors such as maximum size–density lines (Zhang *et al.*, 2005), potential diameter increments (Pretzsch and Biber, 2010), and individual-tree maximum crown widths (Russell and Weiskittel, 2011). Quantile regression is also useful for estimating medians or other quantiles of response variables, which can be helpful for validation purposes (Mäkinen *et al.*, 2008). In ecology, quantile regression is advocated as an effective method for detecting relationships between variables when there is no, or a weak, relationship between means (Cade and Noon, 2003).

14.4.3 Generalized linear regression models

When dependent variables have non-normal distributions, generalized linear regression techniques are useful for ensuring proper parameter estimation. One of the most common generalized linear regression models used in forest growth modeling is the binomial, which is used when response data are binary (0 or 1). For example, tree mortality data are generally binary (1 if dead, 0 if alive). A generalized model consists of three components: a probability distribution, a linear predictor, and a link function. One of the most widely used link functions is the logit:

$$\text{logit}(\rho) = \log\left(\frac{\rho}{1-\rho}\right) \qquad [14.5]$$

where ρ is a number between 0 and 1. A log link is widely used in fitting the Poisson distribution to analyze count data (e.g. Fortin and DeBlois, 2007).

Table 14.5 Confusion matrix used for assessing predictions of a binary event like individual tree mortality.

Hypothesized/predicted	True/observed	
	Positive/alive	Negative/dead
Positive/alive	True positive (Sensitivity)	False positive (1 − specificity)
Negative/dead	False negative (1 − sensitivity)	True negative (Specificity)

Although fitting generalized models is relatively straightforward, validating proper model behavior is more difficult due to the nature of the response variables. Useful techniques for assessing binary models include receiver operator curve (ROC) graphs (Saveland and Neuenschwander, 1990; King, 2003) and cutpoint analysis (Hein and Weiskittel, 2010). To use these techniques, a confusion matrix is first constructed and the cutpoint (threshold between events and non-events) is varied (Table 14.5). A ROC graph is a plot of a model's sensitivity versus one minus the specificity, or false positives, for a range of decision thresholds (Figure 14.4). Estimating the area under the curve (AUC) is a useful metric for assessing prediction performance as well as comparing

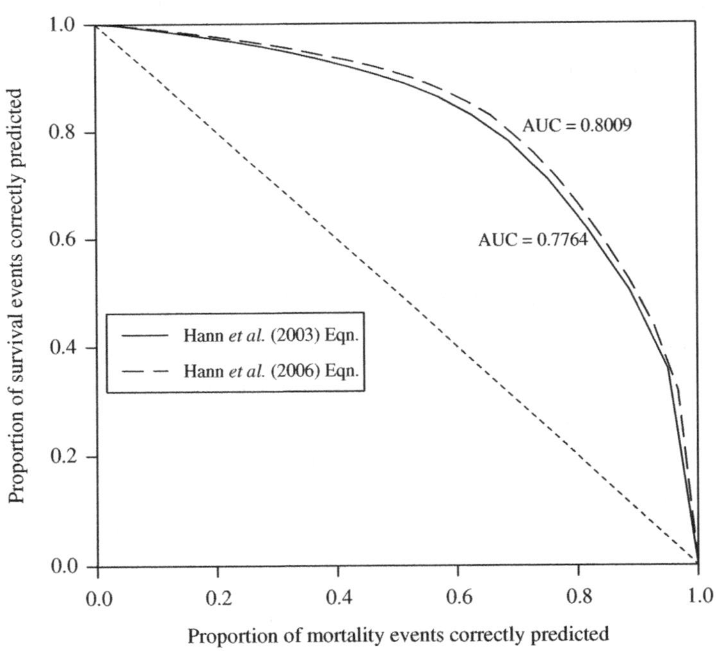

Figure 14.4 Receiver operator curve (ROC) showing the area under the curve (AUC) for two individual-tree mortality equations for coastal Douglas-fir from Hann *et al.* (2003) and Hann *et al.* (2006).

different equations or modeling approaches (King, 2003). Hosmer and Lemeshow (2000) suggest that AUCs between 0.7 and 0.8 provide acceptable discrimination, while AUCs above 0.8 provide excellent discrimination.

14.4.4 Mixed models

The application of mixed models in forestry has seen considerable recent interest. Mixed models have been used for modeling stand basal area (Gregoire *et al.*, 1995) and dominant height growth (Fang and Bailey, 2001; Wang *et al.*, 2007b), stem taper (Garber and Maguire, 2003; Leites and Robinson, 2004), individual tree increment (Calegario *et al.*, 2005; Weiskittel *et al.*, 2007a), cumulative bole volume (Gregoire and Schabenberger, 1996), and HT to DBH (Robinson and Wykoff, 2004; Temesgen *et al.*, 2008).

This interest is due to the ability of mixed models to account for multiple levels of hierarchy within datasets and provide individual specific best linear unbiased predictions (BLUPs) (Figure 14.5). Forestry data are often nested both spatially and temporally. For example, permanent plots contain multiple individual trees that are repeatedly measured

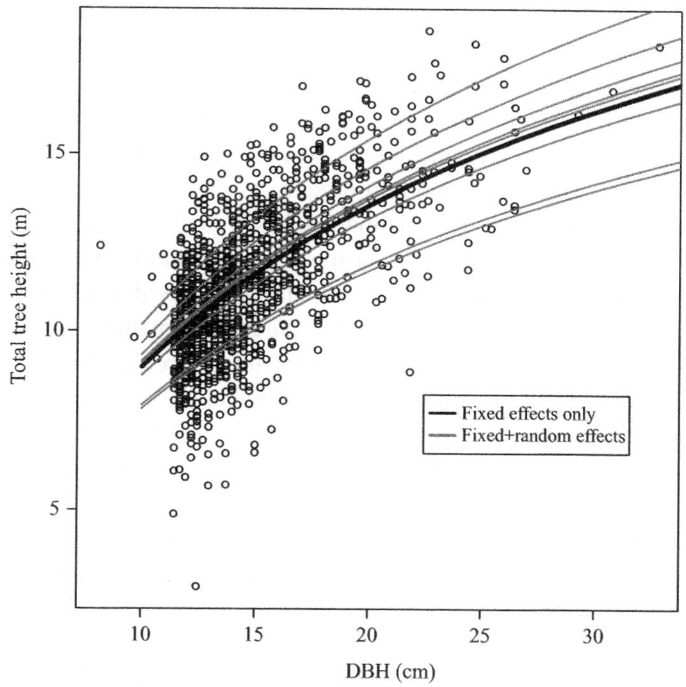

Figure 14.5 Balsam fir total height against diameter at breast height (DBH; cm) fit with and without random effects for different stands in the US Forest Service Penobscot Experiment Forest located in Maine. The solid black line is the overall population trend, while the gray lines depict each stand's deviation away from the population.

over time (e.g. Gregoire *et al.*, 1995). Hence, the assumption of independence is often violated and parameter standard errors cannot be properly estimated.

The mixed-model approach can be used with a variety of parameter estimation techniques including linear and nonlinear (Pinheiro and Bates, 2000), generalized linear (Bolker *et al.*, 2009), and additive regression (Woods, 2006). A mixed model consists of fixed and random effects, which are both estimated using maximum likelihood. The fixed components are population-average values similar to parameters obtained using OLS regression; while the random component is each individual's deviation from the fixed parameter, and these are expected to be independent, identically and normally distributed with a mean of zero. Random effects can be estimated for each hierarchical level in a dataset (multi-level mixed effects) and for multiple parameters in a model. The key decisions in fitting a mixed model are: (1) whether it is necessary; (2) what parameter or parameters should be random; and (3) how to apply the model to members not included in the original dataset.

Fitting mixed models can be computationally demanding and it can be difficult to obtain convergence, particularly with larger datasets. To determine if a mixed model is necessary, the model should be fitted with and without random effects (Pinheiro and Bates, 2000). A likelihood ratio test can be used to assess if the random effect significantly improves model fit. If a mixed model is necessary, selecting the best parameter or parameters that should be random is often difficult. This often involves fitting the model with different parameters selected as being random and using AIC or BIC to select the best model. Again, likelihood ratio tests are effective for comparing models with different numbers of random effects.

The effectiveness of applying a mixed model to individuals not in the original modeling dataset depends upon whether the model form is linear or nonlinear and whether or not a subsample of the response variable is available for the new individuals. If a subsample of the response variable is not available, then the random parameters of the mixed model cannot be estimated and must, therefore, be set to zero. The resulting predictions are unbiased if the model form is linear, but biased if it is nonlinear (Monleon, 2003; Temesgen *et al.*, 2008). Although the predictions may be more biased, the actual bias might be quite small and the mixed model might be more precise in certain situations, particularly when a more complex relationship like diameter increment is being modeled (Weiskittel *et al.*, 2007a).

If a subsample of the response variable for the new individuals is available, then random parameters of both the linear and nonlinear mixed models can be locally calibrated with relatively few observations using a BLUP estimator. The use of this technique for locally calibrating stem taper and tree HT to DBH equations is discussed in Yang *et al.* (2009b) and Temesgen *et al.* (2008), respectively. Although the number and location of samples needed for proper calibration varies, the gain in precision is generally significant.

When only one new observation is available, Meng and Huang (2010) found that mixed models that included a correlated error structure were more effective predictors than those that did not have one. However, the effectiveness of including a correlated error structure with mixed effects was not assessed when new observations are not available for calibration. For example, Weiskittel *et al.* (2007a) found that a mixed model with a correlated error structure and weighting function was a less biased and more precise predictor than a similar model fitted with no random effects, even when the fixed effects were just used for prediction on an independent dataset.

For the dynamic equations in a growth model, it is unlikely that a subsample of the dynamic response variables, such as future diameter increment, height increment, and mortality, will be available for locally calibrating the random parameters of a mixed model to the plot being projected. Therefore, the most likely usage of mixed models in a growth simulator is for predicting missing measurements of static variables, such as HT, HCB and crown width, when subsamples of the static variables are available on the plot to be projected.

14.4.5 Generalized algebraic difference approach

The generalized algebraic difference approach (GADA) is a method for deriving and fitting dynamic growth equations (Cieszewski and Bailey, 2000), which extends the original approach of Bailey and Clutter (1974). GADA has primarily been used to estimate base-age invariant dominant height-age curves (Cieszewski, 2001; Cieszewski and Strub, 2008), but has also been used to model stand basal area growth (Barrio Anta *et al.*, 2006; Castedo-Dorado *et al.*, 2007a) and tree taper (Strub *et al.*, 2005). GADA works by defining a model form that includes a site-specific localization term, solving the equation for the localization term, specifying the localization term as a function of initial conditions, simplifying the equation, and estimating parameters (Krumland and Eng, 2005). This ensures that redundant parameters are eliminated and the model form is robust (Cieszewski and Bailey, 2000). An array of GADA model forms exists, with some of the most common being the Chapman–Richards, logistic, and Schumacher model forms (see Krumland and Eng, 2005).

Multiple methods for estimating GADA model parameters exist. The most common include: (1) dummy variable; (2) iterative; (3) difference; and (4) mixed effects. The mixed-effects approach has been found to be generally inferior to the first three approaches (Wang *et al.*, 2008a; Weiskittel *et al.*, 2009a), so attention is given to the other methods. The dummy variable approach for GADA uses indicator variables (0 or 1) to characterize the localization parameters, but it can be difficult to obtain convergence (Krumland and Eng, 2005). The iterative approach is similar to the dummy variable approach, but first estimates the global parameters and then optimizes site-specific localization parameter estimates, holding the global parameters constant. Krumland and Eng (2005) found that this approach gave good results and significantly reduced model convergence times by up to 500%.

The difference approach avoids estimating the localization parameters by substituting the initial measurements and assuming that measurement error does not drastically bias parameter estimates. The difference approach can be formulated with many different data structures (see Wang *et al.*, 2007a).

The best method for GADA is still debated in the literature (Cieszewski and Strub, 2007; Wang *et al.*, 2008b). Cieszewski and Strub (2007) found that the dummy variable approach was superior, while Wang *et al.* (2007a) found the difference approach worked better when a non-overlapping forward difference structure was used.

14.4.6 System of equations

Growth models utilize a set of equations in order to adequately characterize system dynamics, and often the same data are used to fit the set of equations. Estimating each equation separately using OLS or GLS might not result in efficient parameter estimates and variances (Kmenta, 1997). Inefficiency can arise if there are cross-equation correlations in the residuals or if the response variable of one equation is used as a predictor variable in another equation. There are methods for dealing with these situations, but deciding which method is appropriate is complicated. The following is a brief overview of the situations a growth modeler might face and the possible techniques for addressing the situations.

The literature dealing with systems of equations refers to response variables as endogenous variables, while explanatory variables are considered exogenous variables (e.g. Kmenta, 1997). In addition, the response or endogenous variables are often referred to as the "left side" of the equations and the predictor or exogenous variables are often referred to as the "right side" of the equations. The following techniques are applicable if none of the endogenous variables in a set of equations are also used as exogenous variables.

1. If the residuals between equations are not correlated, then OLS or GLS is appropriate.

2. If the residuals between equations are correlated but the right sides of the equations are identical in form, then OLS or GLS least squares is again appropriate.

3. If the residuals between equations are correlated and the right sides of the equations are not identical in form, then seemingly unrelated regression (SUR) (e.g. Rose Jr. and Lynch, 2001) or maximum likelihood (e.g. Fang *et al.*, 2001; Cao and Strub, 2008) methods are appropriate.

The situation where endogenous variables are also used as exogenous variables is further complicated by differences in the form of the relationships between equations and by the possibility that the system is under, over, or exactly identified (Kmenta, 1997).

Deciding whether or not the system of equations is identifiable is beyond the scope of this book and the reader is referred to an econometrics book such as Kmenta (1997).

In general, systems of equations can be classified as being triangular or integrated. A triangular system occurs if the equations can be arranged in an order such that each endogenous variable is expressed as a function of the exogenous variables and only those endogenous variables which were subjects of previous equations. As a result, the first equation in a triangular system will contain only exogenous variables on the right-hand side; the second equation in the system can contain only exogenous variables and the endogenous variable from the previous equation on the right-hand side, etc. An integrated system of equations occurs if the set of equations cannot be rearranged to form a triangular system.

The following parameter and variance estimation techniques are applicable to triangular systems of equations.

1. If the residuals between equations are not correlated, then the system is called recursive, and OLS or GLS techniques can be applied to each equation to estimate the parameters and variances.

2. If the residuals between equations are correlated, then three-stage least squares (3SLS) regression (e.g. Hasenauer *et al.*, 1998), restricted 3SLS (Borders, 1989), or restricted maximum likelihood (e.g. Fang *et al.*, 2001; Cao and Strub, 2008) are appropriate.

Finally, techniques for estimating the parameters and variances of an integrated system of equations are as follows.

1. If the residuals between equations are not correlated, then two-stage least squares (2SLS) regression is appropriate.

2. If the residuals between equations are correlated, then three-stage least squares (3SLS) regression is appropriate.

These various techniques have been applied to linear systems of growth equations (Borders and Bailey, 1986; Borders, 1989; Hasenauer *et al.*, 1998), nonlinear systems of growth equations (e.g. Zhang *et al.*, 1997b; Huang and Titus, 1999b; Cao and Strub, 2008), and a combination of linear and nonlinear growth equations (Borders, 1989; Fang *et al.*, 2001). LeMay (1990) demonstrated how systems of equations could be corrected for the presence of autocorrelation and heteroscedasticity using GLS techniques. The primary limitation of systems of equations is that missing observations in some equations are not allowed. Thus, simultaneously estimating individual tree diameter and height increment equations when only a subsample of trees has been measured for height increment is restricted to using just the subsampled trees.

If sample size is not an issue, then the system-of-equations approach has been applied to a set of stand-level growth model equations (e.g. Borders and Bailey, 1986;

Borders, 1989; Fang *et al.*, 2001), a set of tree-level growth model equations (e.g. Hasenauer *et al.*, 1998; Cao *et al.*, 2002; Cao and Strub, 2008), and a combined set of stand- and tree-level equations (Borders, 1989; Zhang *et al.*, 1997b). While most of these growth model applications have assumed a system composed of fixed-parameter equations, Fang *et al.* (2001) did demonstrate how system-of-equations methods could be applied to a set of mixed-parameter equations.

An interesting application of SUR is for ensuring additivity of individual equations, which is particularly important in biomass equations (Parresol, 2001). Kant and Yang (2008) also found SUR to be useful for improving predictions in a matrix growth model by accounting for cross-equation correlation and constraining probability predictions to be less than or equal to one.

14.4.7 Bayesian

Although Bayesian techniques have long been available, they have seen relatively little application in growth and yield modeling (e.g. Green and Strawderman, 1996; Bullock and Boone, 2007; Clark *et al.*, 2007; Metcalf *et al.*, 2009a). This is likely because of the high computational requirements of the Bayesian approach (e.g. $>5\,000$ iterations). The Bayesian method differs from other techniques primarily in how the likelihood function is specified.

A Bayesian estimator is dependent on both prior and posterior distributions, with the latter being summarized by random draws from the data. The primary benefit of using a Bayesian approach lies in the ability to generate the full posterior distribution of estimated parameters; thus various statistics (mean, mode, median, etc.) can be easily calculated from generated samples. In addition, the Bayesian approach does not impose additional constraints or assumptions on posterior distributions as in the frequentist maximum likelihood method. Although Li *et al.* (2011) found that classical maximum likelihood and Bayesian approaches resulted in similar parameter estimates, the Bayesian technique offered several advantages.

Other important uses of Bayesian methods are in estimating multi-level/hierarchical models (Gelman and Hill, 2007; Dietze *et al.*, 2008), calibrating models to new observations (Van Oijen *et al.*, 2005), and combining predictions from different models (Radtke and Robinson, 2006). Furthermore, Metcalf *et al.* (2009a) found them useful for overcoming data sparseness.

14.4.8 Nonparametric

Much like Bayesian techniques, nonparametric approaches rely heavily on available data rather than an assumed distribution. Important nonparametric techniques used in growth and yield modeling are artificial neural networks (ANNs; Guan and Gertner, 1991a; Hasenauer and Merkl, 2001), support vectors (King *et al.*, 2000), classification and regression trees (CART; Nigh and Love, 2004; Fan *et al.*, 2006; Räty and Kangas, 2008), generalized additive models (GAMs; Zhang *et al.*, 2008; Aertsen *et al.*, 2010), and k-nearest neighbor methods (k-NN; Temesgen *et al.*, 2003; Sironen *et al.*, 2008).

Nonparametric techniques are generally considered distribution-free techniques, and sometimes referred to as data mining. These techniques are flexible and can often produce accurate predictive models and identify key covariates, but are often difficult to interpret and implement. In addition, nonparametric approaches may overfit data and, consequently, not extrapolate well. Although they are similar in principle, each nonparametric technique has distinctive features. For example, CART simply splits the data into a series of homogeneous groups, while a GAM uses splining techniques to represent the relationship between independent and dependent variables.

Although CART is one of the most frequently used nonparametric techniques in forest modeling (Dobbertin and Biging, 1998; Nigh and Love, 2004; Fan *et al.*, 2006; Räty and Kangas, 2008), the method may not be the best technique to use in all situations. For example, Aertsen *et al.* (2010) compared different nonparametric techniques for predicting site index of three species and found GAMs to outperform the other techniques. Ecologists recommend the use of boosted regression trees (De'ath, 2007), which Aertsen *et al.* (2010) recommend when ecological interpretability is the most important model attribute.

One nonparametric technique not widely used in forest modeling is the random forests method (Breiman, 2001). Recently, Crookston *et al.* (2010) used random forests extensively to develop a climate-sensitive version of FVS. The use of nonparametric techniques will likely continue to grow in the future due to their ease of use, ability to handle complex datasets, and strong predictive power. However, as with any statistical model, extensive testing and validation is suggested prior to the use of any technique.

14.4.9 Annualization

Permanent plot measurements are often conducted periodically, which creates problems when modeling multiple datasets with different remeasurement intervals. The use of annualization techniques is an effective method for addressing this problem. Weiskittel *et al.* (2007a) suggested annualization of increment equations because it provides a finer resolution of stand dynamics, is easier to modify to represent the often nonlinear response to forest silvicultural treatments, and better allows the use of the model hybridization techniques discussed in Chapter 13. Various techniques for parameter annualization are used (see McDill and Amateis, 1993).

Although simple linear interpolation techniques are widely used, they tend to result in biases, and instead a model-based interpolation method should be utilized (McDill and Amateis, 1993). Cao (2000) generalized the approach of McDill and Amateis (1993) by using an iterative summation method. Weiskittel *et al.* (2007a) extended the method of Cao (2000) by incorporating multi-level mixed effects. On an independent dataset, Weiskittel *et al.* (2007a) found that the fixed effects from annualized equations fitted with mixed effects performed better than annualized equations that did not include random effects and similar equations that used a five-year time step.

These methods primarily deal with predicting tree increment because predicting annual tree survival from periodic measurements is more complicated. One of the most common means for predicting annual survival is to assume that the probability is constant for the measured interval (Monserud, 1976). This is often achieved by discounting mortality rates to achieve annual rates (Hamilton, 1986) or by including measurement interval in the logistic model (Monserud, 1976). Flewelling and Monserud (2002) compared different methods for estimating annual survival from variable-length measurement data, and found that the parameter estimates obtained from each method were relatively similar, but actual mortality rates in their fitting dataset were low.

Cao *et al.* (2002) developed a system of equations to predict both annual changes in individual-tree diameter, height, and crown ratio, as well as the probability of tree survival. Nord-Larsen (2006) demonstrated the usefulness of this technique for modeling annual increment and mortality for data with a highly irregular measurement interval. Although this iterative technique works well, there are still some questions on how the explanatory variables should be treated during iterations. Cao and Strub (2008) compared four different approaches and found that using an annual stand-level model to predict the stand-level attributes used as explanatory variables in the annual tree equations worked best. Crecente-Campo *et al.* (2010) used weighted regression to deal with missing observations in developing an annualized system for individual tree growth and mortality.

14.5 Summary

Model development is a very involved and lengthy process. Ideally, model development is an iterative process. Even developing relatively simple models requires making decisions on the type and quality of data needed, the degree of allowable error, the assumptions implicit in different model forms, and the proper statistical techniques for estimating parameters. Permanent plots with periodic individual-tree-level measurements across a wide range of site conditions offer the most flexibility for model development, but are the most expensive type of plot to maintain.

In any model, there are multiple sources of error that are important to deal with. Although multiple model forms exist, a suitable model form is one that is flexible, fits the data well, and properly extrapolates. Model fitting techniques are continually changing, and only a brief overview is presented here. The use of mixed models has seen increased usage where the method is appropriate, but Bayesian and nonparametric techniques will likely see greater use in the future. However, it is important to note that no model fitting technique can overcome poorly collected, erroneous, or nonrepresentative data. Thus, emphasis in model development should always be on the quality of data available.

15
Model evaluation and calibration

15.1 Introduction

Model evaluation is an important part of model construction, and some examination of a model is required at all stages of model design, fitting, and implementation (e.g. Hann, 1980). This illustrates the iterative and ongoing nature of model development, since evaluation is not merely an afterthought or an acceptance trial. Too often, a model is simply taken at face value and projections are discussed with no representation of their validity or degree of certainty. A much safer approach is to assume that all models are wrong and only a few might actually be useful (Box and Draper, 1987).

When developing a growth and yield model that incorporates multiple dynamic and static equations, there will always be numerous alternative equation forms and parameterizations for a given relationship that will exhibit similar fit statistics. Growth and yield models are always developed with modeling datasets that are small fractions of the population. Therefore, an equation that characterized the modeling data well might not provide biologically reasonable predictions when applied beyond the range of the modeling dataset. This is true even if the set of equations in the model are jointly optimized using seemingly unrelated or system-of-equations procedures (see Chapter 14). Very different system behaviors can occur depending upon which combination of the various alternative equation forms and parameterizations is chosen for the model. Therefore, the creation of the best-behaving model will require the examination and testing of numerous combinations of the alternative equation forms and parameterizations. For example, the development of ORGANON (Hann, 2011) required the examination of literally hundreds of combinations and months of evaluation and testing before the final set of equations was decided.

One difficulty in any model evaluation is defining the criteria for determining usefulness. Other difficulties are identifying model components to evaluate, since errors

Forest Growth and Yield Modeling, First Edition. Aaron R. Weiskittel, David W. Hann, John A. Kershaw, Jr. and Jerome K. Vanclay.

from one equation may partially compensate for errors from another (e.g. Mette *et al.*, 2009), and often independent data of a similar nature to the fitting data are needed to properly evaluate any model. A thorough evaluation of a model involves several steps, including two which are often called *verification* and *validation*. In forest growth modeling, these usually denote *qualitative* and *quantitative* tests of the model, respectively. Rather than use the terms of verification and validation, *model criticism* and *benchmarking* is preferred because they are more value-neutral and are not confused with similar terms in mathematics and logic.

Model criticism involves examination of the structure and properties of a model, with or without supplementary data, to confirm that there are no internal inconsistencies and that the model is biologically realistic. Benchmarking requires comparisons with independent data to quantify model performance. Model evaluation criteria are discussed in several places, including Vanclay and Skovsgaard (1997), Stage (2003), Schmidt *et al.* (2006), and Pretzsch (2009).

The focus of this chapter is to provide an overview of some fundamental criteria for model users and to connect the concepts of model evaluation and *calibration*, which refers to the search for adjustments to improve model predictions for a specific locality. It relates to model evaluation because, if benchmark tests reveal deficiencies in a model, then one must determine if and how the same data may be used to re-calibrate the model so that predictions are improved for that locality. Although the techniques discussed in this chapter are applicable to the individual equations within a growth and yield model, the primary focus is on the evaluation and calibration of growth and yield models as an ensemble of equations.

15.2 Model criticism

Model criticism is the first step in any model evaluation process, particularly if the model is going to be used to make any management or policy recommendations. This process requires users to avoid the "black box" syndrome, which occurs when the steps needed to run a model successfully are known but the inner workings of a model are not understood or appreciated. To criticize a model, a user must understand how a model was constructed, the quality of data used, and if predictions are biologically reasonable. Monserud (2003) demonstrates the use of model criticism techniques to evaluate growth and yield models for sustainable forest management.

There are many criteria by which to evaluate a model, but they are often classified into several broad classes. Schmidt *et al.* (2006) modified the model criticism categories of Vanclay and Skovsgaard (1997) into: (1) model form and parameterization; (2) variable selection and model simplicity; (3) biological realism; (4) compatibility; and (5) reliability (Table 15.1). An additional criterion is model adaptability, as suggested by Robinson and Monserud (2003). Each of these criteria is discussed below.

Table 15.1 Model evaluation criteria and key metrics.

Criteria	Key metrics			
Model form and parameterization	Are fit statistics adequate?	Does model form allow proper extrapolation?	Is model form flexible?	If mixed effects are used, do the random effects explain more of the variation than the fixed effects?
Variable selection and model simplicity	Is the model a simple product of a stepwise regression technique?	Are the selected variables logical?	Are any variables redundant?	Is the model overly simple or complex?
Biological realism	Are the signs of parameter estimates logical?	Are the predictions consistent with existing theories of forest growth?	Do growth equations show both acceleration and deceleration when plotted over size or time?	Does the model behave properly when covariates have a value of zero?
Compatibility	Do predictions of growth and yield match?	Is the model time-step invariant?	Are static equations used to represent dynamic processes?	
Reliability	Is the model overly sensitive to a few parameters or inputs?	Was the model fitted to data for which it is intended to represent?		
Adaptability	Is the model portable?	Is the model extendable?	Is the model source code available?	Is the model properly and fully documented?

Criteria adapted from Schmidt *et al.* (2006) and Vanclay and Skovsgaard (1997).

15.2.1 Model form and parameterization

As discussed in Chapter 14, there are many combinations of model forms and statistical techniques available for model parameterization. Each has their own advantages and disadvantages, but the most important, from a model evaluation viewpoint are: (1) model fit; (2) ability to extrapolate; and (3) appropriateness for the data being fitted.

Model fit is often judged by the coefficient of determination (R^2, the portion of original sum of squares in the data explained by the model), the adjusted coefficient of determination (R_a^2, variation explained by the model adjusted for the number of model parameters), the mean residual (average difference between predicted and observed values), and root mean square error (RMSE, the square root of the average squared difference between predicted and observed values). Although R^2 and R_a^2 are helpful and easily understood metrics of model fit, they have two limitations as a criterion of model evaluation: (1) models with the same R^2 or R_a^2 can have vastly different fits; and (2) some data have unexplainable random error. Furthermore, R^2, but not R_a^2 also has the limitation of always increasing as parameters are added to the model. Mean residual and RMSE, on the other hand, are measures of accuracy and precision, respectively, and are preferred metrics for model evaluation. It is often helpful to express the mean residual and RMSE as a percentage of the mean. A model with a high R^2 or R_a^2 and low mean residual and RMSE is generally preferred.

Nonlinear model forms often assure, but do not guarantee, proper extrapolation, whereas linear or polynomial models can have a limited ability to extrapolate and should be treated with caution. Some nonlinear model forms are more flexible than others because they can adopt a range of shapes depending on the nature of the fitting data. Very flexible model forms include the Weibull, modified logistic, Chapman–Richards, exponential, and Schnute (1981) (Huang *et al.*, 1992). Parameterization should follow the type of data being modeled, and maximum likelihood methods are preferred. However, caution is warranted against naïve application of mixed-effects methods, because they can result in overly simplistic model forms and biased predictions (Kershaw *et al.*, 2009).

15.2.2 Variable selection and model simplicity

Models need to be developed using a logical method of construction. Although the use of stepwise regression techniques may provide the best model for any given dataset, their usage is limited to linear model forms and does not guarantee the best fit to the data. Instead, a model should include the relevant variables determined from previous research or logic. For example, either tree age or tree size is widely used in tree-level growth and yield models because they are well correlated with growth. However, age is not easy to interpret because it is often confounded with several other factors. The effectiveness of tree age is particularly limited under periods of long suppression that are common in many naturally regenerated forests or when multiple thinnings, or other silvicultural treatments, are implemented. Consequently, models based on tree size are generally preferred.

In addition to size or age, many tree-level prediction equations, particularly diameter increment, need to include measures of both one- and two-sided competition, some measure of site productivity, and perhaps a measure of vigor. Once these factors are accounted for, additional variables should be added with caution since all models need to balance simplicity with complexity (Kimmins *et al.*, 2008).

There has never been a growth model developed for which the modeling data was not a very small proportion of a very complex population. While complex model forms may improve fit statistics, they may also result in overfitting the peculiarities of the particular modeling dataset and, therefore, not extrapolating to the broader population. For example, Liang's (2010) use of the approach of Stage and Salas (2007) to represent site productivity may be problematic because it requires 11 parameters and explains a limited amount of variation in comparison to the other factors in the equation, such as tree size and competition. Consequently, a model should be as simple as possible, but also as complex as necessary (Kimmins *et al.*, 2008).

15.2.3 Biological realism

Although growth models are an abstraction of reality, they should still provide predictions that are biologically realistic. Qualitative assessment of biological realism is often difficult. The easiest method is to ensure that the signs of parameter estimates are logical and consistent with biological understanding. For example, a parameter estimate for site index in a diameter increment equation should be positive, while competition should be negative.

Graphical techniques are also helpful in assessing biological realism. Leary (1997) used a modified Bakuzis matrix of stand properties to assess model biological realism (Figure 15.1). This matrix uses eight rules or "law-like" relationships such as Sukachev (stands on good sites self-thin faster), Reinke (stands self-thin in a linear way when number of trees and DBH are log transformed), and yield–density (yield is a function of number of trees) to assess model predictions. Leary (1997) stated that every model he has tested violated some aspect of the modified Bakuzis matrix. These rules and others can be useful in assessing model predictions, but often have multiple limitations (Table 15.2).

Sievänen *et al.* (2000) proposed the idea of volume growth and survival graphs, which graph individual tree growth and risk of mortality relative to the largest value in a stand. Mäkelä *et al.* (2000b) found volume growth and survival graphs useful for comparing predictions from four mechanistic models. Another useful graphical technique to assess biological realism of individual-tree growth increment equations is to plot their predictions for an open-grown tree (i.e. no one- or two-sided competition) over tree size and age (when appropriate). The graphs should show a steep increase (juvenile growth), a peak, and then a decline that approaches zero.

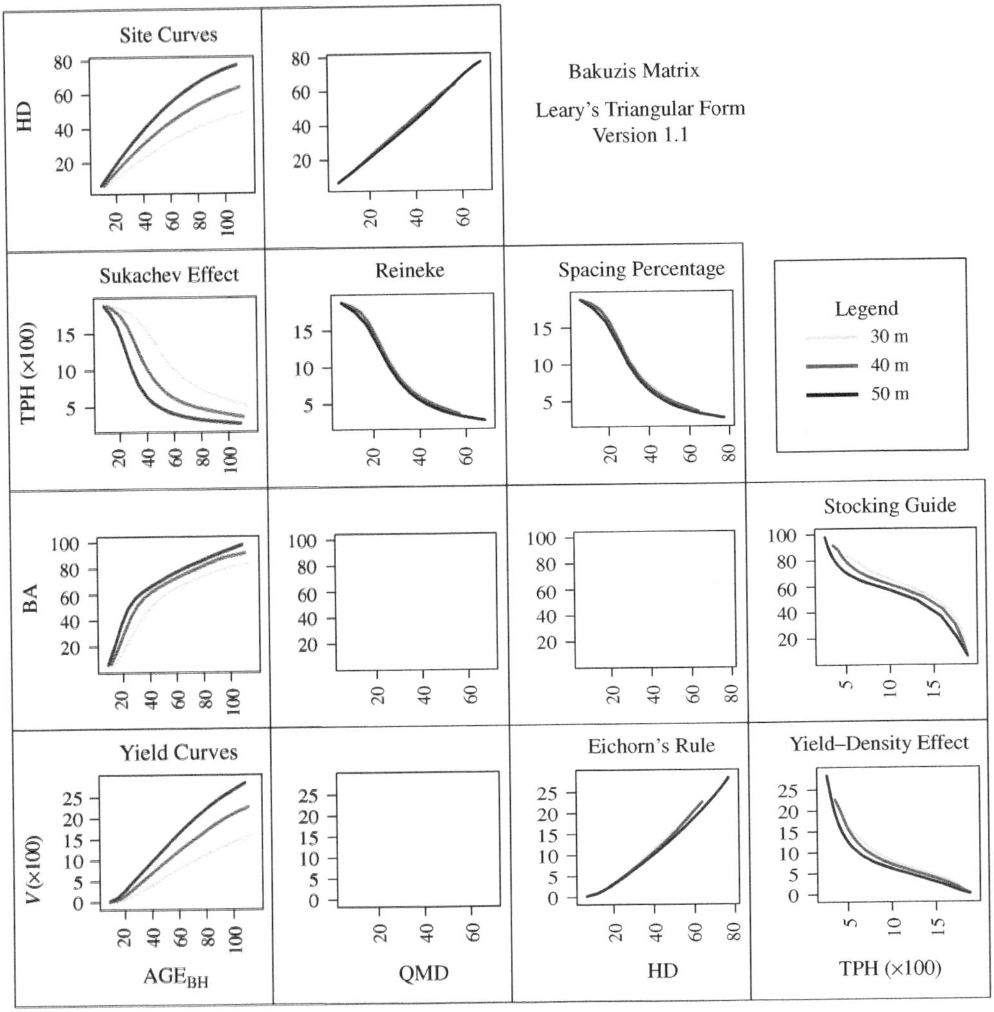

Figure 15.1 Bakuzis matrix for the Stand Management Cooperative (SMC) variant of ORGANON used in verifying model predictions, and constructed using a R package developed by Greg Johnson. This package is available online at: http://www.growthmodel.org/. AGE_{BH} is breast-height age; V is total stand volume ($m^3\ ha^{-1}$). Adapted from Leary (1997).

Finally, Vospernik *et al.* (2010) indicated that assessing predicted development of stand mean height-to-diameter ratios was useful in benchmarking four growth models in Austria, given the importance of this ratio for making management decisions. In general, predicted height-to-diameter ratios should not: (1) exceed that of very dense stands; (2) fall below that of open-grown trees; (3) increase with greater spacings; and (4) be higher in dominant trees when compared to suppressed individuals (Vospernik *et al.*, 2010).

Table 15.2 Important growth and yield principles useful for assessing model predictions.

Principle/theory	Expectation	Limitation	Reference
Yield–age effect	The yield of top height, basal area per unit area, and cubic volume per unit area are sigmoidal in shape over stand age	Applicable only to even-aged stands	—
Eichorn's rule	A given mean height of a stand is matched by the same volume in all site classes	Mean height is not as sensitive to differences in density; differences in past density influence current average crown length; size of crown influences amount of stand volume	Eichorn (1904)
Langsaeter's hypothesis	A stand's cubic volume increment is constant and optimum for a wide range of cubic volume stocking levels on a given site	Not supported in several studies, and the curve may not be perfectly flat over the maximum range	Langsaeter (1941)
Reineke's rule and the $-\frac{3}{2}$ power rule	Stands at their maximum relative density self-thin in a manner that maintains a linear relationship between the logarithm of number of trees and the logarithm of size	Applicable primarily to even-aged stands	Reineke (1933); Yoda *et al.* (1963)
Sukachev effect	Stands on good sites self-thin faster than stands on poor sites	—	Yoda *et al.* (1963)
Size–density trajectory	The trajectory that a stand of a given initial density follows as it approaches the maximum size–density line is independent of site quality, and the shape of the trajectory is independent of initial density	Applicable primarily to even-aged stands	Smith and Hann (1984)

When multiple models exist in the same region, an effective test of biological realism is to compare predictions from the alternative growth models using the same starting conditions. Johnson (2005) illustrated this approach for growth models in the US Pacific Northwest, and found a wide range of differences between the models tested, particularly under various forest silvicultural systems (Figure 15.2). Similarly, Papaik *et al.* (2010) compared predictions from a stand-level model to those from a landscape-level model and identified weaknesses in both modeling approaches.

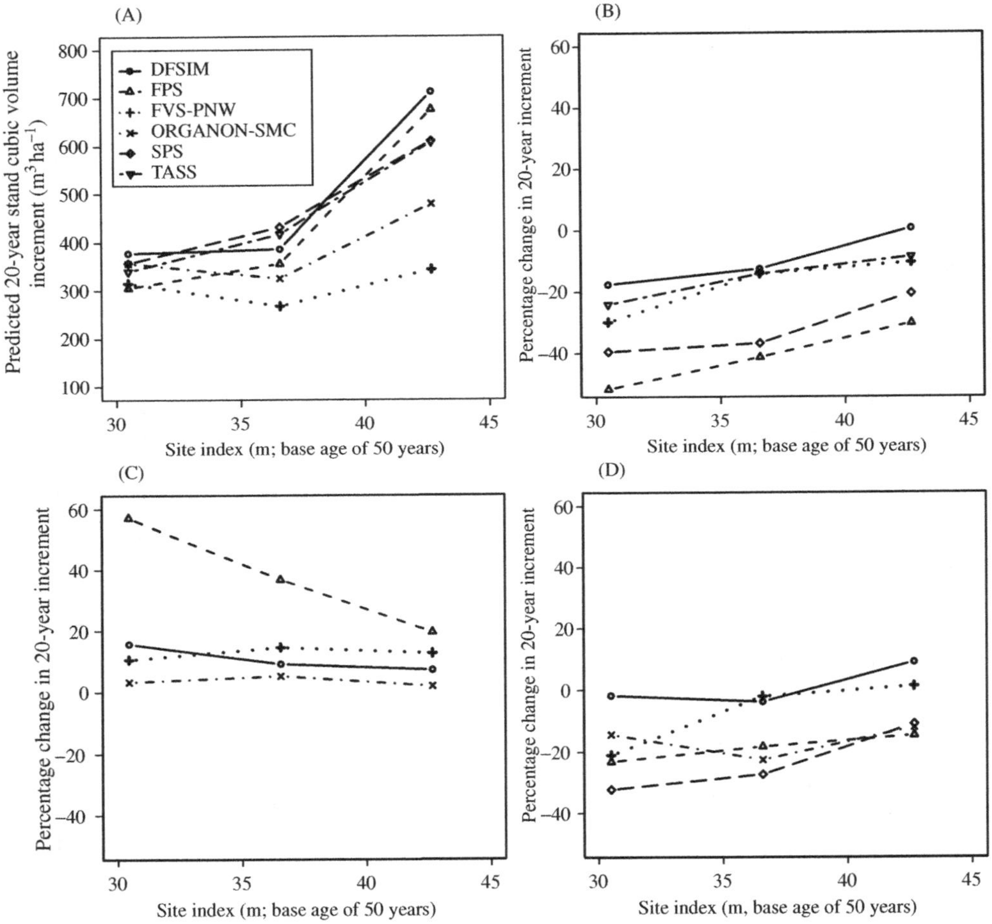

Figure 15.2 Simulated 20-year stand cubic volume increment ($m^3\,ha^{-1}$) and percentage change from the baseline after treatment, for several commonly used growth models in the US Pacific Northwest, across site index and different forest silvicultural systems. The regimes included: no management (A), thinned to 446 stems ha^{-1} (B), fertilized with 224 kg ha^{-1} of nitrogen (C), and thinned to 446 stems ha^{-1} and fertilized with 224 kg ha^{-1} of nitrogen (D). Adapted from Johnson (2005).

15.2.4 Compatibility

Compatibility has multiple meanings in model evaluation (Schmidt *et al.*, 2006). One is that the summation of growth predictions gives the same result as a direct prediction of yield. For example, Buckman (1962) and Clutter (1963) developed stand-level compatible growth and yield equations by using a sigmoidal equation to predict yield and its differential form for predicting growth rate. Compatibility also refers to obtaining the same predictions from models with different temporal and spatial resolutions. For example, Daniels and Burkhart (1988) presented a model that provided compatible results at the individual-tree, size-class, and stand-levels.

As elaborated on in Chapter 10, several other techniques have been used to link models of differing resolutions and to ensure compatible results at each level (Ritchie and Hann, 1997b; Cao, 2006; Yue *et al.*, 2008). Sullivan and Clutter (1972) developed a system of growth equations that were temporally compatible, or step invariant, so that the result from multiple annual projections was the same as the result from a single multi-year long projection. Alternatively, Ochi and Cao (2003) suggested that using annual growth models provides more flexibility and better performance than ones constrained to be compatible.

Deterministic and stochastic predictions should be compatible. Several studies on tree mortality show that this is often the case (Weber *et al.*, 1986; Rathbun *et al.*, 2010). Vanclay (1991b) presented an approach whereby deterministic and stochastic predictions from an individual-tree growth model were compatible, by expressing change as a probabilistic function. Finally, Schmidt *et al.* (2006) suggested that another aspect of compatibility is the dependability of reasonable transformations of model predictions. For example, sometimes height increment is estimated from a static height to diameter equation (e.g. Kershaw *et al.*, 2008), which can result in unrealistic predictions (e.g. Lindner *et al.*, 1997).

15.2.5 Reliability

Reliability is the ability of a model to perform for its stated conditions and provide consistent predictions. Often, independent data and model benchmarking are needed to test reliability. However, other techniques exist when independent data are unavailable. One of the most effective is a sensitivity analysis. The usefulness of this technique has been demonstrated for statistical (e.g. Herring, 2007), hybrid (e.g. Esprey *et al.*, 2004), and mechanistic growth and yield models (e.g. Tatarinov and Cienciala, 2006).

A sensitivity analysis attempts to reveal model parameters and sub-models that cause the greatest fluctuations in model predictions when perturbed. Often this is accomplished by changing each parameter and model input variable by a certain percentage (e.g. 25%) and assessing the degree of change in model output. Consequently, this can be a rather tedious task for complicated models that require multiple simulations. However, the approach is helpful for identifying the most influential parameters or inputs, and establishing model reliability. For example, a 1% change in site index should not result in a 50% increase in stand volume increment.

Although sensitivity analyses are generally done "one factor at a time," this method is highly inadequate, and other methods like regressing output variables against input variables are generally more effective (Saltelli and Annoni, 2010). In addition, Makler-Pick *et al.* (2011) recently suggested the use of generalized boosted models (see Chapter 14) as an effective technique to quantitatively assess model sensitivity.

More sophisticated approaches can be used to test model reliability when equation parameter errors and RMSEs are available. For example, Fortin *et al.* (2009) and Guan *et al.* (1997) used Monte Carlo techniques to assess model reliability in a statistical and mechanistic model, respectively. In contrast, Gertner (1987) used a first-order Taylor expansion to assess model predictions, and found it much more efficient than Monte Carlo techniques. Bayesian methods have also been used to assess model reliability (Green *et al.*, 2000; Radtke *et al.*, 2002).

15.2.6 Adaptability

From a model users' perspective, adaptability is one of the most important model evaluation criteria, since it deals with model infrastructure and interface. Robinson and Monserud (2003) identified the four most important aspects of adaptability to be: (1) portability, (2) extendibility, (3) source code availability, and (4) adequate documentation.

Portability refers to the ease of distributing a model to a variety of users and their platforms, while extendibility is the flexibility of a model to be adapted to novel applications. Both portability and extendibility can require that the model source code be available and adequately documented. Models are commonly made available as executable files and, less commonly, as dynamic link libraries (DLLs). Of the two choices, DLLs are preferred because they allow users to customize a model interface that meets their needs and they can also be easily incorporated into other programs such as R or Microsoft Excel. Executable files and, to a lesser extent, DLLs hinder users from easily extending a model to new applications.

The availability of model source code allows users to change parameter estimates and incorporate additional modifier functions. In addition, model source code availability allows users to identify potential bugs in the code and ensure that a model is being implemented as described in the documentation. However, making the source code available also runs the risk that the code will be modified in a manner that destroys the integrity of the model, a problem avoided by DLLs. Furthermore, most model users will be unable to actually understand the source code in order to modify and recompile it, which again suggests DLLs to be the best approach for making a model available to users. Robinson and Monserud (2003) recommend that a well-documented model is one that includes information on the major sub-models, equation forms used, parameter estimates, standard errors, sample sizes, and goodness-of-fit statistics.

In addition to these criteria, Robinson and Monserud (2003) indicate that cost, model interface, ability to accommodate a variety of sampling schemes, and outputs are other criteria that can be used to assess model adaptability. When all of these criteria were used to assess seven growth and yield models in the US Pacific Northwest, Robinson and Monserud (2003) found that none of the models were fully adaptable.

15.3 Model benchmarking

When independent data are available, model benchmarking is a good option because it gives a quantitative assessment of model performance. The difficulty in model benchmarking is in deciding what level of model error is acceptable, which likely depends on the intended applications of a model. Often, a wide array of statistical tests are used to test model performance (Yang *et al.*, 2004). Another important use of model benchmarking is characterizing model error. Both are discussed below.

15.3.1 Statistical tests

Key statistical tests used in model benchmarking include both parametric and nonparametric techniques. What tests are appropriate depends upon whether a deterministic or stochastic model is being benchmarked. For deterministic models, some of the most commonly used parametric tests in the forestry literature are the paired t-test, chi-square test, and simultaneous F-test, while Kolmogorov–Smirnov test, sign test, and Wilcoxon signed-rank test are the most common nonparametric tests (Yang *et al.*, 2004). Many more tests can be found in the general literature.

The paired t-test is an inappropriate test for model benchmarking because it uses one component of accuracy (precision) to test another – bias (Freese, 1960). Therefore, Freese (1960) developed, and Reynolds (1984) improved, several chi-square tests of overall accuracy. Yang *et al.* (2004) found that the chi-square test had a tendency to reject models too often, required data to be grouped into classes or bins, and often required a subjective level of accuracy. While not a test *per se*, the last point of criticism can be addressed by using the critical error bounds of Gregoire and Reynolds (1988).

Linear regression between observed and predicted values has been widely used to statistically test model performance. Multiple approaches have been employed to determine statistical significance using this technique, including: (1) separate t-tests that the intercept and slope of the linear regression are 0 and 1, respectively; (2) a simultaneous F-test that the intercept and slope of the linear regression are 0 and 1, respectively; and (3) a novel test that evaluates whether that intercept and slope are 0 for a regression between prediction errors and the sum of observed and predicted values (Yang *et al.*, 2004).

However, all of these tests suffer from the same problem that Freese (1960) attributed to the paired t-test; namely using one component of accuracy to test another (Hann, 1980).

The nonparametric Kolmogorov–Smirnov (K–S) test can be used to evaluate whether the prediction errors of a deterministic model are normality distributed with a mean of zero, and the sign test can be used to assess whether the median of the prediction errors is zero. Rather than rely simply on the number of positive and negative prediction errors as is done in the sign test, the Wilcoxon signed-rank test also uses the magnitude of the difference to determine if the predicted and actual values are significantly different. When these statistical tests and several others were applied to nine datasets, Yang *et al.* (2004) found that their usefulness was limited, and cautioned against having them be the sole determinant of model performance.

Reynolds *et al.* (1981) developed six parametric and three nonparametric tests for stochastic models. The idea behind these tests was basically to examine how each observation's response variable in the benchmark dataset fit in with the distribution of the predicted response variables prescribed by the model for that observation. Therefore, multiple random predictions of the response variable are made for each observation in the benchmark dataset, and these are then used to calculate the test statistics. The six parametric tests assume normality and calculate a t-statistic for each observation in the benchmark dataset, and the t-statistics are then transformed and combined to form the tests. The three nonparametric tests calculate the rank of the observed response variable in the set composed of the observed and the multiple randomly predicted response variables, and the ranks are then transformed and combined to form the tests.

One alternative to these traditional statistical tests is the equivalence tests of Robinson and Froese (2004). Rather than rely on a null hypothesis that a model is acceptable, as generally assumed by traditional parametric approaches, the equivalence tests uses a null hypothesis that a model is unacceptable. Equivalence tests have been used to compare means (Robinson and Froese, 2004) or the similarity between individual predictions and observations (Robinson *et al.*, 2005). Since the regression approach to equivalence tests assesses both of these, it is a preferred means for model evaluation (Robinson *et al.*, 2005).

To accomplish this, Robinson *et al.* (2005) suggest that: (1) the mean prediction be subtracted from all predicted values; (2) regions of equivalence be established for the shifted intercept and slope (e.g. $\pm$ 10%); (3) a linear regression be fitted between the observations and adjusted predictions; (4) the intercept for equality be tested by calculating the two one-sided confidence intervals for the intercept and comparing these to the estimated region of equivalence; (5) the slope for equality be tested by calculating the two one-sided confidence intervals for the slope and comparing to the estimated region of equivalence; and (6) the hypothesis of dissimilarity be accepted or rejected based upon these test results. Since selecting the region of equivalence is subjective, Froese and Robinson (2007) also suggest calculating the minimum percentage that would result in rejection of the null hypothesis. When a statistical statement on model performance is needed, the equivalence test is a preferred metric. A R package, 'equivalence', is now available to perform this test. A R package, 'equivalence', is now available to perform this test.

15.3.2 Model error characterization

As noted above, there are several problems with using traditional statistical techniques to determine model performance. A more effective approach may be to use model benchmarking for characterizing the magnitudes and distributions of a model's prediction errors. This is commonly achieved by plotting the prediction errors for all combinations of tree and stand variables and visually assessing them for systematic patterns. These patterns are sometimes difficult to detect, so the addition of a lowess regression line to a plot of the prediction errors is often quite helpful. When representing multiple conditions, matrix residual plots are informative (e.g. Soares *et al.*, 1995). Finally, mapping a model's prediction error is quite helpful in detecting key spatial patterns across a region (e.g. Froese and Robinson, 2007). Time needs to be examined in three contexts: (1) stand age or time since disturbance (e.g. cutting; if known); (2) length of projection; and (3) year of measure. These three aspects of time may reveal different model properties.

When a dataset is limited, re-sampling procedures like bootstrapping and jackknifing may help to better characterize the model's prediction error structure. However, these techniques have their own limitations, and often a large continuous forest inventory with a similar design as used to construct the model is required to fully benchmark a model (Soares *et al.*, 1995).

The error structure and the contribution of each model component to total error may be more revealing than a mere evaluation of total model performance. Thus an error budget of variance components of a model may help identify weaknesses and define priorities for future research. If suitable data are available, error budgets may be compiled by successive simulations using predicted values for only one component and observed data for all other model components. For example, Hann (1980) used several simulation cycles to identify faulty components in his model. His first cycle simulated only increments, and used actual mortality, harvesting, and recruitment, and thus resolved that increment prediction was satisfactory. The second cycle predicted increments and mortality, and employed actual harvesting and recruitment, and so forth, until the final test included all predicted values.

15.4 Model calibration

A decision must be made on whether or not to re-calibrate a model when benchmarking suggests that the model has deficiencies. Several considerations need to be taken into account when making this decision. First, was the benchmarking data of sufficient quality to justify re-calibration? Predictions often differ from reality for multiple reasons, including differences in plot size and design (Hann and Zumrawi, 1991), as well as the presence of measurement error (Gertner and Dzialowy, 1984). Projection length is another important factor influencing model accuracy (Soares *et al.*, 1995). Second, is the data used for benchmarking of a sufficient size, representative of the full population range, and at an appropriate temporal scale? Growth and yield models depict regional averages and do not accurately represent every stand. Consequently, the use of a small dataset with a limited temporal scale or range of conditions makes it difficult to fully benchmark a model and recommend model re-calibration.

Re-calibration can involve an array of approaches, including (1) estimating new parameters for some of the equations; (2) estimating new parameters for all of the equations; and (3) using a simple scaling factor to adjust predictions. The limitation of calibrating only one or some of the equations in a model is that it may alter the behavior of the other equations. Re-calibration can also result in no improvement in the predictions, implying that the model needs to be re-engineered (e.g. Pokharel and Froese, 2008). Consequently, benchmarking is required after calibration. Given that options (1) and (2) are very time and data-demanding, option (3) is generally the preferred solution.

As discussed in Chapter 7, there are multiple methods to estimate a scaling factor. One of the most common is to fit a linear regression through the origin between predicted and observed values (Temesgen *et al.*, 2008). Sophisticated and computationally intensive Bayesian algorithms have also been used to calibrate several mechanistic models (Van Oijen *et al.*, 2005; Svensson *et al.*, 2008). Several statistical growth and yield models have a self-calibration feature (Robinson and Monserud, 2003). For example, the Forest Vegetation Simulator locally calibrates height to diameter and squared diameter increment equations when a certain number of observations are available (Crookston and Dixon, 2005), while ORGANON calibrates height to diameter, height to crown base, and the diameter increment equations when a sufficient number of observations are available (Hann, 2011). Finally, the widespread use of mixed effects makes local calibration easier because the random-effects parameters can be predicted from just a few new observations using a best linear unbiased predictor (BLUP; e.g. Temesgen *et al.*, 2008).

15.5 Summary

Model evaluation is an ongoing process that involves multiple steps (Soares *et al.*, 1995). Models are evaluated in multiple ways, including assessment of data quality and fitting statistics, comparison to accepted forest growth rules, sensitivity analysis, and characterization of errors. In general, there is no single best approach to evaluate a model, and each method provides its own insights into model behavior. Although statistical tests are widely used to evaluate model predictions, the blind use of this approach is strongly cautioned against, as illustrated by the results of Yang *et al.* (2004). However, the equivalence test does shift the burden of proof to the model (Robinson *et al.*, 2005). Model evaluation approaches also depend on the intended use of the model output. For example, Larocque *et al.* (2011) presented an analytical framework for evaluating model output to assist policy decision makers.

When certain model deficiencies are noted, model re-calibration techniques need to be considered and used. The one caveat with any calibration is that it can create more problems than those it was intended to solve. Consequently, all model components need to be carefully assessed after calibration and the option of simply developing a new model rather than fixing the old one should always be considered.

16
Implementation and use

16.1 Introduction

Growth models are tools to be used, whether it is for practical, educational, or for research purposes. As outlined in Chapter 1, growth models have a variety of applications, with some of the most important being comparison of alternative forest management options, updating forest inventories to the present, and forecasting future forest structure. However, for a model to be implemented and used, model developers and users need to deal with several issues. Some of the most important are: (1) collection or generation of appropriate data; (2) temporal scale; (3) spatial scale; (4) interface; (5) visualization; and (6) output, which are each discussed in this chapter. Prior to the use of any growth model, proper model evaluation and re-calibration are strongly encouraged (Chapter 15).

In general, for a model to be useful, it needs to provide information that is:

- *timely:* available when needed, and not outdated when made available;
- *accurate:* a reasonable estimate of reality;
- *complete:* include all the details the user needs to know about the situation;
- *concise:* exclude elements not required by the user;
- *relevant:* directly related to the issues under consideration; and
- *appropriate:* in presentation for the particular audience.

Forest Growth and Yield Modeling, First Edition. Aaron R. Weiskittel, David W. Hann, John A. Kershaw, Jr. and Jerome K. Vanclay.

All of these require that the model is implemented in a flexible manner and has the capability to be linked to other resource databases. This is often a difficult task that requires significant foresight and a well-informed model user community.

16.2 Collection of appropriate data

All models need data to initialize a simulation. Data requirements depend both on the type and on spatial scale of the model being considered. For example, individual-tree models often require both stand-level information on site productivity and a tree list that gives species, diameter at breast height (DBH), expansion factor, total height, and height to crown base (when available) for individual trees. Stand-level models, on the other hand, generally just require a measure of site productivity, two-sided competition and stand age. In both cases, the data need to be error free, representative of the population, and appropriate for the intended application.

Several studies have assessed influences of input errors on model predictions, and generally find that they often are sensitive to key variables. For example, Gertner and Dzialowy (1984) found an individual-tree growth model to be highly sensitive to site index measurements, but not crown ratio. Mowrer and Frayer (1986) indicated that errors in the input data have a larger influence on overall prediction variability than the uncertainty associated with the underlying growth model equations. In other words, a model cannot overcome errors in the input data, and in the words of the computer scientist, George Fuechsel, "garbage in, garbage out."

Both model developers and users can ensure that the most appropriate data are used in simulations. Model developers can develop algorithms that perform error-checking prior to simulations and provide warnings when possible errors are detected. For example, the ORGANON growth and yield model uses this approach to identify fatal errors such as improper species codes, blank records, and unlikely tree heights, and these must be corrected before the model will operate (Hann, 2011). Also, model developers should provide a clear description of how plots within a stand are treated, and allow for different approaches. In the Forest Vegetation Simulator (FVS), the stand is the basic projection unit and the default is to process multiple plots within a stand as one large plot (although some equations use estimates of point rather than stand density) (Crookston and Dixon, 2005).

An alternative approach is to process each plot individually and compile the summary statistics after projections. This approach can be advantageous in stands with diverse internal structures because it maintains the attributes of each individual plot, which allows for the calculation of confidence intervals and reduces bias in estimating stand-level variances (García, 2006). However, use of data from very small plots may result in unrealistic projections due to biases in calculating structural parameters such as basal area of larger trees or other measures of competition (e.g. Ledermann and Eckmüllner, 2004). Finally, model developers should provide flexibility in accommodating a variety of sampling schemes that do not introduce too much measurement error to the predictions (see Chapter 14). Most statistical growth models can accommodate various fixed- and variable-radius sampling schemes, while gap and process-based models are generally less flexible.

To ensure collection of appropriate data, model users should try to: (1) use stratification to ensure the full range of conditions are sampled; (2) collect all necessary data needed by a growth model as accurately as possible (e.g. site index); (3) install multiple plots in a stand; (4) measure diameter on all trees and subsample trees for total height and height to crown base measurements; and (5) use a sampling scheme similar to the one used during model construction. In general, the most useful metrics for stratification are species composition, stand age or density, and site productivity.

As previously mentioned, statistical growth models can be quite sensitive to estimates of site index (e.g. Gertner and Dzialowy, 1984; McRoberts *et al.*, 1994). Consequently, attention should focus on accurate determination of site index, which requires proper selection of site trees and measurement of age and total height (see Chapter 3). Other models, like some variants of FVS, require information on slope, aspect, and elevation. It is best to collect this information in the field rather than try to reconstruct or estimate it in the office.

The number of plots to install in a stand depends on its size, the variability within the stand, and desired levels of precision. For example, 10 plots would likely be sufficient for an even-aged, single-species stand, while two- to three-times more plots may be needed to accurately describe a more complex stand structure. Although several randomly placed plots in a stand may be best for estimating stand-level variances (García, 2006), plots are generally systematically located to ensure full coverage of the stand (Husch *et al.* 2003). Many equations in individual-tree growth models rely on crown ratio, which requires estimates of both total height and height to crown base. Although most growth models use static equations to fill in missing heights and height to crown bases, doing so can introduce an unknown amount of measurement-error prediction bias (see Chapter 14).

Locally calibrating the predictions with a few observations can reduce the measurement-error prediction bias, and, therefore, significantly improve model projection accuracy (Leites *et al.*, 2009). However, models differ in the number of observations needed for calibration. For example, FVS requires three or more records for each species to have measured heights before calibration (Crookston and Dixon, 2005), while ORGANON first tests for a statistical difference between predicted and observed values and only calibrates if a significant difference is noted, regardless of sample size (Hann, 2011). Both Garber *et al.* (2009) and Temesgen *et al.* (2008) suggest that at least four or more trees be measured for height to ensure proper local calibration. A similar or even higher number of sample trees for height to crown base would likely be needed, but work has yet to be conducted on verifying this assumption.

Furthermore, height to crown base should be measured rather than visually estimating crown ratio, as the latter can result in significant bias and higher variability in the estimates

(McRoberts *et al.*, 1994). Also, different definitions of crown base exist (see Chapter 7), and a similar definition to that used in the growth model should be used in the field. However, there are equations to convert from different definitions of height to crown base (Monleon *et al.*, 2004; Randolph, 2010), though this will also introduce measurement error bias to the predictions (see Chapter 14).

Finally, model users should utilize a sampling scheme similar to the one used during model construction. Hann and Zumrawi (1991) found that the accuracy of growth predictions is drastically decreased when using a sampling unit design that is substantially different from the one used to construct the model. Their finding was driven primarily by the various sampling unit designs causing differential error in estimating the one-sided measures of competition (i.e. basal area in larger trees) used in the diameter increment equation (Hann and Zumrawi, 1991). Consequently, larger plot sizes should be favored over smaller ones for ensuring proper estimation of competition (Hann and Zumrawi, 1991; Ledermann and Eckmüllner, 2004; Lappi, 2005).

In comparing fixed- and variable-radius methods for estimating competition for an individual tree, Stage and Ledermann (2008) found that the variable radius methods should be preferred. Regardless of the sampling scheme used, Ledermann and Eckmüllner (2004) suggested that the use of record tripling can significantly improve growth predictions by ensuring a uniform resolution of individual-tree competition. Record tripling is the default in several individual growth models like FVS (Crookston and Dixon, 2005), PROGNAUS (Monserud *et al.*, 1997), and an option in others such as ORGANON (Hann, 2011).

16.3 Generation of appropriate data

When actual sample data are not available to initialize a growth model, developers should ensure that users have the ability to generate appropriate data needed for simulations, especially when models require inputs that are not commonly collected in forest inventories. Biging *et al.* (1994) developed a forest Stand Generator (STAG) for use with the individual-tree growth model CACTOS, which allows users to generate tree lists from simple stand summaries. Likewise, Valentine *et al.* (2000) presented a stand generation approach for a mechanistic growth model. For relatively simple stand structures, the use of parameter recovery techniques for various probability density functions is also an effective technique for generating distance-independent tree lists (see Chapter 10).

More sophisticated approaches may be needed in complex stands. For example, Kershaw *et al.* (2010) used copulas to generate complex, multiple-species stand structures. This technique was also useful in generating tree spatial coordinates as required by distance-dependent individual-tree growth models (Kershaw *et al.*, 2010). Several other methods have been used to generate tree coordinates with varying degrees of success (Hanus *et al.*, 1998; Kokkila *et al.*, 2002; Stamatellos and Panourgias, 2005). The non-simple sequential inhibition process of Hanus *et al.* (1998) was used to generate realistic renditions of natural stands in the stand visualization model VIZ4ST. STRUGEN is an interesting tool as it uses qualitative descriptions like row mixture and single-tree mixture to generate stand structures (Pretzsch, 1997). However, it should be noted that errors incurred with the use of a stand structure generator coupled with the uncertainties in the growth equations can drastically decrease projection accuracies (Goreaud *et al.*, 2006).

Remote sensing technologies are increasingly being used to generate appropriate input data for growth models. Temesgen *et al.* (2003) demonstrated the use of aerial photographs and nearest-neighbor imputation techniques to generate tree lists for a growth model in complex, multi-species stands in British Columbia. Eskelson *et al.* (2009) reviewed the uses and limitations of nearest-neighbor methods for estimating forest inventory data. One technique that will likely see increased application in the future is the use of light detection and ranging (LiDAR) technology. Falkowski *et al.* (2010) demonstrated the use of LiDAR and nearest-neighbor imputation techniques to generate tree lists for FVS across an extensive study area, which produced results comparable to projections made using a network of ground-based plots. However, the technique may have a limited success in stands dominated by small trees (Falkowski *et al.*, 2010) or containing multiple species (Suratno *et al.*, 2009).

Satellites also are quite helpful for improving predictions by hybrid and mechanistic models. For example, Waring *et al.* (2010) showed how predictions by the 3-PG model were improved by using remotely sensed estimates of leaf area index, canopy light interception, and foliage nitrogen content. In fact, estimates of gross primary production (GPP) by NASA's Earth Observing System Moderate Resolution Imaging Spectroradiometer (MODIS) are driven entirely by satellites (Running *et al.*, 2004).

16.4 Temporal scale

Implementing the appropriate model also depends on projection length. Burkhart (2003) suggests that, as projection period increases, model dimensionality or complexity should decrease. For example, Mäkinen *et al.* (2008) suggested the use of a stand-level model for long-term projections (>20 years) and an individual-tree approach for short-term projections (<10 years). This is because prediction errors tend to accumulate and projection accuracy decreases as the number of growth periods increases (Kangas, 1997), particularly in individual-tree growth models (Mäkinen *et al.*, 2008). However, individual-tree growth models are needed when information on long-term forest structure (Mäkinen *et al.*, 2008) and resulting mix of forest products are desired, or when complex stand structures and silvicultural systems are being simulated. In addition, gains in precision by using an individual-tree growth model for short-term projections generally are significant. For example, Härkönen *et al.* (2010a) found a 19% reduction in stand volume root mean square error when estimating 10-year growth for an extensive network of permanent plots in southern Finland.

One drawback to short-term simulations is that influences of annual weather fluctuations often are high, but do tend to average out over longer simulations periods (Kangas, 1998b). Consequently, accuracies of short-term projections benefit significantly with a hybrid approach (see Chapter 13). Regardless of the desired temporal scale of projections, it is important to be cognizant of the high uncertainties associated with any projections. Although complex techniques are available to assess this uncertainty (see Chapter 15), Kangas (1999) presents some relatively simple techniques for incorporating uncertainty into projections.

Finally, it is important to remember that the accuracy of predictions decreases as projections are made outside the data used to develop the model. This is particularly the case when long-term projections (> 100 years) are desired, as that type of data are rarely available. In addition, most models will not warn users when projections are made outside the range of the fitting data. However, the ORGANON growth model (Hann, 2011) does warn users when this occurs and more models should probably also make more explicit reminders like this.

16.5 Spatial scale

Most models discussed in this book use the forest stand as their basic level of operation. Stand-level analyses are often useful for comparing alternative silvicultural systems, but updating inventories or forecasting future forest structures requires projecting multiple stands simultaneously. Even today, processing multiple stands is very computationally intensive, particularly when the goal is to optimize a certain condition or objective. Although landscape-level models like LANDIS-II are available (Scheller *et al.*, 2007), stand-level models are still useful for large-scale projections and offer some advantages. One approach for accomplishing this task is to develop a software system capable of processing multiple stands at once. For example, the Landscape Management System (LMS; McCarter *et al.*, 1998) is a widely used tool that incorporates different growth models such as FVS and ORGANON and can process several hundred stands at a time. In addition, Papaik *et al.* (2010) recently presented a methodology useful for scaling predictions made by an individual-tree growth model to the landscape level, which may prove useful for other future efforts.

A drawback to this type of approach is that optimization, as often required by harvest planning efforts, is difficult. A solution to this is to use a stand-level growth model to generate yield curves for different forest types and silvicultural systems and use a linear programming optimizer like Remsoft's Woodstock (Remsoft Inc., Fredericton, NB) to meet certain objective constraints (e.g. Hennigar *et al.*, 2008). However, if a growth and yield model is available as a dynamic link library (DLL; see below), Woodstock is able to generate yield curves on the fly. The continued development of techniques to extend stand-level growth models to large spatial scale is critical, as many questions and decisions are currently being posed at extensive spatial scales.

16.6 Computer interface

Although growth models differ remarkably in their approach, most users only know a growth model by its computer interface. Providing a computer interface is an important aspect of model development, and various approaches are taken. Some models like 3-PG (Landsberg and Waring, 1997), BIOME-BGC (White *et al.*, 2000), and Mixedwood Growth Model (Bokalo *et al.*, in review) are available as Microsoft Excel spreadsheets. This approach makes the model accessible to a rather wide audience, simplifies programming, and allows the user to take advantage of built-in features like graphs and pivot tables. However, the model is dependent on an external piece of software, "cut and paste" is often required to initialize simulations, and it must be continually updated to ensure compatibility.

Another approach is make the software available as a dynamic link library (DLL) as suggested by Robinson and Monserud (2003). This allows the model to be customized to a user's needs and connected to other software programs, but the bulk of the development effort for the interface is generally shifted to the user. Despite the appeal, only a few models such as ORGANON and CONIFERS are available as a DLL. Ritchie (2009b) demonstrated the usefulness of a DLL with the CONIFERS model by incorporating it into the open-source statistical software R (R Development Core Team, 2009), which provides the ability to perform statistical analysis on output, develop custom figures, and easily implement Monte Carlo simulations. The ORGANON DLL has been used as the growth engine in a number of commercial, proprietary, and free integrated forest management products.

The most common approach for providing a computer interface is to develop a graphical user interface (GUI) using different programming languages. For example, ORGANON is also available as a DOS-based program, while FVS can use the Microsoft Windows-based Suppose GUI (Crookston, 1997). This approach requires an intensive computer programming effort, but forces the model user to adopt an interface with a limited capability for modification. Overall, some of the key features of any interface are flexibility, batch processing capabilities, menu or keyword based, and file-based inputs (Robinson and Monserud, 2003).

One drawback to most growth model interfaces is that they do not allow simulations to be conducted with more than one growth model. As demonstrated by Johnson (2005), the ability to run multiple models on the same stand can be quite illustrative. An effort is currently underway in France to develop a universal growth model interface, which would provide the ability to easily run multiple simulations with different models under the same software platform. The CAPSIS platform currently contains over 50 growth models from around the world and is available on a variety of operating systems (de Coligny *et al.*, 2010).

Another limitation is that most growth model interfaces are not interactive since they generally just process input and provide output. The TRAGIC++ model of Parrott and Lange (2004) attempted to provide a highly interactive interface that promoted experimentation. Finally, it is important for a growth model interface to provide a flexible and easily performed method for implementing forest silvicultural treatments. Ideally, a model interface would allow an array of options for characterizing such treatments, particularly for thinning activities; would provide the option of having a treatment automatically executed when certain stand conditions are met; and would generate output that describes the stand pre- and post-treatment. In addition, the treatments should be additive so that complex silvicultural systems can be represented.

16.7 Visualization

Information generated by forest growth models is generally available in the form of tables and charts, which often are difficult to interpret and visualize. Pictures are often an effective way to convey a lot of information, and one of the best approaches for growth models is to provide an image of potential future conditions as a computer representation of the data (Hauhs *et al.*, 1999; Deutschman *et al.*, 2000). One such system, the Stand Visualization System (SVS) (McGaughey, 1997) generates graphic images depicting stand conditions represented by a list of individual stand components, e.g. trees, shrubs, and down material. It is in wide use and is currently connected to other tools such as the FVS (Crookston and Dixon, 2005), LMS (McCarter *et al.*, 1998) and the NorthEast Decision model (NED; Twery *et al.*, 2005). Besides SVS, several other stand-level visualization tools exist such as TREEVIEW (Pretzsch *et al.*, 2008), Sylview (Scott, 2006), TRAGIC++ (Parrott and Lange, 2004), and the VIZ4ST (Hanus and Hann, 1997). All of these stand-level visualization tools depict stands in varying degrees of complexity. SVS just uses generic geometric shapes to represent tree crowns, while the VIZ4ST incorporates information on individual branches.

At the landscape level, there are several tools available for visualization. These tools are particularly useful for maintaining or protecting scenic views, visualizing the landscape under alternative silvicultural systems, and harvest scheduling. The Environmental Visualization tool (ENVISION) is a very powerful and realistic landscape-level visualization tool (McGaughey, 2003). ENVISION uses an algorithm that allows simulated scenes to be matched with real photographs taken from known locations. UTOOLS and UVIEW are geographic analysis and visualization software tools for watershed-level planning (Ager and McGaughey, 1997). The system uses a database to store spatial information and displays landscape conditions of a forested watershed in a flexible framework. Another similar visualization tool is SmartForest (Orland, 1995), which is also an interactive program to display forest data for the purposes of visualizing effects of various alternative treatments before actually implementing them. The tool has been developed to be compatible with most modern PCs (Uusitalo and Kivinen, 2000). Three additional landscape visualization tools are L-VIS (Pretzsch *et al.*, 2008), VRML (Lim and Honjo, 2003), and SILVISO. Like ENVISION, L-VIS and SILVISO are very highly detailed visualization tools, but are unique in that they are tightly coupled with a forest simulation model (Pretzsch *et al.*, 2008).

Effective visualization of forest and landscapes still remains a significant challenge (Wang *et al.*, 2006). Regardless of scale, Pretzsch *et al.* (2008) identified four tenets that all visualization tools should embody: (1) cover temporal and spatial scales that are suited to human perception capabilities; (2) be data driven; (3) be as realistic as possible; and (4) allow free choice of perspective. Most of the described visualization tools address these tenets, but in different ways. However, Wilson and McGaughey (2000) identified some key issues associated with photo-realistic visualization. Instead, they recommended that a landscape be visualized using several complementary techniques to convey information (Wilson and McGaughey, 2000). Future efforts should continue to refine visualization techniques that make growth model output accessible to a wide audience.

16.8 Output

As discussed in Chapter 1, growth models are used for a wide variety of applications ranging from economic assessments (Latta and Montgomery, 2004) to predicting wildlife habitat availability (Barbour *et al.*, 1997). To be useful to a wide audience, growth models must provide a variety of output. From a forest management perspective, the most important output is information on tree size, biomass/carbon, wood quality, and stand attributes. The ability to estimate carbon sequestration is quickly becoming an important attribute. For example, in the USA, FVS uses an approved protocol for carbon calculation as set by the Chicago Climate Exchange and the California Climate Action Registry. Hoover and Rebain (2008) demonstrated the use of FVS to estimate carbon sequestration for a large area. Like carbon, biomass is becoming an important forest product for use in energy and chemical products, so information on amount and composition is needed.

In addition to altering growth and mortality, forest silvicultural treatments modify wood quality (Maguire *et al.*, 1991). Several growth models such ORGANON and CROBAS (Mäkelä and Makinen, 2003) have the ability to estimate key wood quality attributes. This output, combined with a sawing simulator, is used to assess influences of management on forest product outputs (e.g. Barbour *et al.*, 2003). From a wildlife habitat perspective, information on species and size of both living and dead trees is needed. Time since death is also important because it can be used to predict whether a tree is still standing

(Garber *et al.*, 2005) and its decay class. Finally, economic outputs such as net present value and internal rate of return are important for certain model users.

16.9 Summary

To be valued, models must be useable. This means that they must meet end user needs and requirements (von Teuffel *et al.*, 2006). The features that generally make a model practical are the ability to use data from a variety of sampling schemes, generate data when it is not available, provide a flexible software interface, be connected to a visualization tool, and provide a wide range of outputs, which require the model developer to understand the needs of users. However, to properly implement a model, users must collect accurate and appropriate data to initialize the model and be aware of selecting the best model for the temporal and spatial scales of interest. In comparison to landscape models, using a stand-by-stand growth model offers greater flexibility and the ability to provide information at multiple spatial resolutions (e.g. individual tree, stand).

Integrated software systems like LMS and NED provide the ability to make projections, visualize results, and assess various criteria like carbon sequestration, economic value, and wildlife habitat suitability, for large areas with a single program. These types of tools are effective and efficient at helping users accomplish their multiple objectives in using a model and should serve as a template for future efforts. Overall, proper implementation and use of any growth model is ideally a dynamic interaction between modeler, software developer, and end user.

17
Future directions

Forest growth models are useful tools and have evolved greatly in the last few decades. Since the 1960s, statistical growth models have transitioned from simple yield tables, to compatible growth and yield equations, and eventually to individual-tree simulators. Likewise, ecological gap models have transitioned into complex process models, which are now being modified into hybrid models. However, there is not a model or even modeling approach that is applicable to all situations (e.g. Palosuo *et al.*, 2008). Consequently, there is still significant room for advances in all areas of modeling. In addition to better predictions, other areas for improvement are better input data, model interfaces, and efficient processing of extensive inventories.

17.1 Improving predictions

For improving predictions, one of the most promising future directions for growth models is the hybridization of approaches, whether it be linking models of different spatial resolutions (Chapter 10; e.g. Papaik *et al.*, 2010), bridging distance-dependent and distance-independent approaches (e.g. Perot *et al.*, 2010), or combining the predictions of statistical and mechanistic models (Chapter 13). Like hybrid models, summary models can incorporate current knowledge from a variety of fields to further generalize models like 3-PG, be initialized with standard forest inventory data, and provide realistic short-term predictions of growth across a range of stand conditions (Härkönen *et al.*, 2010b). As discussed throughout this book, each approach has advantages and disadvantages. Combining different approaches allows their strengths to be leveraged, while not significantly increasing input requirements or the number of parameters needing to be estimated.

Forest Growth and Yield Modeling, First Edition. Aaron R. Weiskittel, David W. Hann, John A. Kershaw, Jr. and Jerome K. Vanclay.

Extensive empirical datasets are currently available regionally and even nationally. Large databases can contain millions of observations from hundreds of different species, which can make the use of traditional statistical techniques like linear and nonlinear regression difficult. The strong computational power available today may make other statistical techniques like Bayesian and nonparametric techniques more widely available. Although Li *et al.* (2011) found Bayesian methods to give similar predictions compared to traditional statistical techniques, they suggested that Bayesian techniques offered a significant advantage by being able to predict local parameters on a new dataset relatively easily. Other studies found that Bayesian and nonparametric techniques provided superior predictions compared to traditional statistical techniques (Metcalf *et al.*, 2009a; Aertsen *et al.*, 2010). The benefits of Bayesian techniques are well-exemplified in the work of Clark *et al.* (2007), where they present a hierarchical method for predicting diameter increment, and incorporate various sources of uncertainty. Primarily using nonparametric techniques, Crookston *et al.* (2010) developed a climate-sensitive variant of the Forest Vegetation Simulator (FVS). Crookston *et al.* (2010) found nonparametric techniques advantageous because they allowed full use of extensive datasets, simplified model development, and, most importantly, provided robust predictions.

Detailed spatial data and sophisticated geostatistical techniques (e.g. Smith *et al.*, 2008; Finley *et al.*, 2009) are increasingly becoming available, which may be useful for refining regional predictions. For example, Liang and Zhou (2010) combined a traditional forest matrix model with a geostatistical approach and found improved predictions compared to an uncombined approach. Given the current relative lack of quality climate and soils information, this approach avoids the need for this type of data, while still making use of spatially explicit information. This also further suggests the need to continue the integration of forest growth models into geographic information systems (GIS) (e.g. Zeng *et al.*, 2007; Umeki *et al.*, 2008).

As discussed in Chapter 2, quantitative measures of competition are largely imperfect, and distance-dependent measures have generally offered no additional explanatory power. Existing detailed light-interception models are still computationally demanding (e.g. Brunner, 1998), and, while simplified light-interception models are available, they are not applicable to all stand conditions (e.g. Groot, 2004). However, Hanus (2003) found that a detailed light-interception program, which simulated ambient light (SAL; both direct and diffuse) incident on an individual tree for each growing season, performed significantly better than other indirect spatially implicit and explicit indices of competition. SAL was particularly effective for predicting height increment of both overstory and understory trees in a two-storied, mixed-species stand (Hanus, 2003). This measure combined with a two-sided measure of competition may be the most effective means for comprehensively describing competition in statistical growth models.

Process models offer several theoretical advantages to purely statistical growth models (Chapter 12). However, statistical models are generally more stable in their predictions and are not as sensitive to initial stand conditions or silvicultural treatments (e.g. Matala *et al*., 2003). In addition, the physiological parameters needed for a process model, even for a well-studied species like Douglas-fir, are difficult to obtain (e.g. Weiskittel *et al*., 2010). However, the increased availability of physiological data from eddy-flux towers, and the ability to estimate key parameters from remote sensing data may make process models more widely used in the future. Weiskittel *et al*. (2010) also suggested the need for improvements in representing effects of soil water and nutrient availability on physiological processes, particularly carbon allocation, as well as influences of seasonal drought and forest silvicultural treatments on net photosynthesis. Damour *et al*. (2010) illustrated an approach for combining stomatal conductance models developed for well-watered conditions with ones designed specifically for drought conditions, which should be further incorporated in future process models. Finally, prediction of carbon allocation dynamics will likely remain difficult due to the variety of processes involved, but the theory of optimal co-allocation shows promise (Mäkelä *et al*., 2008).

17.2 Improving input data

As discussed in Chapter 16, models need high-quality input data to make accurate projections. A variety of emerging remote sensing technologies may make it easier to obtain better input data for statistical, process, and hybrid models. Light detection and ranging (LiDAR) technology will likely be the most effective remote sensing tool for improving predictions. For example, Mäkinen *et al*. (2010) found a significant improvement in predictions using single-tree data derived from LiDAR rather than data obtained from stand-level field inventory data. However, improvement of LiDAR processing algorithms still continues, and the technology is not applicable in all stand types. For example, applications of LiDAR in young stands and complex, multi-species stands remains problematic because of the high degree of crown overlap and smaller trees being overtopped by dominant ones.

For process models, data obtained from satellites may be the key for improving predictions. Waring *et al*. (2010) provided an overview of the various potential uses of remote sensing information for improving process model predictions. Some of the most important ones include climate data from passive microwave sensors, leaf area and absorbed radiation determination using the normalized difference vegetation index (NDVI), and estimation of canopy nitrogen and chlorophyll content with albedo derived from MODIS (Waring *et al*., 2010). Satellite data are used to improve estimates of rooting depth and soil water holding capacity (e.g. Ichii *et al*., 2009), which are important factors influencing process model predictions and often are not widely available.

Consequently, modelers need to be aware of these likely future improvements in available input data. This will probably mean the further refinement of both statistical and process individual-tree models, as well as improved distance-dependent measures of competition. However, as demonstrated by Mäkinen *et al.* (2010), any new approach for improving input data needs to be verified and tested to ensure that it does not create different sources of error in growth model predictions.

17.3 Improving software

Software development is a continually evolving field that often is frustrating for both modeler and model users alike. For growth models to be widely used, they need to be user friendly to some degree, which more times than not relates primarily to their computer interface (see Chapter 16). In the future, growth models need to be capable of handling and processing large amounts of data, particularly with the increasing availability of LiDAR data, which will potentially provide information on every tree in a stand, rather than just those on small sample plots. Consequently, growth models will need to be able to easily link to database programs like MySQL. In addition, processing of large databases will need to make full use of the computational power available today, so multiple-core processing and 64-bit compatibility need to be considered.

Common model interface projects like CAPSIS (http://capsis.cirad.fr/) are important and have the potential to make models accessible to a wider audience. This is because model predictions are quickly compared to other models, the learning curve is minimized because the interface is common across models, and the program is maintained by programmers rather than modelers. One attractive alternative is to further the development of growth models based in the statistical package R (R Development Core Team, 2009). Although R has a steep learning curve and can invoke user memories of DOS, it is very flexible, has higher-level computational capabilities like parallel processing and 64-bit compatibility, and, most importantly, has built-in statistical and graphing capabilities. The model CONIFERS was successfully incorporated into R with nothing more than a few R scripts and a dynamic link library (DLL). Models packaged as DLLs will have similar flexibility.

17.4 Summary

This book has attempted to summarize the different types of stand-level forest growth and yield models and provide a detailed description of their different components. Much has changed since one of the first comprehensive books on forest growth modeling, Vanclay (1994), was written. In particular, significant focus has shifted from stand-level to individual-tree models, as predicted by Furnival (1987). Currently, purely statistical approaches are being modified into a variety of hybridized modeling approaches. These two trends will likely continue in the foreseeable future.

Although the ultimate future direction of forest growth models is difficult to predict, the novelty and difficulty of some of the key issues facing forestry today will likely require further development and refinement of growth models. As noted by Furnival (1987), the key to making future progress will be avoiding the mistakes of the past, which can come in many forms. Some of the key issues that growth models will need to address in the future are intensive forest silvicultural systems like clonal forestry and biomass harvesting, climate change, and carbon sequestration. However, given the large uncertainty of these future scenarios, it is interesting to ponder, how precise do model predictions need to be in order to be useful? Regardless of future directions, models need to be developed using current knowledge, be rigorously tested, and made accessible to a wide array of users. These tenets will likely not change with time or modeling approach.

Bibliography

Adame, P., del Río, M., Cañellas, I., 2010a. Modeling individual-tree mortality in Pyrenean oak (*Quercus pyrenaica* Willd.) stands. Annals of Forest Science 67, 810.

Adame, P., Del Rìo, M., Cañellas, I., 2010b. Ingrowth model for Pyrenean oak stands in north-western Spain using continuous forest inventory data. European Journal of Forest Research 129, 669–678.

Adams, D.M., Ek, A.R., 1974. Optimizing the management of uneven-aged forest stands. Canadian Journal of Forest Research 4, 274–287.

Adams, J.P., Matney, T.G., Land Jr., S.B., Belli, K.L., Duzan Jr., H.W., 2006. Incorporating genetic parameters into a loblolly pine growth-and-yield model. Canadian Journal of Forest Research 36, 1959–1967.

Aertsen, W., Kint, V., van Orshoven, J., Özkan, K., Muys, B., 2010. Comparison and ranking of different modelling techniques for prediction of site index in Mediterranean mountain forests. Ecological Modelling 221, 1119–1130.

Affleck, D.L.R., 2006. Poisson mixture models for regression analysis of stand-level mortality. Canadian Journal of Forest Research 36, 2994–3006.

Ager, A.A., McGaughey, R.J., 1997. UTOOLS: Microcomputer software for spatial analysis and landscape visualization. General Technical Report PNW-GTR-397. USDA Forest Service, Pacific Northwest Research Station, Portland, OR.

Akaike, H., 1973. Information theory and an extension of the maximum likelihood principle. In Petrov, B.N., Csaki, F. (Eds.), Proceedings of the 2nd International Symposium of Information Theory, Budapest, pp. 267–281.

Alder, D., 1979. A distance-independent tree model for exotic conifer plantations in east Africa. Forest Science 25, 59–71.

Alder, D., Silva, J.N.M., 2000. An empirical cohort model for management of Terra Firme forests in the Brazilian Amazon. Forest Ecology and Management 130, 141–157.

Alemdag, I.S., 1978. Evaluation of some competition indices for the prediction of diameter increment in planted white spruce. Information Report FMR-X-108. Canadian Department of Environment, Forest Management Institute, Ottawa, Ontario.

Forest Growth and Yield Modeling, First Edition. Aaron R. Weiskittel, David W. Hann, John A. Kershaw, Jr. and Jerome K. Vanclay.

Alenius, V., Hökkä, H., Salminen, H., Jutras, S., 2003. Evaluating estimation methods for logistic regression in modelling individual-tree mortality. In Amaro, A., Reed, D., Soares, P. (Eds.), Modelling Forest Systems. CABI Publishing, Wallingford, UK, pp. 225–236.

Almeida, A.C., Landsberg, J.J., Sands, P.J., 2004. Parameterisation of 3-PG model for fast-growing *Eucalyptus grandis* plantations. Forest Ecology and Management 193, 179–195.

Álvarez González, J.G., Schröder, J., Rodríguez Soalleiro, R., Ruíz González, A.D., 2002. Modelling the effect of thinnings on the diameter distribution of even-aged Maritime pine stands. Forest Ecology and Management 165, 57–65.

Álvarez González, J.G., Castedo Dorado, F., Ruiz González, A.D., López Sánchez, C.A., von Gadow, K., 2004. A two-step mortality model for even-aged stands of *Pinus radiata* D. Don in Galicia (Northwestern Spain). Annals of Forest Science 61, 439–448.

Amateis, R.L., 2000. Modeling response to thinning in loblolly pine plantations. Southern Journal of Applied Forestry 24, 17–22.

Amateis, R., Burkhart, H.E., Walsh, T.A., 1989. Diameter increment and survival equations for loblolly pine trees growing in thinned and unthinned plantations on cutover, site-prepared lands. Southern Journal of Applied Forestry 13, 170–174.

Amateis, R.L., Liu, J., Ducey, M.J., Allen, H.L., 2000. Modeling response to midrotation nitrogen and phosphorus fertilization in loblolly pine plantations. Southern Journal of Applied Forestry 24, 207–212.

Amichev, B.Y., Johnston, M., Van Rees, K.C.J., 2010. Hybrid poplar growth in bioenergy production systems: Biomass prediction with a simple process-based model (3PG). Biomass and Bioenergy 34, 687–702.

Amichev, B.Y., Hangs, R.D., Van Rees, K.C.J., 2011. A novel approach to simulate growth of multi-stem willow in bioenergy production systems using a simple process-based model (3PG). Biomass and Bioenergy 35, 473–488.

Andreassen, K., Tomter, S.M., 2003. Basal area growth models for individual trees of Norway spruce, Scots pine, birch and other broadleaves in Norway. Forest Ecology and Management 180, 11–24.

Andrews, L.S., Perkins, J.P., Thrailkill, J.A., Poage, N.J., Tappeiner II, J.C., 2005. Silvicultural approaches to develop northern spotted owl nesting sites, central Coast Ranges, Oregon. Western Journal of Applied Forestry 20, 13–27.

Arabatzis, A.A., Burkhart, H.E., 1992. An evaluation of sampling methods and model forms for estimating height-diameter relationships in loblolly pine plantations. Forest Science 38, 192–198.

Arney, J.D., 1985. A modeling strategy for the growth projection of managed stands. Canadian Journal of Forest Research 15, 511–518.

Arseneault, J.E., Kershaw, J.A., McCarter, J.B., MacLean, D.A., 2008. Forest Vegetation Simulator Ingrowth Tool: Incorporating ingrowth tree lists into Forest Vegetation Simulator growth projections. Northern Journal of Applied Forestry 25, 158–160.

Assmann, E., 1970. The Principles of Forest Yield Study: Studies in the Organic Production, Structure, Increment and Yield of Forest Stands. Pergamon Press, New York, NY.

Auchmoody, L.R., 1985. Evaluating growth responses to fertilization. Canadian Journal of Forest Research 15, 877–880.

Avery, T.E., Burkhart, H.E., 2002. Forest Measurements 5th edition. McGraw-Hill, New York, NY.

Bailey, R.L., 1980. Individual tree growth derived from diameter distribution models. Forest Science 26, 626–632.

Bailey, R.L., 1994. A compatible volume-taper model based on the Schumacher and Hall generalized constant form factor volume equation. Forest Science 40, 303–313.

Bailey, R.L., Clutter, J.L., 1974. Base-age invariant polymorphic site curves. Forest Science 20, 155–159.

Bailey, R.L., Ware, K.D., 1983. Compatible basal-area growth and yield model for thinned and unthinned stands. Canadian Journal of Forest Research 13, 563–571.

Bailey, R.L., Borders, B.E., Ware, K.D., Jones Jr., E.P., 1985. A compatible model for slash pine plantation survival to density, age, site index, and type and intensity of thinning. Forest Science 31, 180–189.

Bailey, R.L., Burgan, T.M., Jokela, E.J., 1989. Fertilized midrotation-aged slash pine plantations—stand structure and yield prediction models. Southern Journal of Applied Forestry 13, 76–80.

Baldwin Jr., V.C., Feduccia, D.P., 1987. Loblolly pine growth and yield prediction for managed West Gulf plantations. Research Paper SO-236. USDA Forest Service, Southern Forest Experiment Station, New Orleans, LA.

Baldwin Jr., V.C., Peterson, K.D., 1997. Predicting the crown shape of loblolly pine trees. Canadian Journal of Forest Research 27, 102–107.

Baldwin Jr., V.C., Peterson, K.D., Burkhart, H.E., Amateis, R.L., Doughterty, P.M., 1997. Equations for estimating loblolly pine branch and foliage weight and surface area distributions. Canadian Journal of Forest Research 27, 918–927.

Baldwin Jr., V.C., Burkhart, H.E., Westfall, J.A., Peterson, K.D., 2001. Linking growth and yield and process models to estimate impact of environmental changes on growth of loblolly pine. Forest Science 47, 77–82.

Ball, J.T., Woodrow, I.E., Berry, J.A., 1986. A model predicting stomatal conductance and its contribution to the control of photosynthesis under different environmental conditions. In Biggins, J. (Ed.), Progress in Photosynthesis Research, Volume IV. Martinus Nijhoff Publisher, Dordrecht, The Netherlands, pp. 221–224.

Barbour, R.J., Johnston, S., Hayes, J.D., Tucker, G.F., 1997. Simulated stand characteristics and wood product yields from Douglas-fir plantations managed for ecosystem objectives. Forest Ecology and Management 91, 205–219.

Barbour, R.J., Parry, D.L., Punches, J., Forsman, J., Ross, R., 2003. AUTOSAW simulations of lumber recovery for small-diameter Douglas-fir and ponderosa pine from southwestern Oregon. Research Note PNW-RN-543. USDA Forest Service, Pacific Northwest Research Station, Portland, OR.

Barclay, H.J., Trofymow, J.A., Leach, R.I., 2000. Assessing bias from boles in calculating leaf area index in immature Douglas-fir with the LI-COR canopy analyzer. Agricultural and Forest Meteorology 100, 255–260.

Bare, B.B., Hann, D.W., 1981. Applications of ridge regression in forestry. Forest Science 27, 339–348.

Barrio Anta, M., Castedo Dorado, F., Diéguez-Aranda, U., Álvarez González, J.G., Parresol, B.R., Rodríguez Soalleiro, R. 2006. Development of a basal area growth system for maritime pine in northwestern Spain using the generalized algebraic difference approach. Canadian Journal of Forest Research 36, 1461–1474.

Barton, C., 2001. The role of allocation in modelling NEE. In Kirschbaum, M.U.F., Mueller, R. (Eds.), Net Ecosystem Exchange: Proceedings of Cooperative Research Centre for Greenhouse Accounting Workshop. Cooperative Research Centre for Greenhouse Accounting, Canberra, Australia, pp. 43–49.

Basslow, S.L., Ford, E.D., Kiester, A.R., 1990. A critique of carbon-based tree growth models. In Dixon, R.K., Meldahl, R.S., Ruark, G.A., Warren, W.G. (Eds.), Process Modeling of Forest Growth Responses to Environmental Stress. Timber Press, Portland, OR, pp. 50–57.

Batho, A., García, O., 2006. De Perthuis and the origins of site index: A historical note. Forest Biometry, Modelling and Information Sciences 1, 1–10.

Battaglia, M., Sands, P., 1997. Modelling site productivity of *Eucalyptus globulus* in response to climatic and site factors. Australian Journal of Plant Physiology 24, 831–850.

Battaglia, M., Sands, P.J., 1998. Process-based forest productivity models and their application in forest management. Forest Ecology and Management 102, 13–32.

Battaglia, M., Sands, P.J., Candy, S.G., 1999. Hybrid growth model to predict height and volume growth in young *Eucalyptus globulus* plantations. Forest Ecology and Management 120, 193–201.

Battaglia, M., Sands, P., White, D., Mummery, D., 2004. CABALA: A linked carbon, water and nitrogen model of forest growth for silvicultural decision support. Forest Ecology and Management 193, 251–282.

Bechtold, W.A., 2003. Crown-diameter prediction models for 87 species of stand-grown trees in the eastern United States. Southern Journal of Applied Forestry 27, 269–278.

Bechtold, W.A., 2004. Largest-crown-width prediction models for 53 species in the western United States. Western Journal of Applied Forestry 19, 245–251.

Beck, D.E., 1971. Polymorphic site index curves for white pine in the southern Appalachians. Research Paper SE-80. USDA Forest Service, Southeastern Forest Experiment Station, Asheville, NC.

Beckage, B., Lavine, M., Clark, J.S., 2005. Survival of tree seedlings across space and time: Estimates from long-term count data. Journal of Ecology 93, 1177–1184.

Belcher, D.W., Holdaway, M.R., Brand, G.J., 1982. A description of STEMS: The stand and tree evaluation and modeling system. General Technical Report NC-79. USDA Forest Service, North Central Forest Experiment Station, St. Paul, MN.

Bella, I.E., 1971. A new competition model for individual trees. Forest Science 17, 364–372.

Belli, K.L., Ek, A.R., 1988. Growth and survival modeling for planted conifers in the Great Lakes region. Forest Science 34, 458–473.

Benavides, R., Roig, S., Osoro, K., 2009. Potential productivity of forested areas based on a biophysical model. A case study of a mountainous region in northern Spain. Annals of Forest Science 66, 108.

Bernier, P.Y., Raulier, F., Stenberg, P., Ung, C.-H., 2001. Importance of needle age and shoot structure on canopy net photosynthesis of balsam fir (*Abies balsamea*): A spatially inexplicit modeling analysis. Tree Physiology 21, 815–830.

von Bertalanffy, L., 1949. Problems of organic growth. Nature 163, 156–158.

Bettinger, P., Lennette, M., Johnson, K.N., Spies, T.A., 2005. A hierarchical spatial framework for forest landscape planning. Ecological Modelling 182, 25–48.

Bettinger, P., Boston, K., Siry, J., Grebner, D. L., 2009. Forest Management and Planning. Academic Press, Burlington, MA.

Beven, K.J., Kirkby, M.J., 1979. A physically based, variable contributing area model of basin hydrology. Hydrological Sciences Journal 24, 43–69.

Bi, H., 2000. Trigonometric variable-form taper equations for Australian eucalypts. Forest Science 46, 397–409.

Bi, H., 2001. The self-thinning surface. Forest Science 47, 361–370.

Biging, G.S., Dobbertin, M., 1992. A comparison of distance-dependent competition measures for height and basal area growth of individual conifer trees. Forest Science 38, 695–720.

Biging, G.S., Dobbertin, M., 1995. Evaluation of competition indices in individual tree growth models. Forest Science 41, 360–377.

Biging, G.S., Gill, S.J., 1997. Stochastic models for conifer tree crown profiles. Forest Science 43, 25–34.

Biging, G.S., Robards, T.A., Turnblom, E.C., Van Deusen, P.C., 1994. The predictive models and procedures used in the Forest Stand Generator (STAG). Hilgardia 61, 1–36.

Bigler, C., Bugmann, H., 2004a. Predicting the time of tree death using dendrochronological data. Ecological Applications 14, 902–914.

Bigler, C., Bugmann, H., 2004b. Assessing the performance of theoretical and empirical tree mortality models using tree-ring series of Norway spruce. Ecological Modelling 174, 225–239.

Binkley, C.S., 1980. Is succession in hardwood forests a stationary Markov process? Forest Science 26, 566–570.

Birch, K.R., Johnson, K.N., 1992. Stand-level wood-production costs of leaving live, mature trees at regeneration harvest in coastal Douglas-fir stands. Western Journal of Applied Forestry 7, 65–68.

Bokalo, M., Stadt, K., Comeau, P., Titus, S., in prep. The validation and evaluation of the Mixedwood Growth Model (MGM).

Bolker, B.M., Brooks, M.E., Clark, C.J., Geange, S.W., Poulsen, J.R., Stevens, M.H.H., White, J.-S.S., 2009. Generalized linear mixed models: A practical guide for ecology and evolution. Trends in Ecology and Evolution 24, 127–135.

Bollandsårings, O.M., Buongiorno, J., Gobakken, T., 2008. Predicting the growth of stands of trees of mixed species and size: A matrix model for Norway. Scandinavian Journal of Forest Research 23, 167–178.

Bond, B.J., Czarnomski, N.M., Cooper, C., Day, M.E., Greenwood, M.S., 2007. Developmental decline in height growth in Douglas-fir. Tree Physiology 27, 441–453.

Bontemps, J.-D., Hervé, J.C., Dhôte, J.-F., 2009. Long-term changes in forest productivity: A consistent assessment in even-aged stands. Forest Science 55, 549–564.

Borders, B.E., 1989. Systems of equations in forest stand modeling. Forest Science 35, 548–556.

Borders, B.E., Bailey, R.L., 1986. A compatible system of growth and yield equations for slash pine fitted with restricted three-stage least square. Forest Science 32, 185–201.

Borders, B.E., Patterson, W.D., 1990. Projecting stand tables: A comparison of the Weibull diameter distribution method, a percentile-based projection method, and a basal area growth projection method. Forest Science 36, 413–424.

Borer, E.T., Seabloom, E.W., Jones, M.B., Schildhauer, M., 2009. Some simple guidelines for effective data management. Bulletin of the Ecological Society of America 90, 205–214.

Bormann, B.T., Cromack Jr., K., Russell III, W.O., 1994. Influences of red alder on soils and long-term ecosystem productivity. In Hibbs, D.E., DeBell, D.S., Tarrant, R.F. (Eds.), The Biology and Management of Red Alder. Oregon State University Press, Corvallis, OR, pp. 47–56.

Bosch, C.A., 1971. Redwoods: A population model. Science 172, 345–349.

Bossel, H., 1986. Dynamics of forest dieback: Systems analysis and simulation. Ecological Modelling 34, 259–288.

Botkin, D.B., Janak, J.F., Wallis, J.R., 1972a. Rationale, limitations, and assumptions of a northeastern forest growth simulator. IBM Journal of Research and Development 16, 101–116.

Botkin, D.B., Janak, J.F., Wallis, J.R., 1972b. Some ecological consequences of a computer model of forest growth. Journal of Ecology 60, 849–872.

Box, G.E.P., Draper, N.R., 1987. Empirical Model-Building and Response Surfaces. John Wiley & Sons, Inc., Hoboken, NJ.

Bragg, D.C., 2001a. Potential relative increment (PRI): A new method to empirically derive optimal tree diameter growth. Ecological Modelling 137, 77–92.

Bragg, D.C., 2001b. A local basal area adjustment for crown width prediction. Northern Journal of Applied Forestry 18, 22–28.

Bragg, D.C., 2008. Practical extension of a Lake States tree height model. Northern Journal of Applied Forestry 25, 186–194.

Bragg, D.C., Roberts, D.W., Crow, T.R., 2004. A hierarchical approach for simulating northern forest dynamics. Ecological Modelling 173, 31–94.

Bravo, F., Hann, D.W., Maguire, D.A., 2001. Impact of competitor species composition on predicting diameter growth and survival rates of Douglas-fir trees in southwestern Oregon. Canadian Journal of Forest Research 31, 2237–2247.

Bravo-Oviedo, A., Sterba, H., del Río, M., Bravo, F., 2006. Competition-induced mortality for Mediterranean *Pinus pinaster* Ait. and *P. sylvestris* L. Forest Ecology and Management 222, 88–98.

Bréda, N.L.L., 2003. Ground-based measurements of leaf area index: A review of methods, instruments and current controversies. Journal of Experimental Botany 54, 2403–2417.

Breiman, L., 2001. Random forests. Machine Learning 45, 5–32.

Briggs, R.D., 1994. Site classification field guide. CFRU Technical Report 6. Cooperative Forestry Research Unit, University of Maine, Orono, ME.

Brisson, J., 2001. Neighborhood competition and crown asymmetry in *Acer saccharum*. Canadian Journal of Forest Research 31, 2151–2159.

Brown, G.S., 1965. Point density in stems per acre. New Zealand Forest Research Note 38. New Zealand Forestry Service, Forest Research Institute, Rotorua, New Zealand.

Bruce, D., 1981. Consistent height-growth and growth-rate estimates for remeasured plots. Forest Science 27, 711–725.

Bruner, H.D., Moser Jr., J.W., 1973. A Markov chain approach to the prediction of diameter distribution in uneven-aged forest stands. Canadian Journal of Forest Research 3, 409–417.

Brunner, A., 1998. A light model for spatially explicit forest stand models. Forest Ecology and Management 107, 19–46.

Brunner, A., Nigh, G., 2000. Light absorption and bole volume growth of individual Douglas-fir trees. Tree Physiology 20, 323–332.

Buchman, R.G., Pederson, S.P., Walters, N.R., 1983. A tree survival model with application to species of the Great Lakes region. Canadian Journal of Forest Research 13, 601–608.

Buckman, R.E., 1962. Growth and yield of red pine in Minnesota. Technical Bulletin 1272. USDA, Washington, DC.

Buford, M.A., 1989. Mean stem size and total volume development of various loblolly pine seed sources planted at one location. Canadian Journal of Forest Research 19, 396–400.

Buford, M.A., Burkhart, H.E., 1987. Genetic improvement effects on growth and yield of loblolly pine plantations. Forest Science 33, 707–724.

Bugmann, H., 2001. A review of forest gap models. Climatic Change 51, 259–305.

Bugmann, H., Grote, R., Lasch, P., Lindner, M., Suckow, F., 1998. A new forest gap model to study the effects of environmental change on forest structure and functioning. In Mohren, G.M.J., Kramer, K., Sabaté, S. (Eds.), Impacts of Global Change on Tree Physiology and Forest Ecosystems. Kluwer Academic, The Netherlands, pp. 255–261.

Bullock, B.P., Boone, E.L., 2007. Deriving tree diameter distributions using Bayesian model averaging. Forest Ecology and Management 242, 127–132.

Buongiorno, J., Michie, B.R., 1980. A matrix model of uneven-aged forest management. Forest Science 26, 609–625.

Buongiorno, J., Peyron, J.-L., Houllier, F., Bruciamacchie, M., 1995. Growth and management of mixed-species, uneven-aged forests in the French Jura: Implications for economic returns and tree diversity. Forest Science 41, 397–429.

Burk, T.E., Newberry, J.D., 1984. A simple algorithm for moment-based recovery of Weibull distribution parameters. Forest Science 30, 329–332.

Burkhart, H.E., 1977. Cubic-foot volume of loblolly pine to any merchantable top limit. Southern Journal of Applied Forestry 1, 7–9.

Burkhart, H.E., 2003. Suggestions for choosing an appropriate level for modelling forest stands. In Amaro, A., Reed, D., Soares, P. (Eds.), Modelling Forest Systems. CABI Publishing, Wallingford, UK, pp. 3–10.

Burkhart, H.E., Sprinz, P.T., 1984. Compatible cubic volume and basal area projection equations for thinned old-field loblolly pine plantations. Forest Science 30, 86–93.

Burkhart, H.E., Walton, S.B., 1985. Incorporating crown ratio into taper equations for loblolly pine trees. Forest Science 31, 478–484.

Burkhart, H.E., Farrar, R.L., Amateis, R.L., Daniels, R.F., 1987. Simulation of individual tree growth and stand development in loblolly pine plantations on cutover, site-prepared areas. Publication Number FWS-1-87. Virginia Polytechnic Institute and State University, School of Forestry and Wildlife Resources, Blacksburg, VA.

Burkhart, H.E., Farrar, K.D., Amateis, R.L., Daniels, R.F., 1987 revised 2001. Simulation of individual tree growth and stand development in loblolly pine plantations on cutover, site-prepared areas. Publication Number FWS-1-87. Virginia Tech Department of Forestry, Blacksburg, VA.

Burnham, K.P., Anderson, D.R., 2002. Model Selection and Multimodel Inference: A Practical Information-Theoretic Approach, 2nd edition. Springer-Verlag, New York, NY.

Busby, G.M., Montgomery, C.A., Latta, G.S., 2007. The opportunity cost of forest certification on private land in western Oregon. Western Journal of Applied Forestry 22, 55–60.

Cade, B.S., Noon, B.R., 2003. A gentle introduction to quantile regression for ecologists. Frontiers in Ecology and the Environment 1, 412–420.

Cajander, A.K., 1949. Forest types and their significance. Acta Forestalia Fennica 56, 1–71.

Calama, R., Montero, G., 2004. Interregional nonlinear height-diameter model with random coefficients for stone pine in Spain. Canadian Journal of Forest Research 34, 150–163.

Calama, R., Mutke, S., Tomé, J., Gordo, J., Montero, G., Tomé, M., 2011. Modelling spatial and temporal variability in a zero-inflated variable: The case of stone pine (*Pinus pinea* L.) cone production. Ecological Modelling, 222, 606–618.

Calegario, N., Maestri, R., Leal, C.L., Daniels, R.F., 2005. Growth estimate of *Eucalyptus* stands based on nonlinear multilevel mixed-effects model theory. Ciência Florestal 15, 285–292.

Calkin, D.E., Montgomery, C.A., Schumaker, N.H., Polasky, S., Arthur, J.L., Nalle, D.J., 2002. Developing a production possibility set of wildlife species persistence and timber harvest value. Canadian Journal of Forest Research 32, 1329–1342.

Campbell, G.S., Norman, J.M., 1998. An Introduction to Environmental Biophysics, 2nd Edition. Springer, New York, NY.

Campbell, R.G., Ferguson, I.S., Opie, J.E., 1979. Simulating growth and yield of mountain ash stands: A deterministic model. Australian Journal of Forest Research 9, 189–202.

Canavan, S.J., 2002. The presence and characterization of measurement error in forestry. PhD dissertation. Oregon State University, Department of Forestry Resources, Corvallis, OR.

Canavan, S.J., Hann, D.W., 2004. The two-stage method of measurement error characterization. Forest Science 50, 743–756.

Canham, C.D., LePage, P.T., Coates, K.D., 2004. A neighborhood analysis of canopy tree competition: Effects of shading versus crowding. Canadian Journal of Forest Research 34, 778–787.

Cannell, M.G.R., 1989. Physiological basis of wood production: A review. Scandinavian Journal of Forest Research 4, 459–490.

Cannell, M.G.R., Dewar, R.C., 1994. Carbon allocation in trees: A review of concepts for modelling. Advances in Ecological Research 25, 59–104.

Cao, Q.V., 1997. A method to distribute mortality in diameter distribution models. Forest Science 43, 435–442.

Cao, Q.V., 2000. Prediction of annual diameter growth and survival for individual trees from periodic measurements. Forest Science 46, 127–131.

Cao, Q.V., 2004. Predicting parameters of a Weibull function for modeling diameter distribution. Forest Science 50, 682–685.

Cao, Q.V., 2006. Predictions of individual-tree and whole-stand attributes for loblolly pine plantations. Forest Ecology and Management 236, 342–347.

Cao, Q.V., 2007. Incorporating whole-stand and individual-tree models in a stand-table projection system. Forest Science 53, 45–49.

Cao, Q.V., 2009. Calibrating a segmented taper equation with two diameter measurements. Southern Journal of Applied Forestry 33, 58–61.

Cao, Q.V., Baldwin Jr., V.C., 1999. A new algorithm for stand table projection models. Forest Science 45, 506–511.

Cao, Q.V., Burkhart, H.E., 1980. Cubic-foot volume of loblolly pine to any height limit. Southern Journal of Applied Forestry 4, 166–168.

Cao, Q.V., Dean, T.J., 2008. Using segmented regression to model the density–size relationship in direct-seeded slash pine stands. Forest Ecology and Management 255, 948–952.

Cao, Q.V., Pepper, W.D., 1986. Predicting inside bark diameter for shortleaf, loblolly and longleaf pines. Southern Journal of Applied Forestry 10, 220–224.

Cao, Q., Strub, M., 2008. Evaluation of four methods to estimate parameters of an annual tree survival and diameter growth model. Forest Science 54, 617–624.

Cao, Q.V., Burkhart, H.E., Max, T.A., 1980. Evaluation of two methods for cubic-volume prediction of loblolly pine to any merchantable limit. Forest Science 26, 71–80.

Cao, Q.V., Burkhart, H.E., Lemin Jr., R.C., 1982. Diameter distributions and yields of thinned loblolly pine plantations. Publication Number FSW-1-82. Virginia Polytechnic Institute and State University, School of Forestry and Wildlife Resources, Blacksburg, VI.

Cao, Q.V., Dean, T.J., Baldwin Jr., V.C., 2000. Modeling the size-density relationship in direct-seeded slash pine stands. Forest Science 46, 317–321.

Cao, Q.V., Li, S., McDill, M.E., 2002. Developing a system of annual tree growth equations for the loblolly pine-shortleaf pine type in Louisiana. Canadian Journal of Forest Research 32, 2051–2059.

Carlson, C.A., Fox, T.R., Allen, H.L., Albaugh, T.J., 2008. Modeling mid-rotation fertilizer responses using the age-shift approach. Forest Ecology and Management 256, 256–262.

Carmean, W.H., 1975. Forest site quality evaluation in the United States. Advances in Agronomy 27, 209–269.

Carmean, W.H., Hahn, J.T., Jacobs, R.D., 1989. Site index curves for forest tree species in the Eastern United States. North Central General Technical Report NC-128. USDA Forest Service, North Central Forest Experiment Station, St. Paul, MN.

Carroll, R.J., Ruppert, D., Stefanski, L.A., Crainiceanu, C.M. 2006. Measurement Error in Nonlinear Models: A Modern Perspective 2nd edition. Chapman & Hall, Boca Raton, FL.

Carson, S.D., García, O., Hayes, J.D., 1999. Realized gain and prediction of yield with genetically improved *Pinus radiata* in New Zealand. Forest Science 45, 186–200.

Casper, B.B., Jackson, R.B., 1997. Plant competition underground. Annual Review of Ecology and Systematics 28, 545–570.

Cassell, R.F., Moser Jr., J.W., 1974. A programmed Markov model for predicting diameter distributions and species composition in uneven-aged forests. Research Bulletin 915. Purdue University Agricultural Experiment Station, West Lafayette, IN.

Castedo Dorado, F., Diéguez-Aranda, U., Barrio Anta, M., Sánchez Rodríguez, M., von Gadow, K., 2006. A generalized height–diameter model including random components for radiata pine plantations in northwestern Spain. Forest Ecology and Management 229, 202–213.

Castedo-Dorado, F., Diéguez-Aranda, U., Barrio-Anta, M., Álvarez-González, J.G., 2007a. Modelling stand basal area growth for radiata pine plantations in Northwestern Spain using the GADA. Annals of Forest Science 64, 609–619.

Castedo-Dorado, F., Diéguez-Aranda, U., Álvarez-González, J.G., 2007b. A growth model for *Pinus radiata* D. Don stands in north-western Spain. Annals of Forest Science 64, 453–465.

Chan, S.S., McCreight, R.W., Walstad, J.D., Spies, T.A., 1986. Evaluating forest vegetative cover with computerized analysis of fisheye photographs. Forest Science 32, 1085–1091.

Chen, J.-L., Reynolds, J.F., 1997. A coordination model of whole-plant carbon allocation in relation to water stress. Annals of Botany 80, 45–55.

Chen, J.M., Liu, J., Cihlar, J., Goulden, M.L., 1999. Daily canopy photosynthesis model through temporal and spatial scaling for remote sensing applications. Ecological Modelling 124, 99–119.

Chen, J.M., Menges, C.H., Leblanc, S.G., 2005. Global mapping of foliage clumping index using multi-angular satellite data. Remote Sensing of Environment 97, 447–457.

Chikumbo, O., Mareels, I.M.Y., Turner, B.J., 1999. Predicting stand basal area in thinned stands using a dynamical model. Forest Ecology and Management 116, 175–187.

Chisman, H.H., Schumacher, F.X., 1940. On the tree-area ratio and certain of its applications. Journal of Forestry 38, 311–317.

Chuine, I., Cour, P., Rousseau, D.D., 1999. Selecting models to predict the timing of flowering of temperate trees: Implications for tree phenology modelling. Plant, Cell and Environment 22, 1–13.

Cieszewski, C.J., 2001. Three methods of deriving advanced dynamic site equations demonstrated on inland Douglas-fir site curves. Canadian Journal of Forest Research 31, 165–173.

Cieszewski, C.J., Bailey, R.L., 2000. Generalized algebraic difference approach: Theory based derivation of dynamic site equations with polymorphism and variable asymptotes. Forest Science 46, 116–126.

Cieszewski, C.J., Bella, I.E., 1993. Modeling density-related lodgepole pine height growth, using Czarnowski's stand dynamics theory. Canadian Journal of Forest Research 23, 2499–2506.

Cieszewski, C.J., Strub, M.R., 2007. Parameter estimation of base-age invariant site index models: Which data structure to use? – A discussion. Forest Science 53, 552–555.

Cieszewski, C.J., Strub, M.R., 2008. Generalized algebraic difference approach derivation of dynamic site equations with polymorphism and variable asymptotes from exponential and logarithmic functions. Forest Science 54, 303–315.

Cieszewski, C.J., Zasada, M., Strub, M.R., 2006. Analysis of different base models and methods of site model derivation for Scots pine. Forest Science 52, 187–197.

Cieszewski, C.J., Strub, M.R., Zasada, M., 2007. New dynamic site equation that fits best the Schwappach data for Scots pine (*Pinus sylvestris* L.) in Central Europe. Forest Ecology and Management 243, 83–93.

Clark, J.S., Macklin, E., Wood, L., 1998. Stages and spatial scales of recruitment limitation in southern Appalachian forests. Ecological Monographs 68, 213–235.

Clark, J.S., Silman, M., Kern, R., Macklin, E., HilleRisLambers, J., 1999. Seed dispersal near and far: Patterns across temperate and tropical forests. Ecology 80, 1475–1494.

Clark, J.S., Wolosin, M., Dietze, M., Ibáñez, I., LaDeau, S., Welsh, M., Kloeppel, B., 2007. Tree growth inference and prediction from diameter censuses and ring widths. Ecological Applications 17, 1942–1953.

Clark, P.J., Evans, F.C., 1954. Distance to nearest neighbor as a measure of spatial relationships in populations. Ecology 35, 445–453.

Clark III, A.C., Souter, R.A., Schlaegel, B.E., 1991. Stem profile equations for southern tree species. Research Paper SE-282. USDA Forest Service, Southeastern Forest Experiment Station, Asheville, NC.

Cline, S.P., Berg, A.B., Wight, H.M., 1980. Snag characteristics and dynamics in Douglas-fir forests, western Oregon. Journal of Wildlife Management 44, 773–786.

Clutter, J.L., 1963. Compatible growth and yield models for loblolly pine. Forest Science 9, 354–371.

Clutter, J.L., Allison, B.J., 1974. A growth and yield model for *Pinus radiata* in New Zealand. In Fries, J. (Ed.), Growth models for tree and stand simulation. Research Note 30. Department of Forest Yield Research, Royal College of Forestry, Stockholm, Sweden, pp. 136–160.

Clutter, J.L., Jones, E.P., 1980. Prediction of growth after thinning in old field slash pine plantations. Research Paper SE-217. USDA Forest Service, Southeastern Forest Experiment Station, Asheville, NC.

Cluzeau, C., Le Goff, N., Ottorini, J.-M., 1994. Development of primary branches and crown profile for *Fraxinus excelsior*. Canadian Journal of Forest Research 24, 2315–2323.

Coates, K.D., Canham, C.D., Beaudet, M., Sachs, D.L., Messier, C., 2003. Use of a spatially explicit individual-tree model (SORTIE/BC) to explore the implications of patchiness in structurally complex forests. Forest Ecology and Management 186, 297–310.

Coble, D.W., Milner, K.S., Marshall, J.D., 2001. Above- and below-ground production of trees and other vegetation on contrasting aspects in western Montana: A case study. Forest Ecology and Management 142, 231–241.

Cole, D.M., Stage, A.R., 1972. Estimating future diameters of lodgepole pine. Research Paper INT-131. USDA Forest Service, Intermountain Forest and Range Experiment Station, Ogden, UT.

Cole, W.G., Lorimer, C.G., 1994. Predicting tree growth from crown variables in managed northern hardwood stands. Forest Ecology and Management 67, 159–175.

de Coligny, F., Dufour, S., Dreyfus, P., Courbaud, B., Auclair, D., 2010. CAPSIS: Computer-Aided Projection for Strategies In Silviculture. In Falcao, A. O., Rosset, C. (Eds.), Proceedings of the Workshop on Decision Support Systems in Sustainable Forest Management, Lisbon, Portugal, 19–21 April, 2010 Available at: www.fc.ul.pt/dsfm2010/docs/presentations/paper_6.pdf (accessed 17 March, 2011).

Condés, S., Sterba, H., 2005. Derivation of compatible crown width equations for some important tree species of Spain. Forest Ecology and Management 217, 203–218.

Coomes, D.A., Grubb, P.J., 2000. Impacts of root competition in forests and woodlands: A theoretical framework and review of experiments. Ecological Monographs 70, 171–207.

Coops, N.C., Waring, R.H., Landsberg, J.J., 2001. Estimation of potential forest productivity across the Oregon transect using satellite data and monthly weather records. International Journal of Remote Sensing 22, 3797–3812.

Corbeels, M., McMurtrie, R.E., Pepper, D.A., O'Connell, A.M., 2005a. A process-based model of nitrogen cycling in forest plantations. Part I. Structure, calibration, and analysis of the decomposition model. Ecological Modelling 187, 426–448.

Corbeels, M., McMurtrie, R.E., Pepper, D.A., O'Connell, A.M., 2005b. A process-based model of nitrogen cycling in forest plantations. Part II. Simulating growth and nitrogen mineralisation of *Eucalyptus globulus* plantations in south-western Australia. Ecological Modelling 187, 449–474.

Corral Rivas, J.J., Álvarez González, J.G., Aguirre, O., Hernández, F.J., 2005. The effect of competition on individual tree basal area growth in mature stands of *Pinus cooperi* Blanco in Durango (Mexico). European Journal of Forest Research 124, 133–142.

Cosby, B.J., Hornberger, G.M., Clapp, R.B., Ginn, T.R., 1984. A statistical exploration of the relationships of soil moisture characteristics to the physical properties of soils. Water Resources Research 20, 682–690.

Crecente-Campo, F., Marshall, P., LeMay, V., Diéguez-Aranda, U., 2009a. A crown profile model for *Pinus radiata* D. Don in northwestern Spain. Forest Ecology and Management 257, 2370–2379.

Crecente-Campo, F., Marshall, P., Rodríguez-Soalleiro, R., 2009b. Modeling non-catastrophic individual-tree mortality for *Pinus radiata* plantations in northwestern Spain. Forest Ecology and Management 257, 1542–1550.

Crecente-Campo, F., Soares, P., Tomé, M., Diéguez-Aranda, U., 2010. Modelling annual individual-tree growth and mortality of Scots pine with data obtained at irregular measurement intervals and containing missing observations. Forest Ecology and Management 260, 1965–1974.

Crookston, N.L., 1997. Suppose: An interface to the Forest Vegetation Simulator. In Teck, R., Moeur, M., Adams, J. (Eds.), Proceedings: Forest Vegetation Simulator Conference. General Technical Report INT-GTR-373. USDA Forest Service, Intermountain Research Station, Ogden, UT, pp. 7–14.

Crookston, N.L., Dixon, G.E., 2005. The forest vegetation simulator: A review of its structure, content, and applications. Computers and Electronics in Agriculture 49, 60–80.

Crookston, N.L., Colbert, J.J., Thomas, P.W., Sheehan, K.A., Kemp, W.P., 1990. User's guide to the western spruce budworm modeling system. General Technical Report INT-274. USDA Forest Service, Intermountain Research Station, Ogden, UT.

Crookston, N.L., Rehfeldt, G.E., Warwell, M.V., 2007. Using Forest Inventory and Analysis data to model plant-climate relationships. In McRoberts, R.E., Reams, G.A., Van Deusen, P.C., McWilliams, W.H. (Eds.), Proceedings of the seventh annual Forest Inventory and Analysis symposium. General Technical Report WO-77. USDA Forest Service, Washington, DC, pp. 243–250.

Crookston, N.L., Rehfeldt, G.E., Dixon, G.E., Weiskittel, A.R., 2010. Addressing climate change in the forest vegetation simulator to assess impacts on landscape forest dynamics. Forest Ecology and Management 260, 1198–1211.

Curt, T., Bouchaud, M., Agrech, G., 2001. Predicting site index of Douglas-Fir plantations from ecological variables in the Massif Central area of France. Forest Ecology and Management 149, 61–74.

Curtis, J.D., 1964a. Roots of a ponderosa pine. Research Paper INT-9. USDA Forest Service, Intermountain Forest and Range Experiment Station, Ogden, UT.

Curtis, R.O., 1964b. A stem-analysis approach to site-index curves. Forest Science 10, 241–256.

Curtis, R.O., 1967. Height-diameter and height-diameter-age equations for second-growth Douglas-fir. Forest Science 13, 365–375.

Curtis, R.O., 1970. Stand density measures: An interpretation. Forest Science 16, 403–414.

Curtis, R.O., 1972. Yield tables past and present. Journal of Forestry 70, 28–32.

Curtis, R.O., 1982. A simple index of stand density for Douglas-fir. Forest Science 28, 92–94.

Curtis, R.O., 2010. Effect of diameter limits and stand structure on relative density indices: A case study. Western Journal of Applied Forestry 25, 169–175.

Curtis, R.O., Marshall, D.D., 2000. Why quadratic mean diameter? Western Journal of Applied Forestry 15, 137–139.

Curtis, R.O., Marshall, D.D., 2005. Permanent-plot procedures for silvicultural and yield research. General Technical Report PNW-GTR-634. USDA Forest Service, Pacific Northwest Research Station, Portland, OR.

Curtis, R.O., Post, B.W., 1964. Basal area, volume, and diameter related to site index and age in unmanaged even-aged northern hardwoods in the Green Mountains. Journal of Forestry 62, 864–870.

Curtis, R.O., Clendenen, G.W., DeMars, D.J., 1981. A new stand simulator for coast Douglas-fir: DFSIM user's guide. General Technical Report PNW-128. USDA Forest Service, Pacific Northwest Research Station, Portland, OR.

D'Amato, A.W., Puettmann, K.J., 2004. The relative dominance hypothesis explains interaction dynamics in mixed species *Alnus rubra*/*Pseudotsuga menziesii* stands. Journal of Ecology 92, 450–463.

Dahms, W.G., 1983. Growth-simulation model for lodgepole pine in central Oregon. Research Paper PNW-302. USDA Forest Service, Portland, OR.

Damour, G., Simonneau, T., Cochard, H., Urban, L., 2010. An overview of models of stomatal conductance at the leaf level. Plant, Cell and Environment 33, 1419–1438.

Daniels, R.F., 1976. Simple competition indices and their correlation with annual loblolly pine tree growth. Forest Science 22, 454–456.

Daniels, R.F., Burkhart, H.E., 1975. Simulation of individual tree growth and stand development in managed loblolly pine plantations. Publication FWS-5-75. Virginia Polytechnic Institute and State University, Blacksburg, VA.

Daniels, R.F., Burkhart, H.E., 1988. An integrated system of forest stand models. Forest Ecology and Management 23, 159–177.

Daniels, R.F., Burkhart, H.E., Clason, T.R., 1986. A comparison of competition measures for predicting growth of loblolly pine trees. Canadian Journal of Forest Research 16, 1230–1237.

Danjon, F., 1995. Observed selection effects on height growth, diameter and stem form in maritime pine. Silvae Genetica 44, 10–19.

Daubenmire, R., 1976. The use of vegetation in assessing the productivity of forest lands. Botanical Review 42, 115–143.

De'ath, G., 2007. Boosted trees for ecological modeling and prediction. Ecology 88, 243–251.

Dean, T.J., 2001. Potential effect of stand structure on belowground allocation. Forest Science 47, 69–76.

Deckmyn, G., Mali, B., Kraigher, H., Torelli, N., Op de Beeck, M., Ceulemans, R., 2009. Using the process-based stand model ANAFORE including Bayesian optimisation to predict wood quality and quantity and their uncertainty in Slovenian beech. Silva Fennica 43, 523–534.

Deleuze, C., Hervé, J.-C., Colin, F., Ribeyrolles, L., 1996. Modelling crown shape of *Picea abies*: Spacing effects. Canadian Journal of Forest Research 26, 1957–1966.

Deleuze, C., Pain, O., Dhôte, J.-F., Hervé, J.-C., 2004. A flexible radial increment model for individual trees in pure even-aged stands. Annals of Forest Science 61, 327–335.

Dell, T.R., 1979. Potential for using crown ratio in predicting product yield. In Proceedings of 1979 SAF/IUFRO Forest Resource Inventories Workshop. Colorado State University, Fort Collins, CO, pp. 843–851.

Deutschman, D.H., Devine, C., Buttel, L.A., 2000. The role of visualization in understanding a complex forest simulation model. Computer Graphics 34, 51–55.

Dey, D.C., Johnson, P.S., Garrett, H.E., 1996. Modeling the regeneration of oak stands in the Missouri Ozark Highlands. Canadian Journal of Forest Research 26, 573–583.

Dhôte, J.-F., 1994. Hypotheses about competition for light and water in even-aged common beech (*Fagus sylvatica* L.). Forest Ecology and Management 69, 219–232.

Dhrymes, P.J., Howrey, H.P., Hymans, S.H., Kmenta, J., Leamer, E.E., Quandt, R.E., Ransey, J.B., Shapiro, H.T., Zarnowitz, V., 1972. Criteria for evaluation of econometric models. Annals of Economic and Social Measurement 1, 291–325.

Di Lucca, C.M., 1999. TASS/SYLVER/TIPSY: Systems for predicting the impact of silvicultural practices on yield, lumber value, economic return and other benefits. In Bamsey, C.R. (Ed.), Stand Density Management Conference: Using the Planning Tools, Edmonton, AB, pp. 7–16.

Di Lucca, C.M., Polsson, K., Goudie, J., Bogle, T., 2003. Yield implications of variable retention harvesting. Presentation at Western Mensurationist Meeting, July 3, 2003, Victoria, British Columbia.

Diéguez-Aranda, U., Castedo-Dorado, F., Álvarez-González, J.G., Rodríguez-Soalleiro, R., 2005. Modelling mortality of Scots pine (*Pinus sylvestris* L.) plantations in the northwest of Spain. European Journal of Forest Research 124, 143–153.

Diéguez-Aranda, U., Castedo-Dorado, F., Álvarez-González, J.G., Rojo, A., 2006. Compatible taper function for Scots pine plantations in northwestern Spain. Canadian Journal of Forest Research 36, 1190–1205.

Dietze, M.C., Wolosin, M.S., Clark, J.S., 2008. Capturing diversity and interspecific variability in allometries: A hierarchical approach. Forest Ecology and Management 256, 1939–1948.

Dixon, G.E., 1985. Crown ratio modeling using stand density index and the Weibull distribution. Internal Report. USDA Forest Service, Forest Management Service Center, Fort Collins, CO.

Dixon, G.E., White, R., Frank, J., 1995. Northeast (NE) Variant Overview. USDA Forest Service, Forest Management Service Center, Fort Collins, CO.

Dixon, G.E., White, R., Frank, J., 2007. Northeast (NE) Variant Overview. USDA Forest Service, Forest Management Service Center, Fort Collins, CO.

Dobbertin, M., Biging, G.S., 1998. Using the non-parametric classifier CART to model forest tree mortality. Forest Science 44, 507–516.

Dobbertin, M., Brang, P., 2001. Crown defoliation improves tree mortality models. Forest Ecology and Management 141, 271–284.

Dolph, K.L., 1988. Prediction of periodic basal area increment for young-growth mixed conifers in the Sierra Nevada. Research Paper PSW-190. USDA Forest Service, Pacific Southwest Forest and Range Experiment Station, Berkeley, CA.

Dolph, K.L., 1992. Predicting height increment of young-growth red fir in California and southern Oregon. Research Paper PSW-210. USDA Forest Service, Pacific Southwest Research Station, Albany, CA.

Doolittle, W.T., Vimmerstedt, J.P., 1960. Site index curves for natural stands of white pine in the southern Appalachians. Research Note 141. USDA Forest Service, Southeastern Forest Experiment Station, Asheville, NC.

Doruska, P.F., Mays, J.E., 1998. Crown profile modeling of loblolly pine by nonparametric regression analysis. Forest Science 44, 445–453.

Draper, N.R., Smith, H., 1998. Applied Regression Analysis 3rd edition. John Wiley & Sons, Inc., New York, NY.

Drew, T.J., Flewelling, J.W., 1977. Some recent Japanese theories of yield-density relationships and their application to Monterey pine plantations. Forest Science 23, 517–534.

Drew, T.J., Flewelling, J.W., 1979. Stand density management: An alternative approach and its application to Douglas-fir plantations. Forest Science 25, 518–532.

Ducey, M., 2009a. The ratio of additive and traditional stand density indices. Western Journal of Applied Forestry 24, 5–10.

Ducey, M.J., 2009b. Predicting crown size and shape from simple stand variables. Journal of Sustainable Forestry 28, 5–21.

Ducey, M., Allen, H.L., 2001. Nutrient supply and fertilization efficiency in midrotation loblolly pine plantations: A modeling analysis. Forest Science 47, 96–102.

Ducey, M.J., Knapp, R.A., 2010. A stand density index for complex mixed species forests in the northeastern United States. Forest Ecology and Management 260, 1613–1622.

Ducey, M.J., Larson, B.C., 2003. Is there a correct stand density index? An alternative interpretation. Western Journal of Applied Forestry 18, 179–184.

Dudek, A., Ek, A.R., 1980. A bibliography of worldwide literature on individual tree based forest stand growth models. Staff Paper Series 12. University of Minnesota, Department of Forest Resources, St. Paul, MN.

Duerr, W.A., Gevorkiantz, S.R., 1938. Growth prediction and site determination in uneven-aged timber stands. Journal of Agricultural Research 56, 81–98.

Dunbar, A., Dhubhain, A.N., Bulfin, M., 2002. The productivity of Douglas fir in Ireland. Forestry, 75, 537–545.

Duursma, R.A., 2004. A simple process-based model of forest growth and a test for the Priest River Experimental Forest PhD dissertation. University of Idaho, Moscow, ID.

Duursma, R.A., Mäkelä, A., 2007. Summary models for light interception and light-use efficiency of non-homogeneous canopies. Tree Physiology 27, 859–870.

Duursma, R.A., Robinson, A.P., 2003. Bias in the mean tree model as a consequence of Jensens' inequality. Forest Ecology and Management 186, 373–380.

Duursma, R.A., Marshall, J.D., Robinson, A.P., 2003. Leaf area index inferred from solar beam transmission in mixed conifer forests on complex terrain. Agricultural and Forest Meteorology 118, 221–236.

Duzan Jr., H.W., Allen, H.L., Ballard, R., 1982. Predicting fertilizer response in established loblolly pine plantations with basal area and site index. Southern Journal of Applied Forestry 6, 15–19.

Dyer, M.E., Burkhart, H.E., 1987. Compatible crown ratio and crown height models. Canadian Journal of Forest Research 17, 572–574.

Dzierzon, H., Mason, E.G., 2006. Towards a nationwide growth and yield model for radiata pine plantations in New Zealand. Canadian Journal of Forest Research 36, 2533–2543.

Eamus, D., McGuinness, K., Burrows, W., 2000. Review of allometric relationship for estimating woody biomass for Queensland, the Northern Territory and Western Australia. National Carbon Accounting System Technical Report 5A. Australian Greenhouse Office, Canberra, Australia.

Eichorn, F., 1904. Beziehungen zwischen Bestandshöhe und Bestandsmasse. Allgemeine Forst und Jagdzeitung 80, 45–49.

Eid, T., Øyen, B.-H., 2003. Models for prediction of mortality in even-aged forest. Scandinavian Journal of Forest Research 18, 64–77.

Eid, T., Tuhus, E., 2001. Models for individual tree mortality in Norway. Forest Ecology and Management 154, 69–84.

Eis, S., 1972. Root grafts and their silvicultural implications. Canadian Journal of Forest Research 2, 111–120.

Eis, S., 1974. Root system morphology of western hemlock, western red cedar, and Douglas-fir. Canadian Journal of Forest Research 4, 28–38.

Eis, S., 1986. Differential growth of individual components of trees and their interrelationships. Canadian Journal of Forest Research 16, 352–359.

Eis, S., 1987. Root systems of older immature hemlock, cedar, and Douglas-fir. Canadian Journal of Forest Research 17, 1348–1354.

Ek, A.R., 1974a. Nonlinear models for stand table projection in northern hardwood stands. Canadian Journal of Forest Research 4, 23–27.

Ek, A.R., 1974b. Dimensional relationships of forest and open grown trees in Wisconsin. Forestry Research Note 181. University of Wisconsin, Madison, WI.

Ek, A.R., Monserud, R.A., 1974. Trials with program FOREST: Growth and reproduction simulation for mixed species even- or uneven-aged forest stands. In Fries, J. (Ed.), Growth Models for Tree and Stand Simulation. Research Note 30. Department of Forest Yield Research Royal College of Forestry, Stockholm, Sweden, pp. 56–73.

Ek, A.R., Birdsall, E.T., Spears, R.J., 1984. A simple model for estimating total and merchantable tree heights. Research Note NC-309. US Forest Service, North Central Forest Experiment Station, St. Paul, MN.

Ek, A.R., Robinson, A.P., Radtke, P.J., Walters, D.K., 1997. Development and testing of regeneration imputation models for forests in Minnesota. Forest Ecology and Management 94, 129–140.

Emmingham, W.H., 1977. Comparison of selected Douglas-fir seed sources for cambial and leader growth patterns in four western Oregon environments. Canadian Journal of Forest Research 7, 154–164.

Emmingham, W.H., Waring, R.H., 1977. An index of photosynthesis for comparing forest sites in western Oregon. Canadian Journal of Forest Research 7, 165–174.

Eskelson, B.N.I., Temesgen, H., LeMay, V., Barrett, T.M., Crookston, N.L., Hudak, A.T., 2009. The roles of nearest neighbor methods in imputing missing data in forest inventory and monitoring databases. Scandinavian Journal of Forest Research 24, 235–246.

Esprey, L.J., Sands, P.J., Smith, C.W., 2004. Understanding 3-PG using a sensitivity analysis. Forest Ecology and Management 193, 235–250.

Falkowski, M.J., Hudak, A.T., Crookston, N.L., Gessler, P.E., Uebler, E.H., Smith, A.M.S., 2010. Landscape-scale parameterization of a tree-level forest growth model: A k-nearest neighbor imputation approach incorporating LiDAR data. Canadian Journal of Forest Research 40, 184–199.

Fan, Z., Kabrick, J.M., Shifley, S.R., 2006. Classification and regression tree based survival analysis in oak-dominated forests of Missouri's Ozark highlands. Canadian Journal of Forest Research 36, 1740–1748.

Fang, Z., Bailey, R.L., 2001. Nonlinear mixed effects modeling for slash pine dominant height growth following intensive silvicultural treatments. Forest Science 47, 287–300.

Fang, Z., Borders, B.E., Bailey, R.L., 2000. Compatible volume-taper models for loblolly and slash pine based on a system with segmented-stem form factors. Forest Science 46, 1–12.

Fang, Z., Bailey, R.L., Shiver, B.D., 2001. A multivariate simultaneous prediction system for stand growth and yield with fixed and random parameters. Forest Science 47, 550–562.

Farquhar, G.D., von Caemmerer, S., Berry, J.A., 1980. A biochemical model of photosynthetic CO_2 assimilation in leaves of C_3 leaves. Planta 149, 78–90.

Farr, W.A., Harris, A.S., 1979. Site index of Sitka spruce along the Pacific Coast related to latitude and temperatures. Forest Science 25, 145–153.

Farr, W.A., DeMars, D.J., Dealy, J.E., 1989. Height and crown width related to diameter for opengrown western hemlock and Sitka spruce. Canadian Journal of Forest Research 19, 1203–1207.

Farrar Jr., R.M., 1987. Stem-profile functions for predicting multiple-product volumes in natural longleaf pines. Southern Journal of Applied Forestry 11, 161–167.

Ferguson, D.E., Stage, A.R., Boyd, R.J., 1986. Predicting regeneration in the grand fir-cedar-hemlock ecosystem of the northern Rocky Mountains. Forest Science Monograph 26.

Fight, R.D., Bolon, N.A., Cahill, J.M., 1993. Financial analysis of pruning Douglas-fir and ponderosa pine in the Pacific Northwest. Western Journal of Applied Forestry 8, 58–61.

Fight, R.D., Knowles, L., McInnes, I., 1995. Effect of pruning on early growth and stand dynamics in Douglas-fir plantations. In Proceedings of IUFRO XX World Congress, Tampere, Finland.

Filho, A.F., Schaaf, L.B., 1999. Comparison between predicted volumes estimated by taper equations and true volumes obtained by the water displacement technique (xylometer). Canadian Journal of Forest Research 29, 451–461.

Filipescu, C.N., Comeau, P.G., 2007. Competitive interactions between aspen and white spruce vary with stand age in boreal mixedwoods. Forest Ecology and Management 247, 175–184.

Finley, A.O., Banerjee, S., McRoberts, R.E., 2009. Hierarchical spatial models for predicting tree species assemblages across large domains. Annals of Applied Statistics 3, 1052–1079.

Fisher, R.F., Binkley, D., 2000. Ecology and Management of Forest Soils, 3rd Edition. John Wiley & Sons, Inc., New York, NY.

Flewelling, J.W., Monserud, R.A., 2002. Comparing methods for modelling tree mortality. In Crookston, N.L., Havis, R.N. (Eds.), Second Forest Vegetation Simulator Conference, Fort Collins, Colorado, February 12–14, 2002. Proceedings RMRS-P-25. USDA Forest Service, Rocky Mountain Research Station, Fort Collins, CO, pp. 168–177.

Flewelling, J.W., Pienaar, L.V., 1981. Multiplicative regression with lognormal errors. Forest Science 27, 281–289.

Flewelling, J., Collier, R., Gonyea, B., Marshall, D., Turnblom, E., 2001. Height-age curves for planted stands of Douglas-fir, with adjustments for density. Stand Management Cooperative Working Paper 1. University of Washington, College of Forest Resources, Seattle, WA.

Fonseca, T.F., Marques, C.P., Parresol, B.R., 2009. Describing maritime pine diameter distributions with Johnson's S_B distribution using a new all-parameter recovery approach. Forest Science 55, 367–373.

Fontes, L., Tomé, M., Thompson, F., Yeomans, A., Luis, J.S., Savill, P., 2003. Modelling the Douglas-fir (*Pseudotsuga menziesii* (Mirb.) Franco) site index from site factors in Portugal. Forestry 76, 491–507.

Ford, E.D., Diggle, P.J., 1981. Competition for light in a plant monoculture modelled as a spatial stochastic process. Annals of Botany 48, 481–500.

Ford, E.D., Sorrensen, K.A., 1992. Theory and models of inter-plant competition as a spatial process. In DeAngelis, D.L., Gross, L.J. (Eds.), Individual-Based Models and Approaches in Ecology. Chapman & Hall, New York, NY, pp. 363–407.

Fortin, M., DeBlois, J., 2007. Modeling tree recruitment with zero-inflated models: The example of hardwood stands in southern Québec, Canada. Forest Science 53, 529–539.

Fortin, M., DeBlois, J., 2010. A statistical estimator to propagate height prediction errors into a general volume model. Canadian Journal of Forest Research 40, 1930–1939.

Fortin, M., Bédard, S., DeBlois, J., Meunier, S., 2009. Assessing and testing prediction uncertainty for single tree-based models: A case study applied to northern hardwood stands in southern Québec, Canada. Ecological Modelling 220, 2770–2781.

Fox, J.C., Ades, P.K., Bi, H., 2001. Stochastic structure and individual-tree growth models. Forest Ecology and Management 154, 261–276.

Fox, T.R., 2000. Sustained productivity in intensively managed forest plantations. Forest Ecology and Management 138, 187–202.

Fox, T.R., Allen, H.L., Albaugh, T.J., Rubilar, R., Carlson, C.A., 2007. Tree nutrition and forest fertilization of pine plantations in the southern United States. Southern Journal of Applied Forestry 31, 5–11.

Franklin, J.F., Dyrness, C.T., 1973. Natural vegetation of Oregon and Washington. General Technical Report PNW-8. USDA Forest Service, Portland, OR.

Frazer, G.W., Canham, C.D., Lertzman, K.P., 1999. Gap light anayzer (GLA), Version 2.0: Imaging software to extract canopy structure and gap light transmission indices from true-color fisheye photographs: Users manual and program documentation. Simon Fraser University, Burnaby, BC.

Frazier, B.E., Rodgers, T.M., Briggs, C.A., Rupp, R.A., 2009. Remote area soil proxy modeling technique. Soil Survey Horizons 50, 62–67.

Freese, F., 1960. Testing accuracy. Forest Science 6, 139–145.

Fridman, J., Ståhl, G., 2001. A three-step approach for modelling tree mortality in Swedish forests. Scandinavian Journal of Forest Research 16, 455–466.

Friend, A.D., Stevens, A.K., Knox, R.G., Cannell, M.G.R., 1997. A process-based, terrestrial biosphere model of ecosystem dynamics (Hybrid v3.0). Ecological Modelling 95, 249–287.

Fries, J., 1974. Growth models for tree and stand simulation. Research Note 30. Department of Forest Yield Research, Royal College of Forestry, Stockholm, Sweden.

Froese, R.E., Robinson, A.P., 2007. A validation and evaluation of the Prognosis individual-tree basal area increment model. Canadian Journal of Forest Research 37, 1438–1449.

Fuller, W.A., 1987. Measurement Error Models. John Wiley & Sons, Inc., New York, NY.

Furnival, G.M., 1961. An index for comparing equations used in constructing volume tables. Forest Science 7, 337–341.

Furnival, G.M., 1987. Growth and yield prediction: Some criticisms and suggestions. In Chappell, H.N., Maguire, D.A. (Eds.), Predicting forest growth and yield: Current issues, future prospects. Contribution Number 58. Institute of Forest Resources, University of Washington, Seattle, WA, pp. 23–30.

Furnival, G.M., Wilson, R.W., 1971. Systems of equations for predicting forest growth and yield. In Patil, G.P., Pielou, E.C., Walters, W.E. (Eds.), Statistical Ecology, Volume 3 Penn State University Press, University Park, PA, pp. 43–57.

Furnival, G.M., Gregoire, T.G., Valentine, H.T., 1990. An analysis of three methods for fitting site-index curves. Forest Science 36, 464–469.

Garber, S.M., Maguire, D.A., 2003. Modeling stem taper of three central Oregon species using nonlinear mixed effects models and autoregressive error structures. Forest Ecology and Management 179, 507–522.

Garber, S.M., Brown, J.P., Wilson, D.S., Maguire, D.A., Heath, L.S., 2005. Snag longevity under alternative silvicultural regimes in mixed-species forests of central Maine. Canadian Journal of Forest Research 35, 787–796.

Garber, S.M., Monserud, R.A., Maguire, D.A., 2008. Crown recession patterns in three conifer species of the northern Rocky Mountains. Forest Science 54, 633–646.

Garber, S.M., Temesgen, H., Monleon, V.J., Hann, D.W., 2009. Effects of height imputation strategies on stand volume estimation. Canadian Journal of Forest Research 39, 681–690.

García, O., 1974. Sobre modelos matemáticos de rodal [On mathematical stand models]. Informe Ténico Number 48. Instituto Forestal, Santiago, Chile.

García, O., 1984. New class of growth models for even-aged stands: *Pinus radiata* in Golden Downs Forest. New Zealand Journal of Forestry 14, 65–88.

García, O., 1994. The state-space approach in growth modelling. Canadian Journal of Forest Research 24, 1894–1903.

García, O., 1998. Estimating top height with variable plot sizes. Canadian Journal of Forest Research 28, 1509–1517.

García, O., 2005a. Comparing and combining stem analysis and permanent sample plot data in site index models. Forest Science 51, 277–283.

García, O., 2005b. Thinking about time. In Naito, K. (Ed.), The Role of Forests for Coming Generations – Philosophy and Technology for Forest Resource Management. Japan Society of Forest Planning Press, Utsunomiya, Japan, pp. 47–54.

García, O., 2005c. Unifying sigmoid univariate growth equations. Forest Biometry, Modelling and Information Sciences 1, 63–68.

García, O., 2006. Scale and spatial structure effects on tree size distributions: Implications for growth and yield modelling. Canadian Journal of Forest Research 36, 2983–2993.

García, O., 2009. A simple and effective forest stand mortality model. Mathematical and Computational Forestry & Natural-Resources Sciences 1, 1–9.

García, O., 2011. A parsimonious dynamic stand model for interior spruce in British Columbia. Forest Science, in press.

Garman, S.L., Acker, S.A., Ohmann, J.L., Spies, T.A., 1995. Asymptotic height-diameter equations for twenty-four tree species in western Oregon. Research Contribution 10. Oregon State University, Forest Research Laboratory, Corvallis, OR.

Gelman, A., Hill, J., 2007. Data Analysis Using Regression and Multilevel/Hierarchical Models. Cambridge University Press, New York, NY.

Gertner, G., 1987. Approximating precision in simulation projections: An efficient alternative to Monte Carlo methods. Forest Science 33, 230–239.

Gertner, G., 1989. The need to improve models for individual tree mortality. In Rink, G., Budelsky, C.A. (Eds.), Proceedings of the 7th Central Hardwood Conference. General Technical Report NC-132. USDA Forest Service, North Central Forest Experiment Station, St. Paul, MN, pp. 59–61.

Gertner, G.Z., Dzialowy, P.J., 1984. Effects of measurement errors on an individual tree-based growth projection system. Canadian Journal of Forest Research 14, 311–316.

Gessel, S.P., Steinbrenner, E.C., Miller, R.E., 1979. Response of northwest forests to elements other than nitrogen. In Gessel, S.P., Kenady, R.M., Atkinson, W.A. (Eds.), Proceedings of Forest Fertilization Conference. Contribution Number 40. University of Washington Institute of Forest Resources, Seattle, WA, pp. 140–149.

Giardina, C.P., Ryan, M.G., 2002. Total belowground carbon allocation in a fast-growing *Eucalyptus* plantation estimated using a carbon balance approach. Ecosystems 5, 487–499.

Gifford, R.M., 2003. Plant respiration in productivity models: Conceptualisation, representation and issues for global terrestrial carbon-cycle research. Functional Plant Biology 30, 171–186.

Gill, S.J., Biging, G.S., 2002a. Autoregressive moving average models of conifer crown profiles. Journal of Agricultural, Biological, and Environmental Statistics 7, 558–573.

Gill, S.J., Biging, G.S., 2002b. Autoregressive moving average models of crown profiles for two California hardwood species. Ecological Modelling 152, 213–226.

Gill, S.J., Biging, G.S., Murphy, E.C., 2000. Modeling conifer tree crown radius and estimating canopy cover. Forest Ecology and Management 126, 405–416.

Gilmore, D.W., 2001. Equations to describe crown allometry of *Larix* require local validation. Forest Ecology and Management 148, 109–116.

Glover, G.R., Hool, J.N., 1979. A basal area ratio predictor of loblolly pine plantation mortality. Forest Science 25, 275–282.

Godin, C., 2000. Representing and encoding plant architecture: A review. Annals of Forest Science 57, 413–438.

Goelz, J.C.G., Burk, T.E., 1992. Development of a well-behaved site index equation: Jack pine in north central Ontario. Canadian Journal of Forest Research 22, 776–784.

Goelz, J.C.G., Burk, T.E., 1996. Measurement error causes bias in site index equations. Canadian Journal of Forest Research 26, 1585–1593.

Golser, M., Hasenauer, H., 1997. Predicting juvenile tree height growth in uneven-aged mixed species stands in Austria. Forest Ecology and Management 97, 133–146.

Gompertz, B., 1825. On the nature of the function expressive of the law of human mortality, and on a new mode of determining the value of life contingencies. Philosophical Transactions of the Royal Society of London 115, 513–583.

Goodwin, A.N., 2009. A cubic tree taper model. Australian Forestry 72, 87–98.

Goreaud, F., Alvarez, I., Courbaud, B., de Coligny, F., 2006. Long-term influence of the spatial structure of an initial state on the dynamics of a forest growth model: A simulation study using the CAPSIS platform. Simulation 82, 475–495.

Goudie, J.W., Polsson, K.R., Ott, P.K., 2009. An empirical model of crown shyness for lodgepole pine (*Pinus contorta* var. *latifolia* [Engl.] Critch.) in British Columbia. Forest Ecology and Management 257, 321–331.

Goudriaan, J., 1986. A simple and fast numerical method for the computation of daily totals of crop photosynthesis. Agricultural and Forest Meteorology 38, 249–254.

Gould, P.J., Harrington, C.A., 2008. Extending sapwood-leaf area relationships from stems to roots in coast Douglas-fir. Annals of Forest Science 65, 802.

Gould, P.J., Marshall, D.D., 2010. Incorporation of genetic gain into growth projections of Douglas-Fir using ORGANON and the Forest Vegetation Simulator. Western Journal of Applied Forestry 25, 55–61.

Gould, P.J., Steiner, K.C., McDill, M.E., Finley, J.C., 2006. Modeling seed-origin oak regeneration in the central Appalachians. Canadian Journal of Forest Research 36, 833–844.

Gould, P.J., Fei, S., Steiner, K.C., 2007. Modeling sprout-origin oak regeneration in the Central Appalachians. Canadian Journal of Forest Research 37, 170–177.

Gould, P., Johnson, R., Marshall, D., Johnson, G., 2008. Estimation of genetic-gain multipliers for Douglas-fir height and diameter growth. Forest Science 54, 588–596.

Grace, J.C., Pont, D., Goulding, C.J., Rawley, B., 1999. Modelling branch development for forest management. New Zealand Journal of Forestry Science 29, 391–408.

Granberg, H.B., 1988. A forest shading model using bit-mapped graphics. Agricultural and Forest Meteorology 43, 225–234.

Grant, R.F., Zhang, Y., Yuan, F., Wang, S., Hanson, P.J., Gaumont-Guay, D., Chen, J., Black, T.A., Barr, A., Baldocchi, D.D., Arain, A., 2006. Intercomparison of techniques to model water stress effects on CO_2 and energy exchange in temperate and boreal deciduous forests. Ecological Modelling 196, 289–312.

Green, E.J., Strawderman, W.E., 1996. A Bayesian growth and yield model for slash pine plantations. Journal of Applied Statistics 23, 285–300.

Green, E.J., MacFarlane, D.W., Valentine, H.T., 2000. Bayesian synthesis for quantifying uncertainty in predictions from process models. Tree Physiology 20, 415–419.

Greene, D.F., Calogeropoulos, C., 2002. Measuring and modelling seed dispersal of terrestrial plants. In Bullock, J.M., Kenward, R.E., Hails, R.S. (Eds.), Dispersal Ecology: 42nd Symposium of the British Ecological Society. Blackwell Publishing, Oxford, UK, pp. 3–23.

Greene, D.F., Johnson, E.A., 1994. Estimating the mean annual seed production of trees. Ecology 75, 642–647.

Greene, D.F., Johnson, E.A., 2004. Modelling the temporal variation in the seed production of North American trees. Canadian Journal of Forest Research 34, 65–75.

Greene, D.F., Zasada, J.C., Sirois, L., Kneeshaw, D., Morin, H., Charron, I., Simard, M.-J., 1999. A review of the regeneration dynamics of North American boreal forest tree species. Canadian Journal of Forest Research 29, 824–839.

Greene, D.F., Canham, C.D., Coates, K.D., Lepage, P.T., 2004. An evaluation of alternative dispersal functions for trees. Journal of Ecology 92, 758–766.

Gregoire, T.G., 1987. Generalized error structure for forestry yield models. Forest Science 33, 423–444.

Gregoire, T.G., Reynolds, M.R., 1988. Accuracy testing and estimation alternatives. Forest Science 34, 302–320.

Gregoire, T.G., Schabenberger, O., 1996. A non-linear mixed-effects model to predict cumulative bole volume of standing trees. Journal of Applied Statistics 23, 257–271.

Gregoire, T.G., Schabenberger, O., Barrett, J.P., 1995. Linear modelling of irregularly spaced, unbalanced, longitudinal data from permanent-plot measurements. Canadian Journal of Forest Research 25, 137–156.

Gregoire, T.G., Lin, Q.F., Boudreau, J., Nelson, R., 2008. Regression estimation following the square-root transformation of the response. Forest Science 54, 597–606.

Grier, C.C., Waring, R.H., 1974. Conifer foliage mass related to sapwood area. Forest Science 20, 205–206.

Grier, C.C., Lee, K.M., Archibald, R.M., 1984. Effects of urea fertilization on allometric relations in young Douglas-fir trees. Canadian Journal of Forest Research 14, 900–904.

Griffiths, R.P., Bradshaw, G.A., Marks, B., Lienkaemper, G.W., 1996. Spatial distribution of ectomycorrhizal mats in coniferous forests of the Pacific Northwest, USA. Plant and Soil 180, 147–158.

Grigal, D.F., 2009. A soil-based aspen productivity index for Minnesota. Forest Ecology and Management 257, 1465–1473.

Groot, A., 2004. A model to estimate light interception by tree crowns, applied to black spruce. Canadian Journal of Forest Research 34, 788–799.

Groot, A., Saucier, J.-P., 2007. IVY: An individual-tree growth model for complex-structured stands. In Newton, P.F., LeMay, V.M. (Eds.), Complex stand structures and associated dynamics: Measurement indices and modelling approaches. Forest Research Paper 167. Ontario Ministry of Natural Resources, Forest Research Institute, Sault St. Marie, Ontario, pp. 78–79.

Groot, A., Gauthier, S., Bergeron, Y., 2004. Stand dynamics modelling approaches for multicohort management of eastern Canadian boreal forests. Silva Fennica 38, 437–448.

Grote, R., 1998. Integrating dynamic morphological properties into forest growth modelling. II. Allocation and mortality. Forest Ecology and Management 111, 193–210.

Grote, R., Pretzsch, H., 2002. A model for individual tree development based on physiological processes. Plant Biology 4, 167–180.

Gu, L., Pallardy, S.G., Tu, K., Law, B.E., Wullschleger, S.D., 2010. Reliable estimation of biochemical parameters from C_3 leaf photosynthesis-intercellular carbon dioxide response curves. Plant, Cell and Environment 33, 1852–1874.

Guan, B.T., Gertner, G., 1991a. Modeling red pine tree survival with an artificial neural network. Forest Science 37, 1429–1440.

Guan, B.T., Gertner, G., 1991b. Using a parallel distributed processing system to model individual tree mortality. Forest Science 37, 871–885.

Guan, B.T., Gertner, G.Z., Parysow, P., 1997. A framework for uncertainty assessment of mechanistic forest growth models: A neural network example. Ecological Modelling 98, 47–58.

Guiterman, C.H., Seymour, R.S., Weiskittel, A.R., in press. Influence of contrasting thinning regimes on leaf area index of *Pinus strobus*: Long-term trends predicted from allometric equations and litterfall data. Forest Science, in press.

Guo, J.-P., Wang, J.R., 2006. Comparison of height growth and growth intercept models of jack pine plantations and natural stands in northern Ontario. Canadian Journal of Forest Research 36, 2179–2188.

Hahn, J.T., Leary, R.A., 1979. Potential diameter growth functions. In A generalized forest growth projection system applied to the Lake States. General Technical Report NC-49. USDA Forest Service, North Central Research Station, St. Paul, MN, pp. 22–26.

Hamel, B., Bélanger, N., Paré, D., 2004. Productivity of black spruce and jack pine stands in Québec as related to climate, site biological features and soil properties. Forest Ecology and Management 191, 239–251.

Hamilton Jr., D.A., 1974. Event probabilities estimated by regression. Research Paper INT-152. USDA Forest Service, Intermountain Forest and Range Experiment Station, Ogden, UT.

Hamilton Jr., D.A., 1986. A logistic model of mortality in thinned and unthinned mixed conifer stands of northern Idaho. Forest Science 32, 989–1000.

Hamilton Jr., D.A., 1990. Extending the range of applicability of an individual tree mortality model. Canadian Journal of Forest Research 20, 1212–1218.

Hamilton Jr., D.A., Rehfeldt, G.E., 1994. Using individual tree growth projection models to estimate stand-level gains attributable to genetically improved stock. Forest Ecology and Management 68, 189–207.

Hanewinkel, M., Pretzsch, H., 2000. Modelling the conversion from even-aged to uneven-aged stands of Norway spruce (*Picea abies* L. Karst.) with a distance-dependent growth simulator. Forest Ecology and Management 134, 55–70.

Hann, D.W., 1978. The development and evaluation of an even- and uneven-aged ponderosa pine/Arizona fescue stand simulator with managerial decision-making potential. PhD dissertation. University of Washington, Seattle, WA.

Hann, D.W., 1980. Development and evaluation of an even- and uneven-aged ponderosa pine/Arizona fescue stand simulator. Research Paper INT-267. USDA Forest Service, Ogden, UT.

Hann, D.W., 1994. A key to the literature presenting tree volume and taper equations for species in the Pacific Northwest and California. Research Contribution 6. Oregon State University, Forest Research Laboratory, Corvallis, OR.

Hann, D.W., 1997. Equations for predicting the largest crown width of stand-growth trees in Western Oregon. Research Contribution 17. Oregon State University, Forest Research Laboratory, Corvallis, OR.

Hann, D.W., 1998. Extending southwest Oregon's Douglas-fir dominant height growth equation to older ages. Research Contribution 18. Oregon State University, Forest Research Laboratory, Corvallis, OR.

Hann, D.W., 1999. An adjustable predictor of crown profile for stand-grown Douglas-fir trees. Forest Science 45, 217–225.

Hann, D.W., 2011. ORGANON user's manual edition 9.0. Oregon State University, Department of Forest Engineering, Resources, and Management, Corvallis, OR.

Hann, D.W., Bare, B.B., 1979. Uneven-aged forest management: State of the art (or science?). General Technical Report INT-50. USDA Forest Service, Intermountain Forest and Range Experiment Station, Ogden, UT.

Hann, D.W., Brodie, J.D., 1980. Even-aged forest management: Basic managerial questions and available or potential techniques for answering them. General Technical Report INT-83. USDA Forest Service, Intermountain Forest and Range Experiment Station, Ogden, UT.

Hann, D.W., Hanus, M., 2001. Enhanced mortality equations for trees in the mixed conifer zone of southwest Oregon. Research Contribution 34. Oregon State University, Forest Research Laboratory, Corvallis, OR.

Hann, D.W., Hanus, M.L., 2002a. Enhanced diameter-growth-rate equations for undamaged and damaged trees in southwest Oregon. Research Contribution 39. Oregon State University, Forest Research Laboratory, Corvallis, OR.

Hann, D.W., Hanus, M.L., 2002b. Enhanced height-growth-rate for undamaged and damaged trees in southwest Oregon. Research Contribution 41. Oregon State University, Forest Research Laboratory, Corvallis, OR.

Hann, D.W., Hanus, M.L., 2004. Evaluation of nonspatial approaches and equation forms used to predict tree crown recession. Canadian Journal of Forest Research 34, 1993–2003.

Hann, D.W., Larsen, D.R., 1991. Diameter growth equations for fourteen tree species in southwest Oregon Research Bulletin 69. Oregon State University, Forest Research Laboratory, Corvallis, OR.

Hann, D.W., Ritchie, M.W., 1988. Height growth rate of Douglas-fir: A comparison of model forms. Forest Science 34, 165–175.

Hann, D.W., Wang, C.H., 1990. Mortality equations for individual trees in the mixed-conifer zone of southwest Oregon Research Bulletin 67. Oregon State University, Forest Research Laboratory, Corvallis, OR.

Hann, D.W., Weiskittel, A.R., 2010. Evaluation of alternative approaches for predicting individual tree volume increment. Western Journal of Applied Forestry 25, 120–126.

Hann, D.W., Zumrawi, A.-A., 1991. Growth model predictions as affected by alternative sampling-unit designs. Forest Science 37, 1641–1655.

Hann, D.W., Walters, D.K., Scrivani, J.A., 1987. Incorporating crown ratio into prediction equations for Douglas-fir stem volume. Canadian Journal of Forest Research 17, 17–22.

Hann, D.W., Marshall, D.D., Hanus, M.L., 2003. Equations for predicting height-to-crown base, 5-year diameter growth rate, 5-year height growth rate, 5-year mortality rate, and maximum size-density trajectory for Douglas-fir and western hemlock in the coastal region of the Pacific Northwest. Research Contribution 40. Oregon State University, College of Forestry Research Laboratory, Corvallis, OR.

Hann, D.W., Marshall, D.D., Hanus, M.L., 2006. Reanalysis of the SMC-ORGANON equations for diameter-growth rate, height-growth rate, and mortality rate of Douglas-fir. Research Contribution 49. Oregon State University, Forest Research Laboratory, Corvallis, OR.

Hann, D.W., Bluhm, A.A., Hibbs, D.E., 2011. Development and evaluation of the tree-level equations and their combined stand-level behavior in the red alder plantation version of ORGANON. Forest Biometrics Research Note 1. Forest Biometrics Research Paper 1. Oregon State University, Department of Forest Engineering, Resources, and Management, Corvallis, OR.

Hanson, E.J., Azuma, D.L., Hiserote, B.A., 2002. Site index equations and mean annual increment equations for Pacific Northwest Research Station Forest Inventory and Analysis Inventories, 1985–2001. Research Note PNW-RN-533. USDA Forest Service, Pacific Northwest Research Station, Portland, OR.

Hanus, M.L., 2003. Modeling light competition in the forests of Western Oregon: PhD dissertation. Oregon State University, Corvallis, OR.

Hanus, M.L., Hann, D.W., 1997. VIZ4ST: Forest Visualization user's guide, Edition 1.0. Department of Forest Resources, Oregon State University, Corvallis, OR.

Hanus, M.L., Hann, D.W., Marshall, D.D., 1998. Reconstructing the spatial pattern of trees from routine stand examination measurements. Forest Science 44, 125–133.

Hanus, M.L., Hann, D.W., Marshall, D.D., 1999a. Predicting height for undamaged and damaged trees in southwest Oregon. Research Contribution 27. Oregon State University, Forest Research Laboratory, Corvallis, OR.

Hanus, M.L., Marshall, D.D., Hann, D.W., 1999b. Height-diameter equations for six species in the costal regions of the Pacific Northwest. Research Contribution 25. Oregon State University, Forest Research Laboratory, Corvallis, OR.

Hanus, M.L., Hann, D.W., Marshall, D.D., 2000. Predicting height to crown base for undamaged and damaged tress in southwest Oregon. Research Contribution 29. Oregon State University, Forest Research Laboratory, Corvallis, OR.

Härkönen, S., Mäkinen, A., Tokola, T., Rasinmäki, J., Kalliovirta, J., 2010a. Evaluation of forest growth simulators with NFI permanent sample plot data from Finland. Forest Ecology and Management 259, 573–582.

Härkönen, S., Pulkkinen, M., Duursma, R., Mäkelä, A., 2010b. Estimating annual GPP, NPP and stem growth in Finland using summary models. Forest Ecology and Management 259, 524–533.

Harper, J.L., 1977. Population Biology of Plants. Academic Press, New York, NY.

Harrison, W.C., Daniels, R.F., 1988. A new biomathematical model for growth and yield of loblolly pine plantations. In Ek, A.R., Shifley, S.R., Burk, T.E. (Eds.), Forest growth modeling and prediction, Volume I. General Technical Report NC-120. USDA Forest Service, North Central Forest Experimental Station, St. Paul, MN, pp. 293–304.

Hasenauer, H., 1997. Dimensional relationships of open-grown trees in Austria. Forest Ecology and Management 96, 197–206.

Hasenauer, H., 2006. Sustainable Forest Management: Growth Models for Europe. Springer, Berlin, Germany.

Hasenauer, H., Kindermann, G., 2002. Methods for assessing regeneration establishment and height growth in uneven-aged mixed species stands. Forestry 75, 385–394.

Hasenauer, H., Kindermann, G., 2006. Modeling regeneration in even and uneven-aged mixed species forests. In Hasenauer, H. (Ed.), Sustainable Forest Management: Growth Models for Europe. Springer, Berlin, Germany, pp. 167–193.

Hasenauer, H., Merkl, D., 2001. Predicting regeneration establishment within mixed species stands using neural networks. In Koch, B. (Ed.), Spatial technologies for forest inventories. Contributions to the XXI IUFRO World Congress Freiburger Forschungsberichte, Kuala Lumpur, Malaysia, pp. 145–154.

Hasenauer, H., Monserud, R.A., 1996. A crown ratio model for Austrian forests. Forest Ecology and Management 84, 49–60.

Hasenauer, H., Monserud, R.A., 1997. Biased predictions for tree height increment models developed from smoothed 'data'. Ecological Modelling 98, 13–22.

Hasenauer, H., Burkhart, H.E., Sterba, H., 1994. Variation in potential volume yield of loblolly pine plantations. Forest Science 40, 162–176.

Hasenauer, H., Burkhart, H.E., Amateis, R.L., 1997. Basal area development in thinned and unthinned loblolly pine plantations. Canadian Journal of Forest Research 27, 265–271.

Hasenauer, H., Monserud, R.A., Gregoire, T.G., 1998. Using simultaneous regression techniques with individual-tree growth models. Forest Science 44, 87–95.

Hasenauer, H., Merkl, D., Weingartner, M., 2001. Estimating tree mortality of Norway spruce stands with neural networks. Advances in Environmental Research 5, 405–414.

Hassani, B.T., LeMay, V., Marshall, P.L., Temesgen, H., Zumrawi, A.-A., 2004. Regeneration imputation models for complex stands of southeastern British Columbia. Forestry Chronicle 80, 271–278.

Hatch, C.R., 1980. Modelling crown size using inventory data. Mitteilungen der forstlichen Bundesversuchsanstalt Wien 130, 93–97.

Hatch, C.R., Gerrad, D.J., Tappeiner, J.C., 1975. Exposed crown surface area: A mathematical index of individual tree growth potential. Canadian Journal of Forest Research 5, 224–228.

Hauhs, M., Dörwald, W., Kastner-Maresch, A., Lange, H., 1999. The role of visualization in forest growth modelling. In Amaro, A., Tomé, M. (Eds.), Proceedings of IUFRO Congress, Empirical and Process-Based Models for Forest Tree and Stand Growth Simulation, Oeiras, Portugal, September 21–27, 1997, pp. 403–418.

Hawkes, C., 2000. Woody plant mortality algorithms: Description, problems and progress. Ecological Modelling 126, 225–248.

Hayes, J.P., Chan, S.S., Emmingham, W.H., Tappeiner, J.C., Kellogg, L.D., Bailey, J.D., 1997. Wildlife response to thinning young forests in the Pacific Northwest. Journal of Forestry 95, 28–33.

Heath, L.S., Hansen, M., Smith, J.E., Miles, P.D., Smith, B.W., 2009. Investigation into calculating tree biomass and carbon in the FIADB using a biomass expansion factor approach. In McWilliams, W., Moisen, G., Czaplewski, R. (Eds.), Forest Inventory and Analysis (FIA) Symposium 2008. Proceedings RMRS-P-56CD. USDA Forest Service, Rocky Mountain Research Station, Fort Collins, CO.

Hegyi, F., 1974. A simulation model for managing jack-pine stands. In Fries, J. (Ed.), Growth Models for Tree and Stand Simulation. Research Note 30. Royal College of Forestry, Stockholm, Sweden, pp. 74–90.

Hein, S., Weiskittel, A.R., 2010. Cutpoint analysis for models with binary outcomes: A case study on branch mortality. European Journal of Forest Research 129, 585–590.

Hennigar, C.R., MacLean, D.A., Amos-Binks, L.J., 2008. A novel approach to optimize management strategies for carbon stored in both forests and wood products. Forest Ecology and Management 256, 786–797.

Henning, J.G., Burk, T.E., 2004. Improving growth and yield estimates with a process model derived growth index. Canadian Journal of Forest Research 34, 1274–1282.

Herring, N.D., 2007. Sensitivity analysis of the Forest Vegetation Simulator Southern variant (FVS-Sn) for southern Appalachian hardwoods. MS thesis. Virginia Polytechnic Institute and State University, Blacksburg, VA.

Hett, J.M., Loucks, O.L., 1968. Application of life table analyses to tree seedlings in Quetico Provincial Park, Ontario. Forestry Chronicle 44, 29–32.

Hocking, R.R., 1976. The analysis and selection of variables in linear regression. Biometrics 32, 1–49.

Hökkä, H., Groot, A., 1999. An individual-tree basal area growth model for black spruce in second-growth peatland stands. Canadian Journal of Forest Research 29, 621–629.

Hökkä, H., Alenius, V., Penttilä, T., 1997. Individual-tree basal area growth models for Scots pine, pubescent birch and Norway spruce on drained peatlands in Finland. Silva Fennica 31, 161–178.

Holdaway, M.R., 1984. Modeling the effect of competition on tree diameter growth as applied in STEMS. General Technical Report NC-94, USDA Forest Service, North Central Forest Experiment Station, St. Paul, MN.

Holdaway, M.R., 1986. Modelling tree crown ratio. Forestry Chronicle 62, 451–455.

Homann, P.S., McKane, R.B., Sollins, P., 2000. Belowground processes in forest-ecosystem biogeochemical simulation models. Forest Ecology and Management 138, 3–18.

Honer, T.G., 1971. Crown shape in open- and forest-grown balsam fir and black spruce. Canadian Journal of Forest Research 1, 203–207.

Hool, J.N., 1966. A dynamic programming - Markov chain approach to forest production control. Forest Science Monograph 12.

Hoover, C., Rebain, S., 2008. The Kane Experimental Forest carbon inventory: Carbon reporting with FVS. In Harvis, R.N., Crookston, N.L. (Eds.), Third Forest Vegetation Simulator

Conference. Proceedings RMRS-P-54. USDA Forest Service, Rocky Mountain Research Station, Fort Collins, CO, pp. 17–22.

Hopkins, A.D., 1938. Bioclimatics, a science of life and climate relations. Miscellaneous Report 280. USDA, Washington, DC.

Hosmer, D.W., Lemeshow, S., 2000. Applied Logistic Regression, 2nd edition. John Wiley & Sons, Inc., New York, NY.

Hu, Z., García, O., 2010. A height-growth and site-index model for interior spruce in the Sub-Boreal Spruce biogeoclimatic zone of British Columbia. Canadian Journal of Forest Research 40, 1175–1183.

Huang, S., Titus, S.J., 1993. An index of site productivity for uneven-aged or mixed-species stands. Canadian Journal of Forest Research 23, 558–562.

Huang, S., Titus, S.J., 1994. An age-independent individual tree height prediction model for boreal spruce-aspen stands in Alberta. Canadian Journal of Forest Research 24, 1295–1301.

Huang, S., Titus, S.J., 1995. An individual tree diameter increment model for white spruce in Alberta. Canadian Journal of Forest Research 25, 1455–1465.

Huang, S., Titus, S.J., 1999a. An individual tree height increment model for mixed white spruce-aspen stands in Alberta, Canada. Forest Ecology and Management 123, 41–53.

Huang, S., Titus, S.J., 1999b. Estimating a system of nonlinear simultaneous individual tree models for white spruce in boreal mixed-species stands. Canadian Journal of Forest Research 29, 1805–1811.

Huang, S., Titus, S.J., Wiens, D.P., 1992. Comparison of nonlinear height-diameter functions for major Alberta tree species. Canadian Journal of Forest Research 22, 1297–1304.

Huang, S., Price, D., Titus, S.J., 2000. Development of ecoregion-based height–diameter models for white spruce in boreal forests. Forest Ecology and Management 129, 125–141.

Huber, M.O., 2010. Assessing the long-term species composition predicted by PROGNAUS. Forest Ecology and Management 259, 614–623.

Huber, M., Sterba, H., 2009. Development of species composition in long term simulations with an individual-tree growth simulator. Journal of Forest Science 55, 194–200.

Husch, B., Beers, T.W., Kershaw Jr., J.A., 2003. Forest Mensuration 4th edition. John Willey & Sons, Inc., Hoboken, NJ.

Hyink, D.M., Moser Jr., J.W., 1983. A generalized framework for projecting forest yield and stand structure using diameter distributions. Forest Science 29, 85–95.

Hynynen, J., 1995a. Predicting tree crown ratio for unthinned and thinned Scots pine stands. Canadian Journal of Forest Research 25, 57–62.

Hynynen, J., 1995b. Predicting the growth response to thinning for Scots pine stands using individual-tree growth models. Silva Fennica 29, 225–246.

Hynynen, J., Burkhart, H.E., Allen, H.L., 1998. Modeling tree growth in fertilized midrotation loblolly pine plantations. Forest Ecology and Management 107, 213–229.

Hynynen, J., Ojansuu, R., Hökkä, H., Siipilehto, J., Salminen, H., Haapala, P., 2002. Models for predicting stand development in MELA system. Research Paper 835. Finnish Forest Research Institute, Helsinki, Finland.

Hyytiäinen, K., Hari, P., Kokkila, T., Mäkelä, A., Tahvonen, O., Taipale, J., 2004. Connecting a process-based forest growth model to stand-level economic optimization. Canadian Journal of Forest Research 34, 2060–2073.

Ichii, K., Wang, W., Hashimoto, H., Yang, F., Votava, P., Michaelis, A.R., Nemani, R.R., 2009. Refinement of rooting depths using satellite-based evapotranspiration seasonality for ecosystem modeling in California. Agricultural and Forest Meteorology 149, 1907–1918.

Iles, K., 2003. A Sampler of Inventory Topics: A Practical Discussion for Resource Samplers, Concentrating on Forest Inventory Techniques. Kim Iles & Associates, Ltd., Nanaimo, British Columbia.

Ilomäki, S., Nikinmaa, E., Mäkela, A., 2003. Crown rise due to competition drives biomass allocation in silver birch. Canadian Journal of Forest Research 33, 2395–2404.

Inoue, A., 2009. Allometric model of the maximum size–density relationship between stem surface area and stand density. Journal of Forest Research 14, 268–275.

Isebrands, J.G., Rauscher, H.M., Crow, T.R., Dickmann, D.J., 1990. Whole-tree growth process models based on structural-functional relationships. In Dixon, R.K., Meldahl, R.S., Ruark, G.A., Warren, W.G. (Eds.), Forest Growth: Process Modelling of Forest Growth Responses to Environmental Stress. Timber Press, Portland, OR, pp. 96–112.

Iverson, L.R., Dale, M.E., Scott, C.T., Prasad, A., 1997. A GIS-derived integrated moisture index to predict forest composition and productivity of Ohio forests (USA). Landscape Ecology 12, 331–348.

Jack, S.B., Long, J.N., 1996. Linkages between silviculture and ecology: An analysis of density management diagrams. Forest Ecology and Management 86, 205–220.

Jarčuška, B., 2008. Methodical overview to hemispherical photography, demonstrated on an example of the software GLA. Folia Oecologica 35, 66–69.

Jarvis, P.G., 1976. The interpretation of variations in leaf water potential and stomatal conductance found in canopies in the field. Philosophical Transactions of the Royal Society of London B: Biological Sciences 273, 593–610.

Jayawickrama, K.J.S., McKeand, S.E., Jett, J.B., 1998. Phenological variation in height and diameter growth in provenances and families of loblolly pine. New Forests 16, 11–25.

Jenkins, J.C., Chojnacky, D.C., Heath, L.S., Birdsey, R.A., 2004. Comprehensive database of diameter-based biomass regressions for North American tree species. General Technical Report NE-319. USDA Forest Service, Northeastern Research Station, Newtown Square, PA.

Jensen, J.L.W.V., 1906. Sur les fonctions convexes et les inégalités entre les valeurs moyennes. Acta Mathematica 30, 175–193.

Jiang, L., Brooks, J.R., Hobbs, G.R., 2007. Using crown ratio in yellow-poplar compatible taper and volume equations. Northern Journal of Applied Forestry 24, 271–275.

Johansson, T., 2006. Site index conversion equations for *Picea abies* and five broadleaved species in Sweden: *Alnus glutinosa*, *Alnus incana*, *Betula pendula*, *Betula pubescens* and *Populus tremula*. Scandinavian Journal of Forest Research 21, 14–19.

Johnsen, K., Samuelson, L., Teskey, R., McNulty, S., Fox, T., 2001. Process models as tools in forestry and management. Forest Science 47, 2–8.

Johnson, D.L., Johnson, K.N., Hann, D.W., 2007a. The importance of forest stand-level inventory to sustain multiple forest values in the presence of endangered species. In Reynolds, K.M., Thomson, A.J., Kohl, M., Shannon, M.A., Ray, D., Rennolls, K. (Eds.), Sustainable Forestry: From Monitoring and Modelling to Knowledge Management & Policy Science. CABI Publishing, Cambridge, MA, pp. 238–256.

Johnson, G.P., 2005. Growth model runoff II. Growth Model Users Group, Portland, OR.

Johnson, N.K., Bettinger, P., Kline, J.D., Spies, T.A., Lennette, M., Lettman, G., Garber-Yonts, B., Larsen, T., 2007b. Simulating forest structure, timber production, and socioeconomic effects in a multi-owner province. Ecological Applications 17, 34–47.

Johnson, T.S., Wood, G.B., 1987. Simple linear model reliably predicts bark thickness of radiata pine in the Australian capital territory. Forest Ecology and Management 22, 173–183.

Johnston, D.R., Grayson, A.J., Bradley, R.T., 1967. Forest Planning. Faber and Faber, London, UK.

Jokela, E.J., Dougherty, P.M., Martin, T.A., 2004. Production dynamics of intensively managed loblolly pine stands in the southern United States: A synthesis of seven long-term experiments. Forest Ecology and Management 192, 117–130.

Jonckheere, I., Fleck, S., Nackaerts, K., Muys, B., Cooppin, P., Weiss, M., Baret, F., 2004. Review of methods for *in situ* leaf area index determination, Part I. Theories, sensors, and hemispherical photography. Agricultural and Forest Meteorology 121, 19–35.

Jorritsma, I.T.M., van Hees, A.F.M., Mohren, G.M.J., 1999. Forest development in relation to ungulate grazing: A modeling approach. Forest Ecology and Management 120, 23–34.

Jose, S., Williams, R., Zamora, D., 2006. Belowground ecological interactions in mixed-species forest plantations. Forest Ecology and Management 233, 231–239.

Jutras, S., Hökkä, H., Alenius, V., Salminen, H., 2003. Modeling mortality of individual trees in drained peatland sites in Finland. Silva Fennica 37, 235–251.

Kaitaniemi, P., Lintunen, A., 2008. Precision of allometric scaling equations for trees can be improved by including the effect of ecological interactions. Trees 22, 579–584.

Kangas, A., 1997. On the prediction bias and variance in long-term growth projections. Forest Ecology and Management 96, 207–216.

Kangas, A., 1998a. Effect of errors-in-variables on coefficients of a growth model and on prediction of growth. Forest Ecology and Management 102, 203–212.

Kangas, A., 1998b. Uncertainty in growth and yield projections due to annual variation in diameter growth. Forest Ecology and Management 108, 223–230.

Kangas, A.S., 1999. Methods for assessing uncertainty of growth and yield predictions. Canadian Journal of Forest Research 29, 1357–1364.

Kantola, A., Mäkelä, A., 2004. Crown development in Norway spruce (*Picea abies* (L.) Karst.). Trees 18, 408–421.

Karlsson, K., Norell, L., 2005. Modelling survival probability of individual trees in Norway spruce stands under different thinning regimes. Canadian Journal of Forest Research 35, 113–121.

Kaufmann, M.R., Weatherred, J.D., 1982. Determination of potential direct beam solar irradiance. Research Paper RM-242. USDA Forest Service, Rocky Mountain Forest and Range Experiment Station, Fort Collins, CO.

Keane, R.E., Ryan, K.C., Running, S.W., 1996. Simulating effects of fire on northern Rocky Mountain landscapes with the ecological process model FIRE-BGC. Tree Physiology 16, 319–331.

Keane, R.E., Austin, M., Field, C., Huth, A., Lexer, M.J., Peters, D., Solomon, A., Wyckoff, P., 2001. Tree mortality in gap models: Application to climate change. Climatic Change 51, 509–540.

Keen, F.P., 1937. Climatic cycles in eastern Oregon as indicated by tree rings. Monthly Weather Review 65, 175–188.

Keenan, T., Serra, J.M., Lloret, F., Ninyerola, M., Sabate, S., 2011. Predicting the future of forests in the Mediterranean under climate change, with niche- and process-based models: CO_2 matters! Global Change Biology 17, 565–579.

Keith, H., Barrett, D., Keenan, R., 2000. Review of allometric relationships for estimating woody biomass for New South Wales, the Australian Capital Territory, Victoria, Tasmania, and South Australia. National Carbon Accounting System Technical Report 5B. Australian Greenhouse Office, Canberra, Australia.

Kershaw Jr., J.A., Maguire, D.A., 1995. Crown structure in western hemlock, Douglas-fir, and grand fir in western Washington: Trends in branch-level mass and leaf area. Canadian Journal of Forest Research 25, 1897–1912.

Kershaw Jr., J.A., Morrissey, R.C., Jacobs, D.F., Seifert, J.R., McCarter, J.B., 2008. Dominant height-based height-diameter equations for trees in southern Indiana. In Jacobs, D.F., Michler, C. H. (Eds.), Proceedings, 16th Central Hardwood Forest Conference. General Technical Report NRS-P-24. USDA Forest Service, Northern Research Station, Newton Square, PA, pp. 341–355.

Kershaw Jr., J.A., Benjamin, J.G., Weiskittel, A.R., 2009. Approaches for modeling vertical distribution of maximum knot size in black spruce: A comparison of fixed- and mixed-effects nonlinear models. Forest Science 55, 230–237.

Kershaw Jr., J.A., Richards, E.W., McCarter, J.B., Oborn, S., 2010. Spatially correlated stand structures: A simulation approach using copulas. Computers and Electronics in Agriculture 74, 120–128.

Ketterings, Q.M., Coe, R., van Noordwijk, M., Ambagau, Y., Palm, C.A., 2001. Reducing uncertainty in the use of allometric biomass equations for predicting above-ground tree biomass in mixed secondary forests. Forest Ecology and Management 146, 199–209.

Keyes, M.R., Grier, C.C., 1981. Above- and below-ground net production in 40-year old Douglas-fir stands on low and high productivity sites. Canadian Journal of Forest Research 11, 599–605.

Kiernan, D.H., Bevilacqua, E., Nyland, R.D., 2008. Individual-tree diameter growth model for sugar maple trees in uneven-aged northern hardwood stands under selection system. Forest Ecology and Management 256, 1579–1586.

Kiernan, D., Bevilacqua, E., Nyland, R., Zhang, L., 2009. Modeling tree mortality in low- to medium-density uneven-aged hardwood stands under a selection system using generalized estimating equations. Forest Science 55, 343–351.

Kimberley, M.O., Richardson, B., 2004. Importance of seasonal growth patterns in modelling interactions between radiata pine and some common weed species. Canadian Journal of Forest Research 34, 184–194.

Kimberley, M.O., Wang, H., Wilks, P.J., Fisher, C.R., Magesan, G.N., 2004. Economic analysis of growth response from a pine plantation forest applied with biosolids. Forest Ecology and Management 189, 345–351.

Kimmins, J.P., Blanco, J.A., Seely, B., Welham, C., Scoullar, K., 2008. Complexity in modelling forest ecosystems: How much is enough? Forest Ecology and Management 256, 1646–1658.

King, S.L., 2003. Using ROC curves to compare neural networks and logistic regression for modeling individual noncatastrophic tree mortality. In Van Sambeek, J.W., Dawson, J.O., Ponder Jr., F., Loewenstein, E.F., Fralish, J.S. (Eds.), 13th Central Hardwood Forest Conference. General Technical Report NC-234. USDA Forest Service, North Central Research Station, St. Paul, MN, pp. 349–358.

King, S.L., Bennett, K.P., List, S., 2000. Modeling noncatastrophic individual tree mortality using logistic regression, neural networks, and support vector methods. Computers and Electronics in Agriculture 27, 401–406.

Kirschbaum, M.U.F., 1999. CenW, a forest growth model with linked carbon, energy, nutrient and water cycles. Ecological Modelling 118, 17–59.

Kirschbaum, M.U.F., 2000a. CenW: A generic forest growth model. New Zealand Journal of Forestry 45, 15–19.

Kirschbaum, M.U.F., 2000b. Will changes in soil organic carbon act as a positive or negative feedback on global warming? Biogeochemistry 48, 21–51.

Kirschbaum, M.U.F., 2001. The control of ecosystem carbon dynamics by the linkages between above and belowground processes. In Kirschbaum, M.U.F., Mueller, R. (Eds.), Net Ecosystem Exchange: Proceedings of Cooperative Research Centre for Greenhouse Accounting Workshop. Cooperative Research Centre for Greenhouse Accounting, Canberra, pp. 82–88.

Kirschbaum, M.U.F., 2005. A modeling analysis of the interaction between forest age and forest responsiveness to increasing CO_2 concentration. Tree Physiology 25, 953–963.

Kirschbaum, M.U.F., Paul, K.I., 2002. Modelling C and N dynamics in forest soils with a modified version of the CENTURY model. Soil Biology and Biochemistry 34, 341–354.

Kirschbaum, M.U.F., Guo, L.B., Gifford, R.M., 2008. Why does rainfall affect the trend in soil carbon after converting pastures to forests? A possible explanation based on nitrogen dynamics. Forest Ecology and Management 255, 2990–3000.

Kitahara, F., Mizoue, N., Yoshida, S., 2010. Effects of training for inexperienced surveyors on data quality of tree diameter and height measurements. Silva Fennica 44, 657–667.

Klinka, K., Carter, R.E., 1990. Relationships between site index and synoptic environmental factors in immature coastal Douglas-fir stands. Forest Science 36, 815–830.

Klinka, K., Krajina, V.J., Ceska, A., Scagel, A.M., 1989. Indicator plants of coastal British Colombia. University of British Columbia Press, Vancouver, BC.

Kmenta, J., 1997. Elements of Econometrics 2nd edition. University of Michigan, Ann Arbor, MI.

Knoebel, B.R., Burkhart, H.E., 1991. A bivariate distribution approach to modeling forest diameter distributions at two points in time. Biometrics 47, 241–253.

Knowe, S.A., 1994a. Incorporating the effects of interspecific competition and vegetation management treatments into stand table projection models for Douglas-fir saplings. Forest Ecology and Management 67, 87–99.

Knowe, S.A., 1994b. Effect of competition control treatments on height-age and height-diameter relationships in young Douglas-fir plantations. Forest Ecology and Management 67, 101–111.

Knowe, S.A., Foster, G.S., 1989. Application of growth models for simulating genetic gain of loblolly pine. Forest Science 35, 211–228.

Knowe, S.A., Stein, W.I., 1995. Predicting the effects of site preparation and protection on development of young Douglas-fir plantations. Canadian Journal of Forest Research 25, 1538–1547.

Knowe, S.A., Ahrens, G.R., DeBell, D.S., 1997. Comparison of diameter-distribution-prediction, stand-table-projection, and individual-tree-growth modeling approaches for young red alder plantations. Forest Ecology and Management 98, 49–60.

Knowe, S.A., Radosevich, S.R., Shula, R.G., 2005. Basal area and diameter distribution prediction equations for young Douglas-fir plantations with hardwood competition: Coast Ranges. Western Journal of Applied Forestry 20, 77–93.

Kokkila, T., Mäkelä, A., Nikinmaa, E., 2002. A method for generating stand structures using Gibbs marked point process. Silva Fennica 36, 265–277.

Korol, R.L., Running, S.W., Millner, K.S., Hunt Jr., E.R., 1991. Testing a mechanistic carbon balance model against observed tree growth. Canadian Journal of Forest Research 21, 1098–1105.

Korol, R.L., Running, S.W., Milner, K.S., 1995. Incorporating intertree competition into an ecosystem model. Canadian Journal of Forest Research 25, 413–424.

Korol, R.L., Milner, K.S., Running, S.W., 1996. Testing a mechanistic model for predicting stand and tree growth. Forest Science 42, 139–153.

Kozak, A., 1988. A variable-exponent taper equation. Canadian Journal of Forest Research 18, 1363–1368.

Kozak, A., 1997. Effects of multicollinearity and autocorrelation on the variable-exponent taper functions. Canadian Journal of Forestry 27, 619–629.

Kozak, A., 1998. Effects of upper stem measurements on the predictive ability of a variable-exponent taper equation. Canadian Journal of Forest Research 28, 1078–1083.

Kozak, A., 2004. My last words on taper equations. Forestry Chronicle 80, 507–515.

Kozak, A., Kozak, R., 2003. Does cross validation provide additional information in the evaluation of regression models? Canadian Journal of Forest Research 33, 976–987.

Krajicek, J.E., Brinkman, K.A., Gingrich, S.F., 1961. Crown competition: A measure of density. Forest Science, 7, 35–42.

Krumland, B., Eng, H., 2005. Site index systems for major young-growth forest and woodland species in northern California. California Forestry Report 4. Department of Forestry and Fire Protection, State of California Resources Agency, Sacramento, CA.

Krumland, B., Wensel, L.C., 1981. A tree increment model system for north coastal California: Design and implementation. Research Note 15. California Cooperative Redwood Yield Research Project, Department of Forest Conservation, University of California, Berkeley, CA.

Krumland, B.E., Wensel, L.C., 1988. A generalized height-diameter equation for coastal California species. Western Journal of Applied Forestry 3, 113–115.

Kublin, E., Augustin, N.H., Lappi, J., 2008. A flexible regression model for diameter prediction. European Journal of Forest Research 127, 415–428.

Kucharik, C.J., Norman, J.M., Gower, S.T., 1998. Measurements of branch area and adjusting leaf area index indirect measurements. Agricultural and Forest Meteorology 91, 69–88.

Kucharik, C.J., Norman, J.M., Gower, S.T., 1999. Characterization of radiation regimes in nonrandom forest canopies: Theory, measurements, and a simplified modeling approach. Tree Physiology 19, 695–706.

Kurinobu, S., Shingai, Y., 1987. Stand-volume prediction of improved trees based on the realized gain in progeny tests of hinoki (*Chamaecyparis obtusa* Endl.). Silvae Genetica 36, 159–164.

Kuuluvainen, T., Pukkala, T., 1987. Effect of crown shape and tree distribution on the spatial distribution of shade. Agricultural and Forest Meteorology 40, 215–231.

Lacerte, V., Larocque, G.R., Woods, M., Parton, W.J., Penner, M., 2004. Testing the Lake States variant of FVS (Forest Vegetation Simulator) for the main forest types of northern Ontario. Forestry Chronicle 80, 495–506.

Lacointe, A., 2000. Carbon allocation among tree organs: A review of basic processes and representation in functional-structural tree models. Annals of Forest Science 57, 521–533.

Lai, C.-T., Katul, G., Butnor, J., Siqueira, M., Ellsworth, D., Maier, C.A., Johnsen, K., McKeand, S., Oren, R., 2002. Modelling the limits on the response of net carbon exchange to fertilization in a south-eastern pine forest. Plant, Cell and Environment 25, 1095–1119.

Lambeth, C.C., 1980. Juvenile-mature correlations in *Pinaceae* and implications for early selection. Forest Science 26, 571–580.

Landsberg, J.J., 1986. Physiological Ecology Of Forest Production. Academic Press, London, UK.

Landsberg, J.J., 2003. Physiology in forest models: History and the future. Forest Biometry, Modelling and Information Sciences 1, 49–63.

Landsberg, J., Sands, P., 2011. Physiological Ecology of Forest Production: Principles Processes and Models, Volume 4. Academic Press, Maryland Heights, MO.

Landsberg, J.J., Waring, R.H., 1997. A generalized model of forest productivity using simplified concepts of radiation-use efficiency, carbon balance and partitioning. Forest Ecology and Management 95, 209–228.

Landsberg, J.J., Johnsen, K.H., Albaugh, T.J., Allen, H.L., McKeand, S.E., 2001. Applying 3-PG, a simple process-based model designed to produce practical results, to data from loblolly pine experiments. Forest Science 47, 43–51.

Landsberg, J.J., Waring, R.H., Coops, N.C., 2003. Performance of the forest productivity model 3-PG applied to a wide range of forest types. Forest Ecology and Management 172, 199–214.

Landsberg, J., Mäkelä, A., Sievänen, R., Kukkola, M., 2005. Analysis of biomass accumulation and stem size distributions over long periods in managed stands of *Pinus sylvestris* in Finland using the 3-PG model. Tree Physiology 25, 781–792.

Langsaeter, A., 1941. Omtynning i analdret granfurvskog Maddel. Norsite Skogfor Soksvenson 8, 131–216.

Lappi, J., 1991a. Calibration of height and volume equations with random parameters. Forest Science 37, 781–801.

Lappi, J., 1991b. Estimating the distribution of a variable measured with error: Stand densities in a forest inventory. Canadian Journal of Forest Research 21, 469–473.

Lappi, J., 1997. A longitudinal analysis of height/diameter curves. Forest Science 43, 555–570.

Lappi, J., 2005. Plot size related measurement error bias in tree growth models. Canadian Journal of Forest Research 35, 1031–1040.

Lappi, J., 2006. A multivariate, nonparametric stem-curve prediction method. Canadian Journal of Forest Research 36, 1017–1027.

Larocque, G.R., 2002. Examining different concepts for the development of a distance-dependent competition model for red pine diameter growth using long-term stand data differing in initial stand density. Forest Science 48, 24–34.

Larocque, G.R., Marshall, P.L., 1993. Evaluating the impact of competition using relative growth rate in red pine (*Pinus resinosa* Ait.) stands. Forest Ecology and Management 58, 65–83.

Larocque, G.R., Bhatti, J.S., Ascough II, J.C., Liu, J., Luckai, N., Mailly, D., Archambault, L., Gordon, A.M., 2011. An analytical framework to assist decision makers in the use of forest ecosystem model predictions. Environmental Modelling & Software 26, 280–288.

Larsen, D.R., Hann, D.W., 1987. Height-diameter equations for seventeen tree species in southwest Oregon. Research Paper 49. Oregon State University Forest Research Laboratory, Corvallis, OR.

Larsen, D.R., Kershaw Jr., J.A., 1996. Influence of canopy structure assumptions on predictions from Beer's law. A comparison of deterministic and stochastic simulations. Agricultural and Forest Meteorology 81, 61–77.

Larson, P.R., 1963. Stem form development of forest trees. Forest Science Monograph 5, 1–42.

Lassasenaho, J., Melkas, T., Alden, S., 2005. Modelling bark thickness of *Picea abies* with taper curves. Forest Ecology and Management 206, 35–47.

Latta, G., Montgomery, C.A., 2004. Minimizing the cost of stand-level management for older forest structure in western Oregon. Western Journal of Applied Forestry 19, 221–231.

Laubhann, D., Sterba, H., Jan Reinds, G., De Vries, W., 2009. The impact of atmospheric deposition and climate on forest growth in European monitoring plots: An individual tree growth model. Forest Ecology and Management 258, 1751–1761.

Lauer, D.K., Glover, G.R., Gjerstad, D.H., 1993. Comparison of duration and method of herbaceous weed control on loblolly pine response through midrotation. Canadian Journal of Forest Research 23, 2116–2125.

Lavine, M., Beckage, B., Clark, J.S., 2002. Statistical modeling of seedling mortality. Journal of Agricultural, Biological, and Environmental Statistics 7, 21–41.

Le Roux, X., Lacointe, A., Escobar-Gutiérrez, A., Le Dizès, S., 2001. Carbon-based models of individual tree growth: A critical appraisal. Annals of Forest Science 58, 469–506.

Leak, W.B., 1968. Birch regeneration: A stochastic model. Research Note NE-85. USDA Forest Service, Northeastern Forest Experiment Station, Upper Darby, PA.

Leary, R.A., 1979. Design. In A generalized forest growth projection system applied to the Lake States region. General Technical Report NC-49. USDA Forest Service, North Central Forest Experiment Station, St. Paul, MN, pp. 5–15.

Leary, R.A., 1985. A framework for assessing and rewarding a scientist's research productivity. Scientometrics 7, 29–38.

Leary, R.A., 1991. Near-normal, empirical, and identity yield tables for estimating stand growth. Canadian Journal of Forest Research 21, 353–362.

Leary, R.A., 1997. Testing models of unthinned red pine plantation dynamics using a modified Bakuzis matrix of stand properties. Ecological Modelling 98, 35–46.

Leary, R.A., Holdaway, M.R., 1979. Modifier function. In A generalized forest growth projection system applied to the Lake States region. General Technical Report NC-49. USDA Forest Service, North Central Forest Experiment Station, St. Paul, MN, pp. 31–38.

Leary, R.A., Holdaway, M.R., Hahn, J.T., 1979. Diameter growth allocation rule. In A generalized forest growth projection system applied to the Lake States region. General Technical Report NC-49. USDA Forest Service, North Central Forest Experiment Station, St. Paul, MN, pp. 39–46.

Ledermann, T., 2002a. Estimating ingrowth (tree recruitment) using data from the Austrian National Forest Inventory 1981–1996. Austrian Journal of Forest Science 119, 40–77.

Ledermann, T., 2002b. Using logistic regression to model tree selection preferences for harvesting in forests in conversion. In von Gadow, K., Nagel, J., Saborowski, J. (Eds.), Managing Forest Ecosystems: Continuous Cover Forests – Assessment, Analysis, Scenarios. Kluwer Academic Publishers, Dordrecht, The Netherlands, pp. 203–216.

Ledermann, T., 2006. Description of ProgAus for Windows 2.2. In Hasenauer, H. (Ed.), Sustainable Forest Management: Growth Models for Europe. Springer, Berlin, Germany, pp. 71–78.

Ledermann, T., Eckmüllner, O., 2004. A method to attain uniform resolution of the competition variable Basal-Area-in-Larger Trees (BAL) during forest growth projections of small plots. Ecological Modelling 171, 195–206.

Ledermann, T., Stage, A.R., 2001. Effects of competitor spacing in individual-tree indices of competition. Canadian Journal of Forest Research 31, 2143–2150.

LeDuc, D.J., 1987. A comparative analysis of the reduced major axis technique for fitting lines to bivariate data. Canadian Journal of Forest Research 17, 654–659.

Lee, Y.J., 1971. Predicting mortality for even-aged stands of lodgepole pine. Forestry Chronicle 47, 29–32.

Leech, J.W., 1984. Estimating crown width from diameter at breast height for open-grown radiata pine trees in south Australia. Australian Journal of Forest Research 14, 333–337.

Lei, X., Wang, W., Peng, C., 2009. Relationships between stand growth and structural diversity in spruce-dominated forests in New Brunswick, Canada. Canadian Journal of Forest Research 39, 1835–1847.

Leites, L.P., Robinson, A.P., 2004. Improving taper equations of loblolly pine with crown dimensions in a mixed-effects modeling framework. Forest Science 50, 204–212.

Leites, L.P., Robinson, A.P., Crookston, N.L., 2009. Accuracy and equivalence testing of crown ratio models and assessment of their impact on diameter growth and basal area increment predictions of two variants of the Forest Vegetation Simulator. Canadian Journal of Forest Research 39, 655–665.

LeMay, V.M., 1990. MSLS: A linear least squares technique for fitting a simultaneous system of equations with a generalized error structure. Canadian Journal of Forest Research 20, 1830–1839.

Lemmon, P.E., Schumacher, F.X., 1962a. Stocking density around ponderosa pine trees. Forest Science 8, 397–402.

Lemmon, P.E., Schumacher, F.X., 1962b. Volume and diameter growth of ponderosa pine trees as influenced by site index, density, age, and size. Forest Science 8, 236–249.

Leslie, P.H., 1945. On the use of matrices in certain population mathematics. Biometrika 33, 183–212.

Lessard, V.C., McRoberts, R.E., Holdaway, M.R., 2001. Diameter growth models using Minnesota Forest Inventory and Analysis data. Forest Science 47, 301–310.

Leuning, R., 1995. A critical appraisal of a combined stomatal-photosynthesis model for C_3 plants. Plant, Cell and Environment 18, 339–355.

Leuning, R., Kelliher, F.M., DePury, D.G.G., Schulze, E.-D., 1995. Leaf nitrogen, photosynthesis, conductance, and transpiration: Scaling from leaves to canopies. Plant, Cell and Environment 18, 1183–1200.

Lexen, B., 1943. Bole area as an expression of growing stock. Journal of Forestry 41, 883–885.

Lexer, M.J., Hönninger, K., 2001. A modified 3D-patch model for spatially explicit simulation of vegetation composition in heterogeneous landscapes. Forest Ecology and Management 144, 43–65.

Li, R., Weiskittel, A.R., 2010. Comparison of model forms for estimating stem taper and volume in the primary conifer species of the North American Acadian Region. Annals of Forest Science 67, 302.

Li, R., Weiskittel, A.R., 2011. Estimating and predicting bark thickness for seven conifer species in the Acadian Region of North America using a mixed-effects modeling approach: Comparison of model forms and subsampling strategies. European Journal of Forest Research, 130, 219–233.

Li, R., Stewart, B., Weiskittel, A.R., 2011. A Bayesian approach for modeling nonlinear longitudinal/hierarchical data with random effects in forestry. Forestry, in press.

Liang, J., 2010. Dynamics and management of Alaska boreal forest: An all-aged multi-species matrix growth model. Forest Ecology and Management 260, 491–501.

Liang, J., Zhou, M., 2010. A geospatial model of forest dynamics with controlled trend surface. Ecological Modelling 221, 2339–2352.

Liang, J., Buongiorno, J., Monserud, R.A., 2005. Growth and yield of all-aged Douglas-fir – western hemlock forest stands: A matrix model with stand diversity effects. Canadian Journal of Forest Research 35, 2368–2381.

Lichtenstein, M.E., Montgomery, C.A., 2003. Biodiversity and timber in the Coast Range of Oregon: Inside the production possibility frontier. Land Economics 79, 56–73.

Lim, E.-M., Honjo, T., 2003. Three-dimensional visualization forest of landscapes by VRML. Landscape and Urban Planning 63, 175–186.

Lim, K., Treitz, P., Wulder, M., St-Onge, B., Flood, M., 2003. LiDAR remote sensing of forest structure. Progress in Physical Geography 27, 88–106.

Lin, C.-R., Buongiorno, J., 1997. Fixed versus variable-parameter matrix models of forest growth: The case of maple-birch forests. Ecological Modelling 99, 263–274.

Lin, C.-R., Buongiorno, J., Vasievich, M., 1996. A multi-species, density-dependent matrix growth model to predict tree diversity and income in northern hardwood stands. Ecological Modelling 91, 193–211.

Lin, J.Y., 1974. Stand growth simulation models for Douglas-fir and western hemlock in the Northwestern United States. In Fries, J. (Ed.), Growth models for tree and stand simulation. Research Note 30. Royal College of Forestry, Department of Forest Yield Research, Stockholm, Sweden, pp. 102–118.

Lindner, M., Sievänen, R., Pretzsch, H., 1997. Improving the simulation of stand structure in a forest gap model. Forest Ecology and Management 95, 183–195.

Litton, C.M., Raich, J.W., Ryan, M.G., 2007. Carbon allocation in forest ecosystems. Global Change Biology 13, 2089–2109.

Liu, J., Burkhart, H.E., 1994. Modelling inter- and intra-specific competition in loblolly pine (*Pinus taeda* L.) plantations on cutover, site-prepared lands. Annals of Botany 73, 429–435.

Liu, J., Burkhart, H.E., Amateis, R.L., 1995. Projecting crown measures for loblolly pine trees using a generalized thinning response function. Forest Science 41, 43–53.

Loftis, D.L., 1989. Species composition of regeneration after clearcutting southern Appalachian hardwoods. In Miller, J.H. (Ed.), Proceedings of the 5th Biennial Southern Silvicultural Research Conference. General Technical Report SO-74. USDA Forest Service, Southern Forest Experiment Station, New Orleans, LA, pp. 253–257.

Loftis, D.L., 1990. Predicting post-harvest performance of advanced red oak reproduction in the southern Appalachians. Forest Science 36, 908–916.

Long, J.N., 1985. A practical approach to density management. Forestry Chronicle 61, 23–27.

Long, J.N., Daniel, T.W., 1990. Assessment of growing stock in uneven-age stands. Western Journal of Applied Forestry 5, 93–96.

Long, J.N., Shaw, J.D., 2005. A density management diagram for even-aged ponderosa pine stands. Western Journal of Applied Forestry 20, 205–215.

Long, J.N., McCarter, J.B., Jack, S.B., 1988. A modified density management diagram for coastal Douglas-fir. Western Journal of Applied Forestry 3, 88–89.

Longchang, L., Wenkang, H., Guoqing, W., 1991. Study on the method of constructing variable-density yield table. Journal of Forestry Research 2, 95–99.

Lopez Sanchez, C.A., Gorgoro Varela, J., Castedo Dorado, F., Rojo Alboreca, A., Rodriquez Soalleiro, R., Alvarez Gonzalez, J.G., Sanchez Rodriquez, F., 2003. A height-diameter model for *Pinus radiata* D. Don in Galicia (northwestern Spain). Annals of Forest Science 60, 237–245.

Loustau, D., Berbigier, P., Kramer, K., 1997. Sensitivity of the water balance of southwestern France maritime pine forests to climate. In Mohren, G.M.J., Kramer, K., Sabaté, S. (Eds.), Impacts of Global Change on Tree Physiology and Forest Ecosystems. Kluwer Academic, Wageningen, The Netherlands, pp. 193–205.

Lowell, K.E., 1988. Fertilizing radiata pine plantations: Predicting long-term effects. New Zealand Forestry 33, 20–22.

Lowell, K.E., Mitchell, R.J., 1987. Stand growth projection: Simultaneous estimation of growth and mortality using a single probabilistic function. Canadian Journal of Forest Research 17, 1466–1470.

Lynch, D.W., 1958. Effects of stocking on site measurement and yield of second-growth ponderosa pine in the Inland Empire. Research Paper 56. USDA Forest Service, Intermountain Forest and Range Experiment Station, Ogden, UT.

Lynch, T.B., Moser Jr., J.W., 1986. A growth model for mixed species stands. Forest Science 32, 697–706.

MacDonald, B., Morris, D.M., Marshall, P.L., 1990. Assessing components of competition indices for young boreal plantations. Canadian Journal of Forest Research 20, 1060–1068.

MacFarlane, D.W., 2010. Predicting branch to bole volume scaling relationships from varying centroids of tree bole volume. Canadian Journal of Forest Research 40, 2278–2289.

MacFarlane, D.W., Kobe, R.K., 2006. Selecting models for capturing tree-size effects on growth–resource relationships. Canadian Journal of Forest Research 36, 1695–1704.

MacFarlane, D.W., Green, E.J., Burkhart, H.E., 2000. Population density influences assessment and application of site index. Canadian Journal of Forest Research 30, 1472–1475.

MacKinney, A.L., Chaiken, L.E., 1939. Volume, yield and growth of loblolly pine in the mid-Atlantic region. Technical Note 33. USDA Forest Service, Appalachian Forest Experiment Station, Asheville, NC.

MacKinney, A.L., Schumacher, F.X., Chaiken, L.E., 1937. Construction of yield tables for non-normal loblolly pine stands. Journal of Agricultural Research 54, 531–545.

Magnani, F., Mencuccini, M., Grace, J., 2000. Age-related decline in stand productivity: The role of structural acclimation under hydraulic constraints. Plant, Cell and Environment 23, 251–263.

Magnani, F., Grace, J., Borghetti, M., 2002. Adjustment of tree structure in response to the environment under hydraulic constraints. Functional Ecology 16, 385–393.

Magnussen, S., 1999. Effect of plot size on estimates of top height in Douglas-fir. Western Journal of Applied Forestry 14, 17–27.

Maguire, D.A., Batista, J.L.F., 1996. Sapwood taper models and implied sapwood volume and foliage profiles for Coastal Douglas-fir. Canadian Journal of Forest Research 26, 849–863.

Maguire, D.A., Hann, D.W., 1989. The relationship between gross crown dimensions and sapwood area at crown base in Douglas-fir. Canadian Journal of Forest Research 19, 557–565.

Maguire, D.A., Hann, D.W., 1990a. Constructing models for direct prediction of 5-year crown recession in southwest Oregon Douglas-fir. Canadian Journal of Forest Research 20, 1044–1052.

Maguire, D.A., Hann, D.W., 1990b. Bark thickness and bark volume in southwestern Oregon Douglas-fir. Western Journal of Applied Forestry 5, 5–8.

Maguire, D.A., Hann, D.W., 1990c. A sampling strategy for estimating past crown recession on temporary growth plots. Forest Science 36, 549–563.

Maguire, D.A., Kershaw Jr., J.A., Hann, D.W., 1991. Predicting the effects of silvicultural regime on branch size and crown wood core in Douglas-fir. Forest Science 37, 1409–1428.

Maguire, D.A., Brissette, J.C., Gu, L., 1998. Crown structure and growth efficiency of red spruce in uneven-aged, mixed species stand in Maine. Canadian Journal of Forest Research 28, 1233–1240.

Maguire, D.A., Kanaskie, A., Voelker, W., Johnson, R., Johnson, G., 2002. Growth of young Douglas-fir plantations across a gradient in Swiss needle cast severity. Western Journal of Applied Forestry 17, 86–95.

Mailly, D., Gaudreault, M., 2005. Growth intercept models for black spruce, jack pine and balsam fir in Quebec. Forestry Chronicle 81, 104–113.

Mailly, D., Turbis, S., Pothier, D., 2003. Predicting basal area increment in a spatially explicit, individual tree model: A test of competition measures with black spruce. Canadian Journal of Forest Research 33, 435–443.

Mailly, D., Gaudreault, M., Turbis, S., 2004. Modelling individual tree mortality: Tests using the growth efficiency concept with black spruce stands. In Godin, C., Hanan, J., Kurth, W., Lacointe, A., Takenaka, A., Prusinkiewicz, P., DeJong, T., Beveridge, C., Andrieu, B. (Eds.), Proceedings of the 4th International Workshop on Functional-Structural Plant Models. Université de Montpellier, Montpellier, France, p. 328.

Mainwaring, D.B., Maguire, D.A., 2004. The effect of local stand structure on growth and growth efficiency in heterogeneous stands of ponderosa pine and lodgepole pine in central Oregon. Canadian Journal of Forest Research 34, 2217–2229.

Mäkelä, A., 1986. Implications of the pipe model theory on dry matter partitioning and height growth in trees. Journal of Theoretical Biology 123, 103–120.

Mäkelä, A., 1997. A carbon balance model of growth and self-pruning in trees based on structural relationships. Forest Science 43, 7–24.

Mäkelä, A., 2002. Derivation of stem taper from the pipe theory in a carbon balance framework. Tree Physiology 22, 891–905.

Mäkelä, A., Hari, P., 1986. Stand growth model based on carbon uptake and allocation in individual trees. Ecological Modelling 33, 205–229.

Mäkelä, A., Makinen, H., 2003. Generating 3D sawlogs with a process-based growth model. Forest Ecology and Management 184, 337–354.

Mäkelä, A., Valentine, H.T., 2001. The ratio of NPP to GPP: Evidence of change over the course of stand development. Tree Physiology 21, 1015–1030.

Mäkelä, A., Landsberg, J., Ek, A.R., Burk, T.E., Ter-Mikaelian, M., Ågren, G.I., Oliver, C.D., Puttonen, P., 2000a. Process-based models for forest ecosystem management: Current state of the art and challenges for practical implementation. Tree Physiology 20, 289–298.

Mäkelä, A., Sievänen, R., Lindner, M., Lasch, P., 2000b. Application of volume growth and survival graphs in the evaluation of four process-based forest growth models. Tree Physiology 20, 347–355.

Mäkelä, A., Valentine, H.T., Helmisaari, H.-S., 2008. Optimal co-allocation of carbon and nitrogen in a forest stand at steady state. New Phytologist 180, 114–123.

Mäkinen, A., Kangas, A., Kalliovirta, J., Rasinmäki, J., Välimäki, E., 2008. Comparison of treewise and standwise forest simulators by means of quantile regression. Forest Ecology and Management 255, 2709–2717.

Mäkinen, A., Holopainen, M., Kangas, A., Rasinmäki, J., 2010. Propagating the errors of initial forest variables through stand- and tree-level growth simulators. European Journal of Forest Research 129, 887–897.

Mäkinen, H., Ojansuu, R., Sairanen, P., Yli-Kojola, H., 2003. Predicting branch characteristics of Norway spruce (*Picea abies* (L.) Karst.) from simple stand and tree measurements. Forestry 76, 525–546.

Makler-Pick, V., Gal, G., Gorfine, M., Hipsey, M.R., Carmel, Y., 2011. Sensitivity analysis for complex ecological models – A new approach. Environmental Modelling & Software 26, 124–134.

Marshall, D.D., Curtis, R.O., 2002. Levels-of-growing-stock cooperative study in Douglas-fir: Report no. 15 – Hoskins 1963–1998. Research Paper PNW-RP-537. USDA Forest Service, Pacific Northwest Research Station, Portland, OR.

Marshall, D.D., Johnson, D.L., Hann, D.W., 1997. A forest management information system for education, research and operations. Journal of Forestry 95, 27–30.

Marshall, D.D., Johnson, G.P., Hann, D.W., 2003. Crown profile equations for stand-grown western hemlock trees in northwestern Oregon. Canadian Journal of Forest Research 33, 2059–2066.

Martin, G.L., Ek, A.R., 1984. A comparison of competition measures and growth models for predicting plantation red pine diameter and height growth. Forest Science 30, 731–743.

Martin, G.L., Ek, A.R., Monserud, R.A., 1977. Control of plot edge bias in forest stand growth simulation models. Canadian Journal of Forest Research 7, 100–105.

Mason, E.G., Dzierzon, H., 2006. Applications of modeling to vegetation management. Canadian Journal of Forest Research 36, 2505–2514.

Mason, E.G., Milne, P.G., 1999. Effects of weed control, fertilization, and soil cultivation on the growth of *Pinus radiata* at midrotation in Canterbury, New Zealand. Canadian Journal of Forest Research 29, 985–992.

Mason, E.G., Whyte, A.G.D., Woollons, R.C., Richardson, B., 1997. A model of the growth of juvenile radiata pine in the Central North Island of New Zealand: Links with older models and rotation-length analyses of the effects of site preparation. Forest Ecology and Management 97, 187–195.

Mason, E.G., Rose, R.W., Rosner, L.S., 2007. Time vs. light: A potentially useable light sum hybrid model to represent the juvenile growth of Douglas-fir subject to varying levels of competition. Canadian Journal of Forest Research 37, 795–805.

Mason, E.G., Methol, R., Cochrane, H., 2011. Hybrid mensurational and physiological modelling of growth and yield of *Pinus radiata* D Don. using potentially useable radiation sums. Forestry, in press.

Matala, J., Hynynen, J., Miina, J., Ojansuu, R., Peltola, H., Sievänen, R., Väisänen, H., Kellomäki, S., 2003. Comparison of a physiological model and a statistical model for prediction of growth and yield in boreal forests. Ecological Modelling 161, 95–116.

Matney, T.G., Sullivan, A.D., 1982. Compatible stand and stock tables for thinned and unthinned loblolly pine stands. Forest Science 28, 161–171.

Matney, T.G., Belli, K.L., Farrar Jr., R.M., 1990. Parameter-free diameter-distribution recovery. In IUFRO Division 4 World Congress. Publication FWS-2-90. School of Forestry and Wildlife Resources, Virginia Polytechnic Institute and State University, Montréal, Canada, pp. 95–107.

Max, T.A., Burkhart, H.E., 1976. Segmented polynomial regression applied to taper equations. Forest Science 22, 283–289.

McArdle, R.E., Meyer, W.H., Bruce, D., 1961. The yield of Douglas-fir in the Pacific Northwest. Technical Bulletin 201 (rev.). USDA, Washington, DC.

McCarter, J.B., Wilson, J.S., Baker, P.J., Moffett, J.L., Oliver, C.D., 1998. Landscape management through integration of existing tools and emerging technologies. Journal of Forestry 96, 17–23.

McDill, M.E., Amateis, R.L., 1993. Fitting discrete-time dynamic models having any time interval. Forest Science 39, 499–519.

McDowell, N., Barnard, H., Bond, B.J., Hinckley, T.M., Hubbard, R.M., Ishii, H., Köstner, B., Magnani, F., Marshall, J.D., Meinzer, F.C., Phillips, N., Ryan, M.G., Whitehead, D., 2002. The relationship between tree height and leaf area:sapwood ratio. Oecologia 132, 12–20.

McGaughey, R.J., 1997. Visualizing forest stand dynamics using the stand visualization system. In Proceedings of the 1997 ACSM/ASPRS Annual Convention and Exposition. American Society of Photogrammetry and Remote Sensing, Seattle, WA, pp. 248–257.

McGaughey, R.J., 2003. EnVision – Environmental Visualization Software. USDA Forest Service, Pacific Northwest Research Station, Portland, OR.

McNab, W.H., 2010. Effects of landform on site index for two mesophytic tree species in the Appalachian Mountains of North Carolina, USA. International Journal of Forestry Research, doi: 10.1155/2010/298674.

McRoberts, R.E., Hahn, J.T., Hefty, G.J., Van Cleve, J.R., 1994. Variation in forest inventory field measurements. Canadian Journal of Forest Research 24, 1766–1770.

McTague, J.P., Bailey, R.L., 1987. Compatible basal area and diameter distribution models for thinned loblolly pine plantations in Santa Catarina, Brazil. Forest Science 33, 43–51.

McTague, J.P., Stansfield, W.F., 1994. Stand and tree dynamics of uneven-aged ponderosa pine. Forest Science 40, 289–302.

McTague, J.P., Stansfield, W.F., 1995. Stand, species, and tree dynamics of an uneven-aged, mixed conifer forest type. Canadian Journal of Forest Research 25, 803–812.

McTague, J.P., O'Loughlin, D., Roise, J.P., Robison, D.J., Kellison, R.C., 2008. The SOHARC model system for growth and yield of southern hardwoods. Southern Journal of Applied Forestry 32, 173–183.

Mead, R., 1966. A relationship between individual plant-spacing and yield. Annals of Botany 30, 301–309.

Medlyn, B.E., 1998. Physiological basis of the light use efficiency model. Tree Physiology 18, 167–176.

Medlyn, B.E., Dewar, R.C., 1999. Comment on the article by R.H.Waring, J.J. Landsberg and M. Williams relating net primary production to gross primary production. Tree Physiology 19, 137–138.

Medlyn, B., Barrett, D., Landsberg, J., Sands, P., Clement, R., 2003. Conversion of canopy intercepted radiation to photosynthate: A review of modelling approaches for regional scales. Functional Plant Biology 30, 153–169.

Medlyn, B.E., Dreyer, E., Ellsworth, D., Forstreuter, M., Harley, P.C., Kirschbaum, M.U.F., Le Roux, X., Montpied, P., Strassemeyer, J., Walcroft, A., Wang, K., Loustau, D., 2002. Temperature response of parameters of a biochemically based model of photosynthesis. II. A review of experimental data. Plant, Cell and Environment 25, 1167–1179.

Mendoza, G.A., Gumpal, E.C., 1987. Growth projection of a selectively cut-over forest based on residual inventory. Forest Ecology and Management 20, 253–263.

Mendoza, G.A., Setyarso, A., 1986. A transition matrix forest growth model for evaluating alternative harvesting schemes in Indonesia. Forest Ecology and Management 15, 219–228.

Meng, S.X., Huang, S., 2010. Incorporating correlated error structures into mixed forest growth models: Prediction and inference implications. Canadian Journal of Forest Research 40, 977–990.

Mengel, D.L., Roise, J.R., 1990. A diameter-class matrix model for Southeastern U.S. coastal plain bottomland hardwood stands. Southern Journal of Applied Forestry 14, 189–195.

Merganičová, K., Pietsch, S.A., Hasenauer, H., 2005. Testing mechanistic modeling to assess impacts of biomass removal. Forest Ecology and Management 207, 37–57.

Metcalf, C.J.E., McMahon, S.M., Clark, J.S., 2009a. Overcoming data sparseness and parametric constraints in modeling of tree mortality: A new nonparametric Bayesian model. Canadian Journal of Forest Research 39, 1677–1687.

Metcalf, C.J.E., Clark, J.S., Clark, D.A., 2009b. Tree growth inference and prediction when the point of measurement changes: Modelling around buttresses in tropical forests. Journal of Tropical Ecology 25, 1–12.

Mette, T., Albrecht, A., Ammer, C., Biber, P., Kohnle, U., Pretzsch, H., 2009. Evaluation of the forest growth simulator SILVA on dominant trees in mature mixed Silver fir–Norway spruce stands in South-West Germany. Ecological Modelling 220, 1670–1680.

Michie, B.R., Buongiorno, J., 1984. Estimation of a matrix model of forest growth from re-measured permanent plots. Forest Ecology and Management 8, 127–135.

Miehle, P., Battaglia, M., Sands, P.J., Forrester, D.I., Feikema, P.M., Livesley, S.J., Morris, J.D., Arndt, S.K., 2009. A comparison of four process-based models and a statistical regression model to predict growth of *Eucalyptus globulus* plantations. Ecological Modelling 220, 734–746.

Miina, J., Heinonen, J., 2008. Stochastic simulation of forest regeneration establishment using a multilevel multivariate model. Forest Science 54, 206–219.

Miina, J., Saksa, T., 2006. Predicting regeneration establishment in Norway spruce plantations using a multivariate multilevel model. New Forests 32, 265–283.

Miina, J., Eerikäinen, K., Hasenauer, H., 2006. Modelling forest regeneration. In Hasenauer, H. (Ed.), Sustainable Forest Management: Growth Models for Europe. Springer, Berlin, Germany, pp. 93–109.

Miller, R.E., Tarrant, R.F., 1983. Long-term growth response of Douglas-fir to ammonium nitrate fertilizer. Forest Science 29, 127–137.

Miller, R.E., Clendenen, G.W., Bruce, D., 1988. Volume growth and response to thinning and fertilizing of Douglas-fir stands in southwestern Oregon. General Technical Report PNW-GTR-221. USDA Forest Service, Pacific Northwest Research Station, Portland, OR.

Milner, K.S., Coble, D.W., McMahan, A.J., Smith, E.L., 2003. FVSBGC: A hybrid of the physiological model STAND-BGC and the forest vegetation simulator. Canadian Journal of Forest Research 33, 466–479.

Mitchell, K.J., 1969. Simulation of the growth of even-aged stands of white spruce. Bulletin 75. Yale University School of Forestry, New Haven, CT.

Mitchell, K.J., 1975. Dynamics and simulated yield of Douglas-fir. Forest Science Monograph 17, 1–39.

Mitchell, K.J., 1986. Comparison of McArdle, DFSIM, and TASS growth and yield models. In Oliver, C.D., Hanley, D.P., Johnson, J.A. (Eds.), Douglas-fir: Stand management for the future. Contribution Number 55. College of Forest Resources, University of Washington, Seattle, WA, pp. 350–359.

Mitchell, K.J., Cameron, I.R., 1985. Managed stand yield tables for coastal Douglas-fir: Initial density and precommercial thinning. Land Management Report 31. British Columbia Ministry of Forests Research Branch, Victoria, BC.

Mitchell, K.J., Stone, M., Grout, S.E., Di Lucca, C.M., Nigh, G.D., Goudie, J.W., Stone, J.N., Nussbaum, A.J., Yanchuk, A., Stearns-Smith, S., Brockley, R., 2004. TIPSY V4.2. British Columbia Ministry of Forests and Range, Victoria, BC.

Mladenoff, D.J., 2004. LANDIS and forest landscape models. Ecological Modelling 180, 7–19.

Moeur, M., 1981. Crown width and foliage weight of northern Rocky Mountain conifers. Research Paper INT-283. USDA Forest Service, Intermountain Research Station, Ogden, UT.

Moles, A.T., Westoby, M., 2004. Seedling survival and seed size: A synthesis of the literature. Journal of Ecology 92, 372–383.

Moles, A.T., Falster, D.S., Leishman, M.R., Westoby, M., 2004. Small-seeded species produce more seeds per square metre of canopy per year, but not per individual per lifetime. Journal of Ecology 92, 384–396.

Monleon, V.J., 2003. A hierarchical linear model for tree height prediction. Proceedings of the 2003 Meeting of the American Statistical Association, Section on Statistics and the Environment. Alexandria, VA, pp. 2865–2869.

Monleon, V.J., Azuma, D., Gedney, D., 2004. Equations for predicting uncompacted crown ratio based on compacted crown ratio and tree attributes. Western Journal of Applied Forestry 19, 260–267.

Monserud, R.A., 1975. Methodology for simulating Wisconsin northern hardwood stand dynamics. PhD dissertation. University of Wisconsin, Madison, WI.

Monserud, R.A., 1976. Simulation of forest tree mortality. Forest Science 22, 438–444.

Monserud, R.A., 1984. Height growth and site index curves for inland Douglas-fir based on stem analysis data and forest habitat type. Forest Science 30, 943–965.

Monserud, R.A., 1985. Comparison of Douglas-fir site index and height growth curves in the Pacific Northwest. Canadian Journal of Forestry 15, 673–679.

Monserud, R.A., 1986. Time-series analyses of tree-ring chronologies. Forest Science 32, 349–372.

Monserud, R.A., 2003. Evaluating forest models in a sustainable forest management context. Forest Biometry, Modelling and Information Sciences 1, 35–47.

Monserud, R.A., Ek, A.R., 1977. Prediction of understory tree height growth in northern hardwood stands. Forest Science 23, 391–400.

Monserud, R.A., Rehfeldt, G.E., 1990. Genetic and environmental components of variation of site index in inland Douglas-fir. Forest Science 36, 1–9.

Monserud, R.A., Sterba, H., 1996. A basal area increment model for individual trees growing in even- and uneven-aged forest stands in Austria. Forest Ecology and Management 80, 57–80.

Monserud, R.A., Sterba, H., 1999. Modeling individual tree mortality for Austrian forest species. Forest Ecology and Management 113, 109–123.

Monserud, R.A., Moody, U., Breuer, D.W., 1990. A soil-site study for inland Douglas-fir. Canadian Journal of Forest Research 20, 686–695.

Monserud, R.A., Sterba, H., Hasenauer, H., 1997. The single-tree stand growth simulator PROGNAUS. In Teck, R., Moeur, M., Adams, J. (Eds.), Proceedings of the Forest Vegetation Simulator Conference. General Technical Report INT-373. USDA Forest Service, Intermountain Research Station, Ogden, UT, pp. 50–56.

Monserud, R.A., Ledermann, T., Sterba, H., 2005. Are self-thinning constraints needed in a tree-specific mortality model? Forest Science 50, 848–858.

Monserud, R.A., Huang, S.M., Yang, Y.Q., 2006. Predicting lodgepole pine site index from climatic parameters in Alberta. Forestry Chronicle 82, 562–571.

Monsi, M., Saeki, T., 1953. Über den Lichtfakator in den Pflanzengesellschaften und seine Bedeutung für die Stoffproduktion. Japanese Journal of Botany 14, 22–52.

Monteith, J.L., 1965. Evaporation and environment. Symposium of the Society of Experimental Biology 19, 205–234.

Monteith, J.L., 1972. Solar radiation and productivity in tropical ecosystems. Journal of Applied Ecology 9, 747–766.

Monteith, J.L., 1977. Climate and the efficiency of crop production in Britain. Philosophical Transactions of the Royal Society of London B: Biological Sciences 281, 277–294.

Moore, A.D., 1989. On the maximum growth equation used in forest gap simulation models. Ecological Modelling 45, 63–67.

Moore, J.A., Budelsky, C.A., Schlesinger, R.C., 1973. A new index representing individual tree competitive status. Canadian Journal of Forest Research 3, 495–500.

Moore, J.A., Zhang, L., Newberry, J.D., 1994. Effects of intermediate silvicultural treatments on the distribution of within-stand growth. Canadian Journal of Forest Research 24, 398–404.

Morzuch, B.J., Ruark, G.A., 1991. Principal components regression to mitigate the effects of multicollinearity. Forest Science 37, 191–199.

Moser Jr., J.W., Hall, O.F., 1969. Deriving growth and yield functions for uneven-aged forest stands. Forest Science 15, 183–188.

Mowrer, H.T., Frayer, W.E., 1986. Variance propagation in growth and yield projections. Canadian Journal of Forest Research 16, 1196–1200.

Muhairwe, C.K., LeMay, V.M., Kozak, A., 1994. Effects of adding tree, stand, and site variables to Kozak's variable-exponent taper equation. Canadian Journal of Forest Research 24, 252–259.

Munro, D.D., 1974. Forest growth-models: A prognosis. In Fries, J. (Ed.), Growth models for tree and stand simulation. Research Note 30. Royal College of Forestry, Stockholm, Sweden, pp. 7–21.

Murray D.M., von Gadow, K. 1991. Relationships between the diameter distributions before and after thinning. Forest Science 37, 552–559.

Muukkonen, P., 2007. Generalized allometric volume and biomass equations for some tree species in Europe. European Journal of Forest Research 126, 157–166.

Myers, B.J., 1988. Water stress integral – a link between short-term stress and long-term growth. Tree Physiology 4, 315–323.

Myers, C.A., 1977. A computer program for variable density yield tables for loblolly pine plantations. General Technical Report SO-11. USDA Forest Service, Southern Forest Experiment Station, New Orleans, LA.

Nachtmann, G., 2006. Height increment models for individual trees in Austria depending on site and competition. Austrian Journal of Forest Science 123, 199–222.

Nalle, D.J., Montgomery, C.A., Arthur, J.L., Polasky, S., Schumaker, N.H., 2004. Modeling joint production of wildlife and timber. Journal of Environmental Economics and Management 48, 997–1017.

Nance, W.L., Wells, O.O., 1981. Site index models for height growth of planted loblolly pine (*Pinus taeda* L.) seed sources. In Proceedings of 16th Southern Forest Tree Improvement Conference. Virginia Polytechnic Institute and State University, Blacksburg, Virginia, pp. 86–96.

Nance, W.L., Grissom, J.R., Smith, W.R., 1988. A new competition index based on weighted and constrained area potentially available. In Ek, A.R., Shifley, S.R., Burk, T.E. (Eds.), Forest growth modelling and prediction, Volume I. General Technical Report NC-120. USDA Forest Service, North Central Forest Experiment Station, St. Paul, MN, pp. 134–142.

Nanos, N., Larson, K., Millerón, M., Sjöstedt-de Luna, S., 2010. Inverse modeling for effective dispersal: Do we need tree size to estimate fecundity? Ecological Modelling 221, 2415–2424.

Nathan, R., Safriel, U.N., Noy-Meir, I., 2001. Field validation and sensitivity analysis of a mechanistic model for tree seed dispersal by wind. Ecology 82, 374–388.

Nepal, S.K., Somers, G.L., 1992. A generalized approach to stand table projection. Forest Science 38, 120–133.

Nepal, S.K., Somers, G.L., Caudill, S.H., 1996. A stochastic frontier model for fitting tree crown shape in loblolly pine (*Pinus taeda* L.). Journal of Agricultural, Biological, and Environmental Statistics 1, 336–353.

Newnham, R.M., Smith, J.H.G., 1964. Development and testing of stand models for Douglas-fir and lodgepole pine. Forestry Chronicle 40, 494–504.

Newton, P.F., 2009. Development of an integrated decision-support model for density management within jack pine stand-types. Ecological Modelling 220, 3301–3324.

Newton, P.F., Amponsah, I.G., 2007. Comparative evaluation of five height–diameter models developed for black spruce and jack pine stand-types in terms of goodness-of-fit, lack-of-fit and predictive ability. Forest Ecology and Management 247, 149–166.

Nigh, G.D., 1995a. The geometric mean regression line: A method for developing site index conversion equations for species in mixed stands. Forest Science 41, 84–98.

Nigh, G.D., 1995b. Growth intercept models and tables for British Columbia coastal species. Land Management Handbook Field Guide. Ministry of Forests, Research Branch, Victoria, BC.

Nigh, G.D., 1998. A system for estimating height and site index of western hemlock in the interior of British Columbia. Forestry Chronicle 74, 588–596.

Nigh, G., 2002. Site index conversion equations for mixed trembling aspen and white spruce stands in northern British Columbia. Silva Fennica 36, 789–797.

Nigh, G.D., Love, B.A., 1999. How well can we select undamaged site trees for estimating site index? Canadian Journal of Forest Research 29, 1989–1992.

Nigh, G.D., Love, B.A., 2004. Predicting crown class in three western conifer species. Canadian Journal of Forest Research 34, 592–599.

Nightingale, J.M., Hill, M.J., Phinn, S.R., Davies, I.D., Held, A.A., Erskine, P.D., 2008. Use of 3-PG and 3-PGS to simulate forest growth dynamics of Australian tropical rainforests: I. Parameterisation and calibration for old-growth, regenerating and plantation forests. Forest Ecology and Management 254, 107–121.

Nilsson, U., Allen, H.L., 2003. Short- and long-term effects of site preparation, fertilization and vegetation control on growth and stand development of planted loblolly pine. Forest Ecology and Management 175, 367–377.

Nix, H.A., Gillison, A.N., 1985. Towards an operational framework for habitat and wildlife management. In Kikkawa, J. (Ed.), Wildlife Management in the Forests and Forestry-Controlled Lands in the Tropics and Southern Hemisphere. University of Queensland, Brisbane, Australia, pp. 39–55.

Nolet, P., Domon, G., Bergeron, Y., 1995. Potentials and limitations of ecological classification as a tool for forest management: A case study of disturbed deciduous forests in Québec. Forest Ecology and Management 78, 85–98.

Nord-Larsen, T., 2006. Modeling individual-tree growth from data with highly irregular measurement intervals. Forest Science 52, 198–208.

Nord-Larsen, T., Johannsen, V.K., 2007. A state-space approach to stand growth modelling of European beech. Annals of Forest Science 64, 365–374.

Nothdurft, A., Kublin, E., Lappi, J., 2006. A non-linear hierarchical mixed model to describe tree height growth. European Journal of Forest Research 125, 281–289.

Nunifu, T.K., 2009. Compatible diameter and height increment models for lodgepole pine, trembling aspen, and white spruce. Canadian Journal of Forest Research 39, 180–192.

Nyström, K., Kexi, M., 1997. Individual tree basal area growth models for young stands of Norway spruce in Sweden. Forest Ecology and Management 97, 173–185.

O'Hara, K.L., Oliver, C.D., 1988. Three-dimensional representation of Douglas-fir volume growth: Comparison of growth and yield models with stand data. Forest Science 34, 724–743.

Ochi, N., Cao, Q.V., 2003. A comparison of compatible and annual growth models. Forest Science 49, 285–290.

Ogée, J., Brunet, Y., Loustau, D., Berbigier, P., Delzon, S., 2003. MuSICA, a CO_2, water, and energy multilayer, multileaf pine forest model: Evaluation from hourly to yearly time scales and sensitivity analysis. Global Change Biology 9, 697–717.

Oliver, C.W., Larson, B.C., 1996. Forest stand dynamics: Update edition. John Wiley & Sons, Inc., New York, NY.

Ollinger, S.V., Aber, J.D., Federer, C.A., 1998. Estimating regional forest productivity and water yield using an ecosystem model linked to a GIS. Landscape Ecology 13, 323–334.

Opalach, D., Heath, L.S., 1988. Evaluation of long-term fertilizer response. In Ek, A.R., Shifley, S.R., Burk, T.E. (Eds.), Forest growth modeling and prediction. General Technical Report NC-120. USDA Forest Service, North Central Forest Experimental Station, St. Paul, MN, pp. 555–561.

Opie, J.E., 1968. Predictability of individual tree growth using various definitions of competing basal area. Forest Science 14, 314–323.

Orland, B., 1995. SMARTFOREST: A 3-D interactive forest visualization and analysis system. In Power, J.M., Strome, M., Daniel, T.C. (Eds.), Proceedings of the Decision Support – 2001 Conference, Toronto, Ontario, pp. 181–190.

Ottorini, J.-M., 1991. Growth and development of individual Douglas-fir in stands for applications to simulation in silviculture. Annals of Forest Science 48, 651–666.

Ottorini, J.-M., Le Goff, N., Cluzeau, C., 1996. Relationships between crown dimensions and stem development in *Fraxinus excelsior.* Canadian Journal of Forest Research 26, 394–401.

Ouzennous, H., Pothier, D., Raulier, F., 2008. Adjustment of the age-height relationship for uneven-aged black spruce stands. Canadian Journal of Forest Research 38, 2003–2012.

Pacala, S.W., Canham, C.D., Silander Jr., J.A., 1993. Forest models defined by field measurements: I. The design of a northeastern forest simulator. Canadian Journal of Forest Research 23, 1980–1988.

Pacala, S.W., Canham, C.D., Saponara, J., Silander Jr., J.A., Kobe, R.K., Ribbens, E., 1996. Forest models defined by field measurements. II. Estimation, error analysis, and dynamics. Ecological Monographs 66, 1–43.

Paine, D.P., Hann, D.W., 1982. Maximum crown-width equations for southwestern Oregon tree species. Research Paper 46. Oregon State University, Forestry Research Laboratory, Corvallis, OR.

Pairon, M., Jonard, M., Jacquemart, A.-L., 2006. Modeling seed dispersal of black cherry, an invasive forest tree: How microsatellites may help? Canadian Journal of Forest Research 36, 1385–1394.

Palosuo, T., Peltoniemi, M., Mikhailov, A., Komarov, A., Faubert, P., Thürig, E., Lindner, M., 2008. Projecting effects of intensified biomass extraction with alternative modelling approaches. Forest Ecology and Management 255, 1423–1433.

Papaik, M.J., Fall, A., Sturtevant, B., Kneeshaw, D., Messier, C., Fortin, M.-J., Simon, N., 2010. Forest processes from stands to landscapes: Exploring model forecast uncertainties using cross-scale model comparison. Canadian Journal of Forest Research 40, 2345–2359.

Parresol, B.R., 2001. Additivity of nonlinear biomass equations. Canadian Journal of Forest Research 31, 865–878.

Parresol, B.R., Thomas, C.E., 1996. A simultaneous density-integral system for estimating stem profile and biomass: Slash pine and willow oak. Canadian Journal of Forest Research 26, 773–781.

Parrott, L., Lange, H., 2004. Use of interactive forest growth simulation to characterise spatial stand structure. Forest Ecology and Management 194, 29–47.

Paterson, S.S., 1956. The forest area of the world and its potential productivity. Royal University of Göteborg, Göteborg, Sweden.

Paul, K.I., Polglase, P.J., O'Connell, A.M., Carlyle, J.C., Smethurst, P.J., Khanna, P.K., 2002. Soil nitrogen availability predictor (SNAP): A simple model for predicting mineralisation of nitrogen in forest soils. Australian Journal of Soil Research 40, 1011–1026.

Paul, K.I., Polglase, P.J., O'Connell, A.M., Carlyle, J.C., Smethurst, P.J., Khanna, P.K., Worledge, D., 2003. Soil water under forests (SWUF): A model of water flow and soil water content under a range of forest types. Forest Ecology and Management 182, 195–211.

Payandeh, B., Haig, R.A., 1991. A management-oriented model for assessing early stand establishment. Forestry Canada, Frontline Technical Note 5.

Peet, R.K., 1976. Successional patterns and tree population structure in Rocky Mountain forests. Bulletin of the Ecological Society of America 57, 60.

Peet, R.K., 1981. Forest vegetation of the Colorado Front Range. Vegetatio 45, 3–75.

Peltoniemi, M., Mäkipää, R., 2011. Quantifying distance-independent tree competition for predicting Norway spruce mortality in unmanaged forests. Forest Ecology and Management 261, 30–42.

Pelz, D.R., 1978. Estimating individual tree growth with tree polygons. In Fries, J., Burkhart, H.E., Max, T.A. (Eds.), Growth models for long term forecasting of timber yields. Publication FWS-1-78. Virginia Polytechnic Institute and State University, Blacksburg, VA, pp. 172–178.

Peng, C., 2000. Growth and yield models for uneven-aged stands: Past, present, and future. Forest Ecology and Management 132, 259–279.

Peng, C., Zhang, L., Liu, J., 2001. Developing and validating nonlinear height-diameter models for major tree species of Ontario's boreal forests. Northern Journal of Applied Forestry 18, 87–94.

Peng, C., Liu, J., Dang, Q., Apps, M.J., Jiang, H., 2002. TRIPLEX: A generic hybrid model for predicting forest growth and carbon and nitrogen dynamics. Ecological Modelling 153, 109–130.

Penman, H.L., 1948. Natural evaporation from open water, bare soil and grass. Proceedings of the Royal Society of London Series A, Mathematical and Physical Sciences 193, 120–145.

Perot, T., Goreaud, F., Ginisty, C., Dhôte, J.-F., 2010. A model bridging distance-dependent and distance-independent tree models to simulate the growth of mixed forests. Annals of Forest Science 67, 502.

Perry, D.A., 1985. The competition process in forest stands. In Cannell, M.G.R., Jackson, J.E. (Eds.), Attributes of trees as crop plants. Institute of Terrestrial Ecology, Huntingdon, UK, pp. 481–506.

Perry, D.A., Oren, R., Hart, S.C., 2008. Forest Ecosystems 2nd edition. Johns Hopkins University Press, Baltimore, MD.

Peterson, C.E., Hazard, J.W., 1990. Regional variation in growth response of coastal Douglas-fir to nitrogen fertilizer in the Pacific Northwest. Forest Science 36, 625–640.

Peterson, C.E., Webster, S.R., Barker, P.R., Miller, R.E., 1986. Using nitrogen fertilizers in management of coast Douglas-fir: II. Future informational needs. In Oliver, C.D., Hanley, D. P., Johnson, J.A. (Eds.), Douglas-fir: Stand management for the future. Contribution Number 55. University of Washington, Institute of Forest Resources, Seattle, WA, pp. 304–309.

Petersson, H., 1999. A segmented stem profile model for *Pinus sylvestris*. Forest Ecology and Management 124, 13–26.

Petritsch, R., Hasenauer, H., Pietsch, S.A., 2007. Incorporating forest growth response to thinning within Biome-BGC. Forest Ecology and Management 242, 324–336.

Pfister, R.D., Arno, S.F., 1980. Classifying forest habitat types based on potential climax vegetation. Forest Science 26, 52–70.

Picard, N., Bar-Hen, A., Franc, A., 2002. Modeling forest dynamics with a combined matrix/individual-based model. Forest Science 48, 643–652.

Picard, N., Banak, L.N., Namkosserena, S., Yalibanda, Y., 2009. The stock recovery rate in a Central African rainforest: An index of sustainability based on projection matrix models. Canadian Journal of Forest Research 39, 2138–2152.

Picard, N., Ouédraogo, D., Bar-Hen, A., 2010. Choosing classes for size projection matrix models. Ecological Modelling 221, 2270–2279.

Pienaar, L.V., Harrison, W.M., 1988. A stand table projection approach to yield prediction in unthinned even-aged stands. Forest Science 34, 804–808.

Pienaar, L.V., Rheney, J.W., 1993. Yield predictions for mechanically site-prepared slash pine plantations in the southeastern Coastal Plain. Southern Journal of Applied Forestry 17, 163–173.

Pienaar, L.V., Rheney, J.W., 1995. Modeling stand-level growth and yield response to silvicultural treatments. Forest Science 41, 629–638.

Pienaar, L.V., Shiver, B.D., 1981. Survival functions for site-prepared slash pine plantations in the flatwoods of Georgia and northern Florida. Southern Journal of Applied Forestry 5, 59–62.

Pienaar, L.V., Shiver, B.D., 1986. Basal area prediction and projection equations for pine plantations. Forest Science 32, 626–633.

Pienaar, L.V., Shiver, B.D., Grider, G.E., 1985. Predicting basal area growth in thinned slash pine plantations. Forest Science 31, 731–741.

Pienaar, L.V., Page, H.H., Rheney, J.W., 1990. Yield prediction for mechanically site-prepared slash pine plantations. Southern Journal of Applied Forestry 14, 104–109.

Pierce, L.L., Running, S.W., 1988. Rapid estimation of coniferous forest leaf area index using a portable integrating radiometer. Ecology 69, 1762–1767.

Pilli, R., Anfodillo, T., Carrer, M., 2006. Towards a functional and simplified allometry for estimating forest biomass. Forest Ecology and Management 237, 583–593.

Pinheiro, J.C., Bates, D.M., 2000. Mixed-Effects Models in S and S-Plus. Springer-Verlag, New York, NY.

Pinjuv, G., Mason, E.G., Watt, M., 2006. Quantitative validation and comparison of a range of forest growth model types. Forest Ecology and Management 236, 37–46.

Pinkard, E.A., Battaliga, M., Bruce, J., Leriche, A., Kriticos, D.J., 2010. Process-based modelling of the severity and impact of foliar pest attack on eucalypt plantation productivity under current and future climates. Forest Ecology and Management 259, 839–847.

Plonski, W.L., 1974. Normal yield tables (metric) for major forest species of Ontario. Ontario Ministry of Natural Resources, Division of Forestry, Toronto, ON.

Pokharel, B., Froese, R.E., 2008. Evaluating alternative implementations of the Lake States FVS diameter increment model. Forest Ecology and Management 255, 1759–1771.

Pokharel, B., Froese, R.E., 2009. Representing site productivity in the basal area increment model for FVS-Ontario. Forest Ecology and Management 258, 657–666.

Pommerening, A., Stoyan, D., 2006. Edge-correction needs in estimating indices of spatial forest structure. Canadian Journal of Forest Research 36, 1723–1739.

Porté, A., Bartelink, H.H., 2002. Modelling mixed forest growth: A review of models for forest management. Ecological Modelling 150, 141–188.

Powers, R.F., Oliver, W.W., 1978. Site classification of ponderosa pine stands under stocking control in California. Research Paper PSW-128. USDA Forest Service, Pacific Southwest Forest and Range Experiment Station, Berkeley, CA.

Pregitzer, K.S., Goebel, P.C., Wigley, T.B., 2001. Evaluating forestland classification schemes as tools for maintaining biodiversity. Journal of Forestry 99, 33–40.

Pretzsch, H., 1997. Analysis and modeling of spatial stand structures. Methodological considerations based on mixed beech-larch stands in Lower Saxony. Forest Ecology and Management 97, 237–253.

Pretzsch, H., 2002. Application and evaluation of the growth simulator SILVA 2.2 for forest stands, forest estates and large regions. Forstwissenschaftliches Centralblatt 121, 28–51.

Pretzsch, H., 2009. Forest Dynamics, Growth and Yield: From Measurement to Model. Springer-Verlag, Berlin, Germany.

Pretzsch, H., Biber, P., 2005. A re-evaluation of the Reineke's rule and stand density index. Forest Science 51, 304–320.

Pretzsch, H., Biber, P., 2010. Size-symmetric versus size-asymmetric competition and growth partitioning among trees in forest stands along an ecological gradient in Central Europe. Canadian Journal of Forest Research 40, 370–384.

Pretzsch, H., Biber, P., Ďurský, J. 2002. The single tree-based stand simulator SILVA: Construction, application and evaluation. Forest Ecology and Management 162, 3–21.

Pretzsch, H., Biber, P., Ïurský, J., Sodtke, R. 2006. The individual-tree-based stand simulator SILVA. In Hasenauer, H. (Ed.), Sustainable Forest Management: Growth Models for Europe. Springer, Berlin, Germany, pp. 78–84.

Pretzsch, H., Grote, R., Reineking, B., Rötzer, T.H., Seifert, S.T., 2008. Models for forest ecosystem management: A European perspective. Annals of Botany 101, 1065–1087.

Price, D.T., Zimmermann, N.E., van der Meer, P.J., Lexer, M.J., Leadly, P., Jorritsma, I.T.M., Schaber, J., Clark, D.F., Lasch, P., McNulty, S., Wu, J., Smith, B., 2001. Regeneration in gap models: Priority issues for studying forest responses to climate change. Climatic Change 51, 475–508.

Priestley, C.H.B., Taylor, R.J., 1972. On the assessment of surface heat flux and evaporation using large-scale parameters. Monthly Weather Review 100, 81–92.

Puettmann, K.J., Hann, D.W., Hibbs, D.E., 1993. Evaluation of the size-density relationships for pure red alder and Douglas-fir stands. Forest Science 39, 7–27.

Puettmann, K.J., Hibbs, D.E., Hann, D.W., 1992. The dynamics of mixed stands of *Alnus rubra* and *Pseudotsuga menziesii*: Extension of size-density analysis to species mixture. Journal of Ecology 80, 449–458.

Pukkala, T., 1989. Predicting diameter growth in even-aged Scots pine stands with a spatial and non-spatial model. Silva Fennica 23, 101–116.

Pukkala, T., Kolström, T., 1991. Effect of spatial pattern of trees on the growth of a Norway spruce stand: A simulation model. Silva Fennica 25, 117–131.

Pukkala, T., Lähde, E., Laiho, O., 2009. Growth and yield models for uneven-sized forest stands in Finland. Forest Ecology and Management 258, 207–216.

Qin, J., Cao, Q.V., 2006. Using disaggregation to link individual-tree and whole-stand growth models. Canadian Journal of Forest Research 36, 953–960.

Qin, J., Cao, Q.V., Blouin, D.C., 2007. Projection of a diameter distribution through time. Canadian Journal of Forest Research 37, 188–194.

Quero, J.L., Herrero, A., Zamora, R., 2011. Linking stochasticity to determinism of woody plant recruitment in a mosaic landscape: A spatially explicit approach. Basic and Applied Ecology 12, 161–171.

Quicke, H.E., Meldahl, R.S., Kush, J.S., 1994. Basal area growth of individual trees: A model derived from a regional longleaf pine growth study. Forest Science 40, 528–542.

Quicke, H., Glover, G., Meldahl, R.S., 1999. Loblolly pine growth response to herbaceous vegetation control at different planting densities. Canadian Journal of Forest Research 29, 960–967.

R Development Core Team, 2009. R: A language and environment for statistical computing. Reference Index. R Foundation for Statistical Computing, Vienna, Austria.

Radtke, P.J., Robinson, A.P., 2006. A Bayesian strategy for combining predictions from empirical and process-based models. Ecological Modelling 190, 287–298.

Radtke, P.J., Burk, T.E., Bolstad, P.V., 2002. Bayesian melding of a forest ecosystem model with correlated inputs. Forest Science 48, 701–711.

Randolph, K.C., 2010. Equations relating compacted and uncompacted live crown ratio for common tree species in the South. Southern Journal of Applied Forestry 34, 118–123.

Ranger, J., Gelhaye, D., 2001. Belowground biomass and nutrient content in a 47-year-old Douglas-fir plantation. Annals of Forest Science 58, 423–430.

Rathbun, L.C., LeMay, V., Smith, N., 2010. Modeling mortality in mixed-species stands of coastal British Columbia. Canadian Journal of Forest Research 40, 1517–1528.

Räty, M., Kangas, A., 2008. Localizing general models with classification and regression trees. Scandinavian Journal of Forest Research 23, 419–430.

Raulier, F., Lambert, M.-C., Pothier, D., Ung, C.-H., 2003. Impact of dominant tree dynamics on site index curves. Forest Ecology and Management 184, 65–78.

Rautiainen, M., Stenberg, P., 2005. Simplified tree crown model using standard forest mensuration data for Scots pine. Agricultural and Forest Meteorology 128, 123–129.

Rautiainen, M., Mõttus, M., Stenberg, P., Ervasti, S., 2008. Crown envelope shape measurements and models. Silva Fennica 42, 19–33.

Rawls, W.J., Ahuja, L.R., Brakensiek, D.L., 1992. Estimating soil hydraulic properties from soil data. In van Genucthen, M., Leij, F.J., Lund, L.J. (Eds.), Indirect Methods for Estimating Hydraulic Properties of Unsaturated Soils. University of California Riverside Press, Riverside, CA, pp. 329–340.

Rehfeldt, G.E., 2006. A spline model of climate for the Western United States. General Technical Report RMRS-165. USDA Forest Service, Rocky Mountain Research Station, Fort Collins, CO.

Rehfeldt, G.E., Wykoff, W.R., Hoff, R.J., Steinhoff, R.J., 1991. Genetic gains in growth and simulated yield of *Pinus monticola*. Forest Science 37, 326–342.

Reineke, L.H., 1927. A modification of Bruce's method of preparing timber yield tables. Journal of Agricultural Research 35, 843–856.

Reineke, L.H., 1933. Perfecting a stand-density index for even-aged forests. Journal of Agricultural Research 46, 627–638.

Reynolds, J.H., Ford, E.D., 2005. Improving competition representation in theoretical models of self-thinning: A critical review. Journal of Ecology 93, 362–372.

Reynolds Jr., M.R., 1984. Estimating the error in model predictions. Forest Science 30, 454–469.

Reynolds Jr., M.R., Burkhart, H.E., Daniels, R.F., 1981. Procedures for statistical validation of stochastic simulation models. Forest Science 27, 349–364.

Ribbens, E., Silander Jr., J.A., Pacala, S.W., 1994. Seedling recruitment in forests: Calibrating models to predict patterns of tree seedling dispersion. Ecology 75, 1794–1806.

Rich, P.M., Wood, J., Vieglais, D.A., Burek, K., Webb, N., 1999. Guide to HemiView: Software for analysis of hemispherical photography. Manual. Delta–T Devices, Ltd, Cambridge, UK.

Richards, F.J., 1959. A flexible growth equation for empirical use. Journal of Experimental Botany 10, 290–300.

Richardson, B., Kimberely, M.O., Ray, J.W., Coker, G.W., 1999. Indices of interspecific plant competition for *Pinus radiata* in the Central North Island of New Zealand. Canadian Journal of Forest Research 29, 898–905.

Richardson, B., Watt, M.S., Mason, E.G., Kriticos, D.J., 2006. Advances in modelling and decision support systems for vegetation management in young forest plantations. Forestry 79, 29–42.

Ritchie, M.W., 2009a. CONIFERS: A young stand model for Northern California and Southern Oregon. USDA Forest Service, Pacific Southwest Research Station, Albany, CA.

Ritchie, M.W., 2009b. R package for CONIFERS. Growth Model User's Group presentation, Portland, OR.

Ritchie, M.W., Hamann, J.D., 2006. Modeling dynamics of competing vegetation in young conifer plantations of northern California and southern Oregon, USA. Canadian Journal of Forest Research 36, 2523–2532.

Ritchie, M.W., Hamann, J.D., 2008. Individual-tree height-, diameter- and crown-width increment equations for young Douglas-fir plantations. New Forests 35, 173–186.

Ritchie, M.W., Hann, D.W., 1985. Equations for predicting basal area increment in Douglas-fir and grand fir. Research Bulletin 51. Oregon State University, Forestry Research Laboratory, Corvallis, OR.

Ritchie, M.W., Hann, D.W., 1986. Development of a tree height growth model for Douglas-fir. Forest Ecology and Management 15, 135–145.

Ritchie, M.W., Hann, D.W., 1987. Equations for predicting height to crown base for fourteen tree species in southwest Oregon. Research Bulletin 50. Oregon State University, Forestry Research Laboratory, Corvallis, OR.

Ritchie, M.W., Hann, D.W., 1990. Equations for predicting the 5-year height growth of six conifer species in southwest Oregon. Research Paper 54. Oregon State University, Forest Research Laboratory, Corvallis, OR.

Ritchie, M.W., Hann, D.W., 1997a. Implications of disaggregation in forest growth and yield modeling. Forest Science 43, 223–233.

Ritchie, M.W., Hann, D.W., 1997b. Evaluation of individual-tree and disaggregative prediction methods for Douglas-fir stands in western Oregon. Canadian Journal of Forest Research 27, 207–216.

Ritchie, M.W., Powers, R.F., 1993. User's guide for SYSTUM-1 (Version 2.0): A simulator of growth trends in young stands under management in California and Oregon. General Technical Report PSW-GTR-147. USDA Forest Service, Pacific Southwest Research Station, Albany, CA.

Ritz, C., Streibig, J.C., 2008. Nonlinear Regression with R. Springer, New York, NY.

Robinson, A.P., Froese, R.E., 2004. Model validation using equivalence tests. Ecological Modelling 176, 349–358.

Robinson, A.P., Monserud, R.A., 2003. Criteria for comparing the adaptability of forest growth models. Forest Ecology and Management 172, 53–67.

Robinson, A.P., Wykoff, W.R., 2004. Imputing missing height measurements using a mixed-effects modeling strategy. Canadian Journal of Forest Research 34, 2492–2500.

Robinson, A.P., Duursma, R.A., Marshall, J.D., 2005. A regression-based equivalence test for model validation: Shifting the burden of proof. Tree Physiology 25, 903–913.

Rocha, A.V., Su, H.-B., Vogel, C.S., Schmid, H.P., Curtis, P.S., 2004. Photosynthetic and water use efficiency responses to diffuse radiation by an aspen-dominated northern hardwood forest. Forest Science 50, 793–801.

Roeh, R.L., Maguire, D.A., 1997. Crown profile models based on branch attributes in coastal Douglas-fir. Forest Ecology and Management 96, 77–100.

Rogers, R., Johnson, P.S., 1998. Approaches to modeling natural regeneration in oak-dominated forests. Forest Ecology and Management 106, 45–54.

Rojo, A., Perales, X., Sánchez-Rodrígues, F., Álvarez-González, J.G., von Gadow, K., 2005. Stem taper functions for maritime pine (*Pinus pinaster* Ait.) in Galicia (Northwestern Spain). European Journal of Forest Research 124, 177–186.

Rondon, X.J., Gorchov, D.L., Noble Jr., R.B., 2009. Projection of tree growth and timber volume following strip clear-cutting in the Peruvian Amazon. Forest Ecology and Management 257, 588–599.

Rose Jr., C.E., Lynch, T.B., 2001. Estimating parameters for tree basal area growth with a system of equations and seemingly unrelated regressions. Forest Ecology and Management 148, 51–61.

Ross, J., 1981. The Radiation Regime and Architecture of Plant Stands. Dr. W. Junk Publishers, The Hague, The Netherlands.

Rötzer, T., Grote, R., Pretzsch, H., 2005. Effects of environmental changes on the vitality of forest stands. European Journal of Forest Research 124, 349–362.

Rötzer, T., Dieler, J., Mette, T., Moshammer, R., Pretzsch, H., 2010. Productivity and carbon dynamics in managed Central European forests depending on site conditions and thinning regimes. Forestry 83, 483–496.

Rudolf, P.O., 1930. A comparison of several of the growth per cent methods of predicting growth. Journal of Forestry 28, 28–33.

Running, S.W., 1994. Testing FOREST-BGC ecosystem process simulations across a climatic gradient in Oregon. Ecological Applications 4, 238–247.

Running, S.W., Coughlan, J.C., 1988. A general model of forest ecosystem processes for regional applications. I. Hydrologic balance, canopy gas exchange and primary production processes. Ecological Modelling 42, 125–154.

Running, S.W., Gower, S.T., 1991. FOREST-BGC, a general model of forest ecosystem processes for regional applications. II. Dynamic carbon allocation and nitrogen budgets. Tree Physiology 9, 147–160.

Running, S.W., Nemani, R.R., Hungerford, R.D., 1987. Extrapolation of synoptic meteorological data in mountainous terrain and its use for simulating forest evapotranspiration and photosynthesis. Canadian Journal of Forest Research 17, 472–483.

Running, S.W., Nemani, R.R., Heinsch, F.A., Zhao, M., Reeves, M., Hashimoto, H., 2004. A continuous satellite-derived measure of global terrestrial primary production. BioScience 54, 547–560.

Russell, M.B., Weiskittel, A.R., 2011. Maximum and stand-level crown width equations for the major tree species in the Acadian Region. Northern Journal of Applied Forestry, in press.

Russell, M.B., Amateis, R.L., Burkhart, H.E., 2010. Implementing regional locale and thinning response in the loblolly pine height-diameter relationship. Southern Journal of Applied Forestry 34, 21–27.

Russell, M.B., Weiskittel, A.R., Kershaw, J.A., in press. Assessing model performance in forecasting long-term individual tree diameter versus basal area increment for the primary Acadian tree species. Canadian Journal of Forest Research, in press.

Ryan, M.G., 1991. The effect of climate change on plant respiration. Ecological Applications 1, 157–167.

Ryan, M.G., 1990. Growth and maintenance respiration in stems of *Pinus contorta* and *Picea englemannii.* Canadian Journal of Forest Resesarch 20, 48–57.

Ryan, M.G., Yoder, B.J., 1997. Hydraulic limits to tree height and tree growth: What keeps trees from growing beyond a certain height. BioScience 47, 235–242.

Ryan, M.G., Hunt Jr., E.R., McMurtrie, R.E., Ågren, G.I., Aber, J.D., Friend, A.D., Rastetter, E.B., Pulliam, W.M., Raison, R.J., Linder, S., 1996a. Comparing models of ecosystem function for temperate conifer forests. I. Model description and validation. In Melillo, J.M., Ågren, G.I., Breymeyer, A.I. (Eds.), Global Change: Effects on Coniferous Forests and Grasslands. John Wiley & Sons, Inc., New York, NY, pp. 313–362.

Ryan, M.G., Hubbard, R.M., Pongracic, S., Raison, R.J., McMurtrie, R.E., 1996b. Foliage, fine-root, woody-tissue, and stand respiration in *Pinus radiata* in relation to nitrogen status. Tree Physiology 16, 333–343.

Ryan, M.G., Phillips, N., Bond, B.J., 2006. The hydraulic limitation hypothesis revisited. Plant, Cell and Environment 29, 367–381.

Saint-André, L., Nouvellon, Y., Laclau, J.P., Ranger, J., Bouillet, J.P., Nzila, J.D., Deleporte, P., Gonçalves, J.L.M., 2004. Modelling nutrient cycling and integrating nutrient cycling into growth models. In Nambiar, E.K.S., Ranger, J., Tiarks, A.E., Toma, T. (Eds.), Site Management and Productivity in Tropical Plantation Forest. Center for International Forestry Research, Bogor, Indonesia, pp. 171–184.

Salas, C., Stage, A.R., Robinson, A.P., 2008. Modeling effects of overstory density and competing vegetation on tree height growth. Forest Science 54, 107–122.

Salminen, H., Jalkanen, R., 2005. Modelling the effect of temperature on height increment of Scots pine at high latitudes. Silva Fennica 39, 497–508.

Saltelli, A., Annoni, P., 2010. How to avoid a perfunctory sensitivity analysis. Environmental Modelling & Software 25, 1508–1517.

Sands, P.J., 1995. Modelling canopy production. II. From single-leaf photosynthesis parameters to daily canopy photosynthesis. Australian Journal of Plant Physiology 22, 603–614.

Saunders, M.R., Wagner, R.G., 2008. Height-diameter models with random coefficients and site variables for tree species of Central Maine. Annals of Forest Science 65, 203.

Saunders, M.R., Wagner, R.G., Seymour, R.S., 2007. Thinning regimes for spruce-fir stands in the Northeastern United States and Eastern Canada. In Cooperative Forestry Research Unit Annual Report. University of Maine, School of Forest Resources, Orono, ME, pp. 27–32.

Saveland, J.M., Neuenschwander, L.F., 1990. A signal detection framework to evaluate models of tree mortality following fire damage. Forest Science 36, 66–76.

Saxton, K.E., Rawls, W.J., 2006. Soil water characteristic estimates by texture and organic matter for hydrologic solutions. Soil Science Society of America Journal 70, 1569–1578.

Scanlin, D.C., Loewenstein, H., 1979. Response of inland Douglas-fir and grand fir to thinning and nitrogen fertilization in northern Idaho. In Gessel, S.P., Kenady, R.M., Atkinson, W.A. (Eds.), Proceedings of Forest Fertilization Conference. Contribution Number 40. University of Washington Institute of Forest Resources, Seattle, WA, pp. 82–88.

Scheller, R.M., Domingo, J.B., Sturtevant, B.R., Williams, J.S., Rudy, A., Gustafson, E.J., Mladenoff, D.J., 2007. Design, development, and application of LANDIS-II, a spatial landscape simulation model with flexible temporal and spatial resolution. Ecological Modelling 201, 409–419.

Schmid, S., Zingg, A., Biber, P., Bugmann, H., 2006a. Evaluation of the forest growth model SILVA along an elevational gradient in Switzerland. European Journal of Forest Research 125, 43–55.

Schmid, S., Thürig, E., Kaufmann, E., Lischke, H., Bugmann, H., 2006b. Effect of forest management on future carbon pools and fluxes: A model comparison. Forest Ecology and Management 237, 65–82.

Schmidt, M., Nagel, J., Skovsgaard, J.P., 2006. Evaluating individual tree growth models. In Hasenauer, H. (Ed.), Sustainable Forest Management: Growth Models for Europe. Springer, Berlin, Germany, pp. 151–163.

Schmoldt, D.L., Martin, G.L., Bockheim, J.G., 1985. Yield-based measures of northern hardwood site quality and their correlation with soil-site factors. Forest Science 31, 209–219.

Schneider, R., Berninger, F., Ung, C.-H., Bernier, P.Y., Swift, D.E., Zhang, S.Y., 2008. Calibrating jack pine allometric relationships with simultaneous regressions. Canadian Journal of Forest Research 38, 2566–2578.

Schnur, G.L., 1937. Yield, stand, and volume tables for even-aged upland oak forests. Technical Bulletin 560. USDA, Washington, DC.

Schnute, J., 1981. A versatile growth model with statistically stable parameters. Canadian. Journal of Fisheries and Aquatic Sciences 38, 1128–1140.

Scholts, F.E., Lloyd, W.J., Deardorff, C.E., 1956. Some soil characteristics which affect root penetration and timber site quality of Douglas-fir in western Washington. Soil Science Society of America Journal 20, 101–105.

Schreuder, H.T., Williams, M.S., 1995. Assessing measures of tree competition using fixed-area, variable-radius, horizontal-line sampling. Canadian Journal of Forest Research 25, 1578–1582.

Schröder, J., von Gadow, K., 1999. Testing a new competition index for maritime pine in northwestern Spain. Canadian Journal of Forest Research 29, 280–283.

Schröder, J., Rodríguez Soalleiro, R., Vega Alonso, G., 2002. An age-independent basal area increment model for maritime pine trees in northwestern Spain. Forest Ecology and Management 157, 55–64.

Schumacher, F.X., 1939. A new growth curve and its application to timber yield studies. Journal of Forestry 37, 819–820.

Schumacher, F.X., Coile, T.S., 1960. Growth and Yield of Natural Stands of the Southern Pines. T.S. Coile, Inc., Durham, NC.

Schwalm, C.R., Ek, A.R., 2001. Climate change and site: Relevant mechanisms and modeling techniques. Forest Ecology and Management 150, 241–257.

Schwalm, C.R., Ek, A.R., 2004. A process-based model of forest ecosystems driven by meteorology. Ecological Modelling 179, 317–348.

Schwarz, G., 1978. Estimating the dimension of a model. Annals of Statistics 6, 461–464.

Schwärzel, K., Feger, K.-H., Häntzschel, J., Menzer, A., Spank, U., Clausnitzer, F., Köstner, B., Bernhofer, C., 2009. A novel approach in model-based mapping of soil water conditions at forest sites. Forest Ecology and Management 258, 2163–2174.

Schweiger, J., Sterba, H., 1997. A model describing natural regeneration recruitment of Norway spruce (*Picea abies* (L.) Karst.) in Austria. Forest Ecology and Management 97, 107–118.

Schwinning, S., Weiner, J., 1998. Mechanisms determining the degree of size asymmetry in competition among plants. Oecologia 113, 447–455.

Scott, I.R., 2006. Sylview: A visualization system for forest management. MS thesis. University of Missouri, Columbia, MO.

Seber, G.A.F., Wild, C.J., 2003. Nonlinear Regression. John Wiley & Sons, Inc., Hoboken, NJ.

Sellers, P.J., 1985. Canopy reflectance, photosynthesis and transpiration. International Journal of Remote Sensing 6, 1335–1372.

Sessions, J., Johnson, D., Ross, J., Sharer, B., 2000. The Blodgett plan: An active-management approach to developing mature forest habitat. Journal of Forestry. 98, 29–33.

Sessions, J., Bettinger, P., Buckman, R., Newton, M., Hamann, J., 2004. Hastening the return of complex forests following fire: The consequences of delay. Journal of Forestry 102, 38–45.

Seymour, R.S., Fajvan, M.A., 2001. Influence of prior growth suppression and soil on red spruce site index. Northern Journal of Applied Forestry 18, 55–62.

Shaffi, B., Moore, J.A., Newberry, J.D., 1990. Individual-tree diameter growth models for quantifying within-stand response to nitrogen fertilization. Canadian Journal of Forest Research 20, 1149–1155.

Sharma, M., Parton, J., 2007. Height-diameter equations for boreal tree species in Ontario using a mixed-effects modeling approach. Forest Ecology and Management 249, 187–198.

Sharma, M., Parton, J., 2009. Modeling stand density effects on taper for jack pine and black spruce plantations using dimensional analysis. Forest Science 55, 268–282.

Sharma, M., Zhang, S.Y., 2004. Height-diameter models using stand characteristics for *Pinus banksiana* and *Picea mariana*. Scandinavian Journal of Forest Research 19, 442–451.

Sharma, M., Amateis, R.L., Burkhart, H.E., 2002. Top height definition and its effect on site index determination in thinned and unthinned loblolly pine plantations. Forest Ecology and Management 168, 163–175.

Sharma, M., Smith, M., Burkhart, H.E., Amateis, R.L., 2006. Modeling the impact of thinning on height development of dominant and codominant loblolly pine trees. Annals of Forest Science 63, 349–354.

Shaw, D.J., Meldahl, R.S., Kush, J.S., Somers, G.L., 2003. A tree taper model based on similar triangles and use of crown ratio as a measure of form in taper equations for longleaf pine. General Technical Report SRS-66. USDA Forest Service, Asheville, NC.

Shaw, J.D., 2000. Application of stand density index to irregularly structured stands. Western Journal of Applied Forestry 15, 40–42.

Shen, G., Moore, J.A., Hatch, C.R., 2000. The effect of habitat type and rock type on individual tree basal area growth response to nitrogen fertilization. Canadian Journal of Forest Research 30, 613–623.

Shen, G., Moore, J.A., Hatch, C.R., 2001. The effect of nitrogen fertilization, rock type, and habitat type on individual tree mortality. Forest Science 47, 203–213.

Shifley, S.R., Ek, A.R., Burk, T.E., 1993. A generalized methodology for estimating forest ingrowth at multiple threshold diameters. Forest Science 39, 776–798.

Shillinger, R.R., Adams, D.M., Latta, G.S., van Nalts, A.K., 2003. An analysis of future private timber supply potential in western Oregon. Western Journal of Applied Forestry 18, 166–174.

Shinozaki, K., Yoda, K., Hozumi, K., Kira, T., 1964. A quantitative analysis of plant form: The pipe model theory. II. Further evidence of the theory and its application in forest ecology. Japanese Journal of Ecology 14, 133–139.

Short III, E.A., Burkhart, H.E., 1992. Predicting crown-height increment for thinned and unthinned loblolly pine plantations. Forest Science 38, 594–610.

Shugart, H.H., 2002. Forest gap models. In Mooney, H.A., Canadell, J.G. (Eds.), The Earth System: Biological and Ecological Dimensions of Global Environmental Change. John Wiley & Sons, Ltd, Chichester, UK, pp. 316–323.

Shula, R.G., Knowe, S.A., Radosevich, S.R., 1998. Regional Vegetation Management Model (RVMM): Version for the Coast Range Mountains of Oregon and Washington. Oregon State University, Department of Forest Science, Corvallis, OR.

Sievänen, R., Lindner, M., Mäkelä, A., Lasch, P., 2000. Volume growth and survival graphs: A method for evaluating process-based forest growth models. Tree Physiology 20, 357–365.

Silbernagel, J., Moeur, M., 2001. Modeling canopy openness and understory gap patterns based on image analysis and mapped tree data. Forest Ecology and Management 149, 217–233.

Simioni, G., Ritson, P., McGrath, J., Kirschbaum, M.U.F., Copeland, B., Dumbrell, I., 2008. Predicting wood production and net ecosystem carbon exchange of *Pinus radiata* plantations in south-western Australia: Application of a process-based model. Forest Ecology and Management 255, 901–912.

Simioni, G., Ritson, P., Kirschbaum, M.U.F., McGrath, J., Dumbrell, I., Copeland, B., 2009. The carbon budget of *Pinus radiata* plantations in south-western Australia under four climate change scenarios. Tree Physiology 29, 1081–1093.

Sironen, S., Kangas, A., Maltamo, M., Kalliovirta, J., 2008. Localization of growth estimates using non-parametric imputation methods. Forest Ecology and Management 256, 674–684.

Sit, V., Poulin-Costello, M., 1994. Catalogue of Curves for Curve Fitting. Ministry of Forests, Forest Science Research Branch, Victoria, British Columbia.

Skovsgaard, J.P., Vanclay, J.K., 2008. Forest site productivity: A review of the evolution of dendrometric concepts for even-aged stands. Forestry 81, 13–31.

Smith, B.J., Yan, J., Cowles, M.K., 2008. Unified geostatistical modeling for data fusion and spatial heteroskedasticity with R Package ramps. Journal of Statistical Software 25, 1–21.

Smith, J.H.G., Bailey, G.R., 1964. Influence of stocking and stand density on crown widths of Douglas-fir and lodgepole pine. Commonwealth Forestry Review 43, 243–246.

Smith, M.-L., Ollinger, S.V., Martin, M.E., Aber, J.D., Hallett, R.A., Goodale, C.L., 2002. Direct estimation of aboveground forest productivity through hyperspectral remote sensing of canopy nitrogen. Ecological Applications 12, 1286–1302.

Smith, N.J., 1993. Estimating leaf area index and light extinction coefficients in stands of Douglas-fir (*Pseudotsuga menziesii*). Canadian Journal of Forest Research 23, 317–321.

Smith, N.J., Hann, D.W., 1984. A new analytical model based on the -3/2 power rule of self-thinning. Canadian Journal of Forest Research 14, 605–609.

Smith, N.J., Hann, D.W., 1986. A growth model based on the self-thinning rule. Canadian Journal of Forest Research 16, 330–334.

Smith, T.M., Urban, D.L., 1988. Scale and resolution of forest structural pattern. Vegetatio 74, 143–150.

Smith, W.B., 1985. Factors and equations to estimate forest biomass in the North Central region. Research Paper NC-268. USDA Forest Service, North Central Forest Experiment Station, St. Paul, MN.

Snowdon, P., 1987. Predicting foliar biomass of *Pinus radiata* from basal area increment. Australian Journal of Forest Research 17, 277–281.

Snowdon, P., 2001. Short-term predictions of growth of *Pinus radiata* with models incorporating indices of annual climatic variation. Forest Ecology and Management 152, 1–11.

Snowdon, P., 2002. Modeling Type 1 and Type 2 growth responses in plantations after application of fertilizer or other silvicultural treatments. Forest Ecology and Management 163, 229–244.

Snowdon, P., Woollons, R.C., Benson, M.L., 1998. Incorporation of annual climatic indices into models of *Pinus radiata* in a spacing experiment. New Forests 16, 101–123.

Soares, P., Tomé, M., 1999. Distance-dependent competition measures for eucalyptus plantations in Portugal. Annals of Forest Science 56, 307–319.

Soares, P., Tomé, M., 2001. A tree crown ratio prediction equation for eucalypt plantations. Annals of Forest Science 58, 193–202.

Soares, P., Tomé, M., 2002. Height-diameter equation for first rotation eucalypt plantations in Portugal. Forest Ecology and Management 166, 99–109.

Soares, P., Tomé, M., 2003. GLOBTREE: An individual tree growth model for *Eucalyptus globulus* in Portugal. In Amaro, A., Reed, D., Soares, P. (Eds.), Modelling Forest Systems. CABI Publishing, Wallingford, UK, pp. 97–110.

Soares, P., Tomé, M., Skovsgaard, J.P., Vanclay, J.K., 1995. Evaluating a growth model for forest management using continuous forest inventory data. Forest Ecology and Management 71, 251–265.

Söderbergh, I., Ledermann, T., 2003. Algorithms for simulating thinning and harvesting in five European individual-tree growth simulators: A review. Computers and Electronics in Agriculture 39, 115–140.

Solomon, D.S., Zhang, L., 2002. Maximum size-density relationships for mixed softwoods in the northeastern USA. Forest Ecology and Management 155, 163–170.

Solomon, D.S., Hosmer, R.A., Hayslett Jr., H.T., 1986. A two-stage matrix model for predicting growth of forest stands in the Northeast. Canadian Journal of Forest Research 16, 521–528.

Solomon, D.S., Herman, D.A., Leak, W.B., 1995. FIBER 3.0: An ecological growth model for northeastern forest types. General Technical Report NE-204. USDA Forest Service, Northeastern Forest Experiment Station, Radnor, PA.

Somers, G.L., Nepal, S.K., 1994. Linking individual-tree and stand-level growth models. Forest Ecology and Management 69, 233–243.

Sonmez, T., Keles, S., Tilki, F., 2007. Effect of aspect, tree age and tree diameter on bark thickness of *Picea orientalis*. Scandinavian Journal of Forest Research 22, 193–197.

South, D.B., Miller, J.H., 2007. Growth response analysis after early control of woody competition for 14 loblolly pine plantations in the southern U.S. Forest Ecology and Management 242, 569–577.

Sprinz, P.T., Talbert, C.B., Strub, M.R., 1989. Height-age trends from an Arkansas seed source study. Forest Science 35, 677–691.

Spurr, S.H., 1962. A measure of point density. Forest Science 8, 85–96.

Stadt, K.J., Lieffers, V.J., 2000. MIXLIGHT: A flexible light transmission model for mixed-species forest stands. Agricultural and Forest Meteorology 102, 235–252.

Stage, A.R., 1963. A mathematical approach to polymorphic site index curves for grand fir. Forest Science 9, 167–180.

Stage, A.R., 1968. A tree-by-tree measure of site utilization for grand fir related to stand density index. Research Note INT-77. USDA Forest Service, Intermountain Forest and Range Experiment Station, Ogden, UT.

Stage, A.R., 1973. Prognosis model for stand development. Research Paper INT-137. USDA Forest Service, Intermountain Forest and Range Experiment Station, Ogden, UT.

Stage, A.R., 1975. Prediction of height increment for models of forest growth. Research Paper INT-164. USDA Forest Service Intermountain Forest and Range Experimental Station, Ogden, UT.

Stage, A.R., 1976. An expression for the effect of aspect, slope, and habitat type on tree growth. Forest Science 22, 457–460.

Stage, A.R., 2003. How forest models are connected to reality: Evaluation criteria for their use in decision support. Canadian Journal of Forest Research 33, 410–421.

Stage, A.R., Ledermann, T., 2008. Effects of competitor spacing in a new class of individual-tree indices of competition: Semi-distance-independent indices computed for Bitterlich versus fixed-area plots. Canadian Journal of Forest Research 38, 890–898.

Stage, A.R., Salas, C., 2007. Interactions of elevation, aspect, and slope in models of forest species composition and productivity. Forest Science 53, 486–492.

Stage, A.R., Wykoff, W.R., 1993. Calibrating a model of stochastic effects on diameter increment for individual-tree simulations of stand dynamics. Forest Science 39, 692–705.

Stage, A.R., Wykoff, W.R., 1998. Adapting distance-independent forest growth models to represent spatial variability: Effects of sampling design on model coefficients. Forest Science 44, 224–238.

Stage, A.R., Crookston, N.L., Monserud, R.A., 1993. An aggregation algorithm for increasing the efficiency of population models. Ecological Modelling 68, 257–271.

Stamatellos, G., Panourgias, G., 2005. Simulating spatial distributions of forest trees by using data from fixed area plots. Forestry 78, 305–312.

Stape, J.L., Binkley, D., Jacob, W.S., Takahashi, E.N., 2006. A twin-plot approach to determine nutrient limitation and potential productivity in *Eucalyptus* plantations at landscape scales in Brazil. Forest Ecology and Management 223, 358–362.

Steinbrenner, E.C., 1981. Forest soil productivity relationships. In Heilman, P.E., Anderson, H.W., Baumgartner, D.M. (Eds.), Forest Soils of the Douglas-fir Region. Washington State University, Pullman, WA, pp. 199–229.

Sterba, H., 1987. Estimating potential density from thinning experiment and inventory data. Forest Science 33, 1022–1034.

Sterba, H., 2002. Forest inventories and growth models to examine management strategies for forests in transition. Forestry 75, 411–418.

Sterba, H., Ledermann, T., 2005. Inventory and modelling for forests in transition from even-aged to uneven-aged management. Forest Ecology and Management 224, 278–285.

Sterba, H., Monserud, R.A., 1993. The maximum density concept applied to uneven-aged mixed-species stands. Forest Science 39, 432–452.

Sterba, H., Monserud, R.A., 1997. Applicability of the forest stand growth simulator PROGNAUS for the Austrian part of the Bohemian Massif. Ecological Modelling 98, 23–34.

Sterba, H., Korol, N., Rössler, G., 2001. Ein Ansatz zur Evaluierung eines Einzelbaumwachstumsimulators für Fichtenreinbestände (Preliminary evaluation of an individual tree growth model for Norway spruce stands) Forstwissenschaftliches Centralblatt 120, 406–421.

Sterba, H., Blab, A., Katzensteiner, K., 2002. Adapting an individual tree growth model for Norway spruce (*Picea abies* L. Karst.) in pure and mixed species stands. Forest Ecology and Management 159, 101–110.

Strub, M., Cieszewski, C., Hyink, D., 2005. Self-referencing taper curves for loblolly pine. In McRoberts, R.E., Reams, G.A., Van Deusen, P.C., McWilliams, W.H., Cieszewski, C.J. (Eds.), Proceedings of the Fourth Annual Forest Inventory and Analysis Symposium. General Technical Report NC-252. USDA Forest Service, North Central Research Station, St. Paul, MN, pp. 59–64.

Suárez-Mínguez, J.C., 2010. An analysis of the consequences of stand variability in Sitka spruce plantations in Britain using a combination of airborne LiDAR analysis and models. PhD dissertation. The University of Sheffield, Sheffield, UK.

Subedi, N., Sharma, M., 2010. Evaluating height-age determination methods for jack pine and black spruce plantations using stem analysis data. Northern Journal of Applied Forestry 27, 50–55.

Sullivan, A.D., Clutter, J.L., 1972. A simultaneous growth and yield model for loblolly pine. Forest Science 18, 76–86.

Sun, H.-G., Zhang, J.-G., Duan, A.-G., He, C.-Y., 2007. A review of stand basal area growth models. Forestry Studies in China 9, 85–94.

Suratno, A., Seielstad, C., Queen, L., 2009. Tree species identification in mixed coniferous forest using airborne laser scanning. ISPRS Journal of Photogrammetry and Remote Sensing 64, 683–693.

Svensson, M., Jansson, P.-E., Gustafsson, D., Berggren Kleja, D., Langvall, O., Lindroth, A., 2008. Bayesian calibration of a model describing carbon, water and heat fluxes for a Swedish boreal forest stand. Ecological Modelling 213, 331–344.

Swenson, J.J., Waring, R.H., Fan, W., Coops, N., 2005. Predicting site index with a physiologically based growth model across Oregon, USA. Canadian Journal of Forest Research 35, 1697–1707.

Talbert, C.B., Hyink, D.M., 1988. Implications of genetic improvement for the growth and yield of trees and stands. In Ek, A.R., Shifley, S.R., Burk, T.E. (Eds.), Forest Growth Modeling and Prediction. General Technical Report NC-120. USDA Forest Service, North Central Forest Experimental Station, St. Paul, MN, pp. 562–571.

Tan, C.S., Black, T.A., 1976. Factors affecting the canopy resistance of a Douglas-fir forest. Boundary-Layer Meteorology 10, 475–488.

Tang, S., Wang, Y., Zhang, L., Meng, C.-H., 1997. A distribution-independent approach to predicting stand diameter distribution. Forest Science 43, 491–500.

Tappeiner II, J.C., Maguire, D.A., Harrington, T.B., 2007. Silviculture and Ecology and Western U.S. Forests. Oregon State University, Corvallis, OR.

Tatarinov, F.A., Cienciala, E., 2006. Application of BIOME-BGC model to managed forests. 1. Sensitivity analysis. Forest Ecology and Management 237, 267–279.

Taylor, A.R., Chen, H.Y.H., VanDamme, L., 2009. A review of forest succession models and their suitability for forest management planning. Forest Science 55, 23–36.

Temesgen, H., LeMay, V.M., 1999. Examination of tree height and diameter increment models fitted for PROGNOSIS^{BC}. Report prepared for the British Columbia Ministry of Forests. University of British Columbia, Vancouver, BC.

Temesgen, H., Mitchell, S.J., 2005. An individual-tree mortality model for complex stands of southeastern British Columbia. Western Journal of Applied Forestry 20, 101–109.

Temesgen, H., von Gadow, K., 2004. Generalized height-diameter models – an application for major tree species in complex stands of interior British Columbia. European Journal of Forest Research 123, 45–51.

Temesgen, H., LeMay, V.M., Froese, K.L., Marshall, P.L., 2003. Imputing tree-lists from aerial attributes for complex stands of south-eastern British Columbia. Forest Ecology and Management 177, 277–285.

Temesgen, H., LeMay, V.M., Mitchell, S.J., 2005. Tree crown ratio models for multi-species and multi-layered stands of southeastern British Columbia. Forestry Chronicle 81, 133–141.

Temesgen, H., Hann, D.W., Monleon, V.J., 2007. Regional height-diameter equations for major tree species of southwest Oregon. Western Journal of Applied Forestry 22, 213–219.

Temesgen, H., Monleon, V.J., Hann, D.W., 2008. Analysis and comparison of nonlinear tree height prediction strategies for Douglas-fir forests. Canadian Journal of Forest Research 38, 553–565.

Ter-Mikaelian, M.T., Korzukhin, M.D., 1997. Biomass equations for sixty-five North American tree species. Forest Ecology and Management 97, 1–24.

von Teuffel, K., Hein, S., Kotar, M., Pinto Preuhsler, E., Puumalainen, J., Weinfurter, P., 2006. End user needs and requirements. In Hasenauer, H. (Ed.), Sustainable Forest Management: Growth Models for Europe. Springer, Berlin, Germany, pp. 19–38.

Thomas, C.E., Parresol, B.R., 1991. Simple, flexible, trigonometric taper equations. Canadian Journal of Forest Research 21, 1132–1137.

Thornley, J.H.M., 1991. A transport-resistance model of forest growth and partitioning. Annals of Botany 68, 211–226.

Thornton, P.E., Running, S.W., White, M.A., 1997. Generating surfaces of daily meteorological variables over large regions of complex terrain. Journal of Hydrology 190, 214–251.

Thorpe, H.C., Vanderwel, M.C., Fuller, M.M., Thomas, S.C., Caspersen, J.P., 2010. Modelling stand development after partial harvests: An empirically based, spatially explicit analysis for lowland black spruce. Ecological Modelling 221, 256–267.

Tomé, J., Tomé, M., Barreiro, S., Paulo, J.A., 2006. Age-independent difference equations for modelling tree and stand growth. Canadian Journal of Forest Research 36, 1621–1630.

Tomé, M., Burkhart, H.E., 1989. Distance-dependent competition measures for predicting growth of individual trees. Forest Science 35, 816–831.

Trasobares, A., Pukkala, T., 2004. Using past growth to improve individual-tree diameter growth models for uneven-aged mixtures of *Pinus sylvestris* L. and *Pinus nigra* Arn. in Catalonia, north-east Spain. Annals of Forest Science 61, 409–417.

Trincado, G., Burkhart, H.E., 2006. A generalized approach for modeling and localizing stem profile curves. Forest Science 52, 670–682.

Trincado, G., Burkhart, H.E., 2009. A framework for modeling the dynamics of first-order branches and spatial distribution of knots in loblolly pine trees. Canadian Journal of Forest Research 39, 566–579.

Trincado, V.G., Quezada, R., von Gadow, K., 2003. A comparison of two stand table projection methods for young *Eucalyptus nitens* (Maiden) plantations in Chile. Forest Ecology and Management 180, 443–451.

Turnblom, E.C., Burk, T.E., 2000. Modeling self-thinning of unthinned Lake States red pine stands using nonlinear simultaneous differential equations. Canadian Journal of Forest Research 30, 1410–1418.

Turner, J., Thompson, C.H., Turvey, N.D., Hopmans, P., Ryan, P.J., 1990. A soil technical classification for *Pinus radiata* (D. Don) plantations. I. Development. Australian Journal of Soil Research 28, 797–811.

Twery, M.J., Knopp, P.D., Thomasma, S.A., Rauscher, H.M., Nute, D.E., Potter, W.D., Maier, F., Wang, J., Dass, M., Uchiyama, H., Glende, A., Hoffman, R.F., 2005. NED-2: A decision support system for integrated forest ecosystem management. Computers and Electronics in Agriculture 49, 24–43.

Tyler, A.L., MacMillan, D.C., Dutch J., 1996. Models to predict the General Yield Class of Douglas-fir, Japanese larch and Scots pine on better quality land in Scotland. Forestry 69, 13–24.

Umeki, K., Lim, E.-M., Honjo, T., 2008. A GIS-based simulation program to predict multi-species size-structure dynamics for natural forests in Hokkaido, northern Japan Ecological Informatics 3, 218–227.

Ung, C.-H., Bernier, P.Y., Raulier, F., Fournier, R.A., Lambert, M.-C. Régnière, J., 2001. Biophysical site indices for shade tolerant and intolerant boreal species. Forest Science 47, 83–95.

Usher, M.B., 1966. A matrix approach to the management of renewable resources, with special reference to selection forests. Journal of Applied Ecology 3, 355–367.

Usher, M.B., 1969. A matrix model for forest management. Biometrics 25, 309–315.

Uusitalo, J., Kivinen, V.-P., 2000. Implementing SmartForest forest visualization tool on PC environment. In Salminen, H., Saarikko, J., Virtanen, E. (Ed.), Proceedings of Resource Technology '98 Nordic, International Symposium on Advanced Technology in Environmental and Natural Resources. Research Paper 791. Finnish Forest Resource Institute, Helsinki, Finland, pp. 1–7.

Uzoh, F.C.C., Oliver, W.W., 2006. Individual tree height increment model for managed even-aged stands of ponderosa pine throughout the western United States using linear mixed effects models. Forest Ecology and Management 221, 147–154.

Uzoh, F.C.C., Oliver, W.W., 2008. Individual tree diameter increment model for managed even-aged stands of ponderosa pine throughout the western United States using a multilevel linear mixed effects model. Forest Ecology and Management 256, 438–445.

Valenti, M.A., Cao, Q.V., 1986. Use of crown ratio to improve loblolly pine taper equations. Canadian Journal of Forest Research 16, 1141–1145.

Valentine, H.T., 1985. Tree-growth models: Derivations employing the pipe-model theory. Journal of Theoretical Biology 117, 579–585.

Valentine, H.T., 1988. A carbon-balance model of stand growth: A derivation employing pipe-model theory and the self-thinning rule. Annals of Botany 62, 389–396.

Valentine, H.T., 1997. Height growth, site index, and carbon metabolism. Silva Fennica 31, 251–263.

Valentine, H.T., Gregoire, T.G., 2001. A switching model of bole taper. Canadian Journal of Forest Research 31, 1400–1409.

Valentine, H.T., Mäkelä, A., 2005. Bridging process-based and empirical approaches to modeling tree growth. Tree Physiology 25, 769–779.

Valentine, H.T., Ludlow, A.R., Furnival, G.M., 1994. Modeling crown rise in even-aged stands of Sitka spruce or loblolly pine. Forest Ecology and Management 69, 189–197.

Valentine, H.T., Herman, D.A., Gove, J.H., Hollinger, D.Y., Solomon, D.S., 2000. Initializing a model stand for process-based projection. Tree Physiology 20, 393–398.

van der Meer, P.J., Jorritsma, I.T.M., Kramer, K., 2002. Assessing climate change effects on long-term forest development: Adjusting growth, phenology, and seed production in a gap model. Forest Ecology and Management 162, 39–52.

Van Deusen, P.C., Biging, G.S., 1985. STAG: A STAnd Generator for mixed species stands, version 2.0. Research Note Number 11. Northern California Forest Yield Cooperative, Department of Forestry and Management, University of California, Berkeley, CA.

Van Oijen, M., Rougier, J., Smith, R., 2005. Bayesian calibration of process-based forest models: Bridging the gap between models and data. Tree Physiology 25, 915–927.

Van Pelt, R., North, M.P., 1996. Analyzing canopy structure in Pacific Northwest old-growth forests with a stand-scale crown model. Northwest Science 70, 15–30.

Van Wijk, M.T., Dekker, S.C., Bouten, W., Bosveld, F.C., Kohsiek, W., Kramer, K., Mohren, G.M.J., 2000. Modeling daily gas exchange of a Douglas-fir forest: Comparison of three stomatal conductance models with and without a soil water stress function. Tree Physiology 20, 115–122.

Vanclay, J.K., 1988. PLATIPUS physiology: Design of a plantation growth and conversion simulator incorporating silviculture and wood quality. In Ek, A.R., Shifley, S.R., Burk, T.E. (Eds.), Forest Growth Modeling and Prediction. General Technical Report NC-120. USDA Forest Service, North Central Forest Experiment Station, St. Paul, MN, pp. 998–1005.

Vanclay, J.K., 1989a. A growth model for north Queensland rainforests. Forest Ecology and Management 27, 245–271.

Vanclay, J.K., 1989b. Site productivity assessment in rainforests: An objective approach using indicator species. In Mohd, W.R., Chan, H.T., Appanah, S. (Eds.), Proceedings of the Seminar on Growth and Yield in Tropical Mixed/Moist Forests, 20–24 June 1988, Kuala Lumpur. Forest Research Institute, Kuala Lumpur, Malaysia, pp. 225–241.

Vanclay, J.K., 1989c. Modelling selection harvesting in tropical rain forests. Journal of Tropical Forest Science 1, 280–294.

Vanclay, J.K., 1991a. Aggregating tree species to develop diameter increment equations for tropical rainforests. Forest Ecology and Management 42, 143–168.

Vanclay, J.K., 1991b. Compatible deterministic and stochastic predictions by probabilistic modeling of individual trees. Forest Science 37, 1656–1663.

Vanclay, J.K., 1992. Modelling regeneration and recruitment in a tropical rain forest. Canadian Journal of Forest Research 22, 1235–1248.

Vanclay, J.K., 1994. Modelling Forest Growth and Yield: Applications to Mixed Tropical Forests. CAB International, Wallingford, UK.

Vanclay, J.K., Henry, N.B., 1988. Assessing site productivity of indigenous cypress pine forest in southern Queensland. Commonwealth Forestry Review 67, 53–64.

Vanclay, J.K., Sands, P.J., 2009. Calibrating the self-thinning frontier. Forest Ecology and Management 259, 81–85.

Vanclay, J.K., Skovsgaard, J.P., 1997. Evaluating forest growth models. Ecological Modelling 98, 1–12.

VanderSchaaf, C.L., Burkhart, H.E., 2008. Using segmented regression to estimate stages and phases of stand development. Forest Science 54, 167–175.

Vanninen, P., Mäkelä, A., 1999. Fine root biomass of Scots pine stands differing in age and soil fertility in southern Finland. Tree Physiology 19, 823–830.

Vargas-Larreta, B., Castedo-Dorado, F., Álvarez-González, J.G., Barrio-Anta, M., Cruz-Cobos, F., 2009. A generalized height-diameter model with random coefficients for uneven-aged stands in El Salto, Durango (Mexico). Forestry 82, 445–462.

Vaughn, N., 2007. An individual-tree model to predict annual growth of young stands of Douglas-fir (*Pseudotsuga menziesii* (Mirbel)) in the Pacific Northwest. MS thesis. University of Washington, Seattle, WA.

Vaze, J., Teng, J., Spencer, G., 2010. Impact of DEM accuracy and resolution on topographic indices. Environmental Modelling & Software 25, 1086–1098.

Verburg, P.S.J., Johnson, D.W., Harrison, R., 2001. Long-term nutrient cycling patterns in Douglas-fir and red alder stands: A simulation study. Forest Ecology and Management 145, 203–217.

Vertregt, N., Penning de Vries, F.W.T., 1987. A rapid method for determining the efficiency of biosynthesis of plant biomass. Journal of Theoretical Biology 128, 109–119.

Vettenranta, J., 1999. Distance-dependent models for predicting the development of mixed coniferous forests in Finland. Silva Fennica 33, 51–72.

Vezina, P.E., 1962. Crown width-d.b.h. relationships for open-crown balsam fir and white spruce in Quebec. Forestry Chronicle 38, 463–473.

Vickers, L.A., Fox, T.R., Loftis, D.L., Boucugnani, D.A., 2011. Predicting forest regeneration in the Central Appalachians using the REGEN expert system. Journal of Sustainable Forestry, in press.

Vieilledent, G., Courbaud, B., Kunstler, G., Dhôte, J.-F., Clark, J.S., 2009. Biases in the estimation of size-dependent mortality models: Advantages of a semiparametric approach. Canadian Journal of Forest Research 39, 1430–1443.

Vose, J.M., Ryan, M.G., 2002. Seasonal respiration of foliage, fine roots, and woody tissues in relation to growth, tissue N, and photosynthesis. Global Change Biology 8, 182–193.

Vospernik, S., 2004. Models for predicting log grade and stem damage. PhD dissertation. University of Natural Resources and Applied Life Sciences, Vienna, Austria.

Vospernik, S., Bokalo, M., Reimoser, F., Sterba, H., 2007. Evaluation of a vegetation simulator for roe deer habitat predictions. Ecological Modelling 202, 265–280.

Vospernik, S., Monserud, R.A., Sterba, H., 2010. Do individual-tree growth models correctly represent height:diameter ratios of Norway spruce and Scots pine? Forest Ecology and Management 260, 1735–1753.

Vuokila, Y., 1965. Functions for variable density yield tables of pine based on temporary sample plots. Communicationes Instituti Forestalls Fenniae 60, 1–86.

Wagner, R.G., Little, K.M., Richardson, B., McNabb, K., 2006. The role of vegetation management for enhancing productivity of the world's forest. Forestry 79, 57–79.

Wahlenberg, W.G., 1941. Methods of forecasting timber growth in irregular stands. Technical Bulletin 796. USDA, Washington, DC.

Walters, D.K., Hann, D.W., 1986. Taper equations for six conifer species in southwest Oregon. Research Bulletin 56. Oregon State University, Forest Research Laboratory, Corvallis, OR.

Wang, C.-H., 1990. Analysis methods for modeling tree and stand growth responses to fertilization and thinning. PhD dissertation. Oregon State University, Corvallis, OR.

Wang, C.-H., Hann, D.W., 1988. Height-diameter equations for sixteen tree species in the central Williamette Valley of Oregon. Research Paper 51. Oregon State University, Forest Research Laboratory, Corvallis, OR.

Wang, G.G., 1998a. An ecologically based model for site index conversion among species. Canadian Journal of Forest Research 28, 234–238.

Wang, G.G., 1998b. Is height of dominant trees at a reference diameter an adequate measure of site quality? Forest Ecology and Management 112, 49–54.

Wang, G.G., Huang, S., Monserud, R.A., Klos, R.J., 2004. Lodgepole pine site index in relation to synoptic measures of climate, soil moisture and soil nutrients. Forestry Chronicle 80, 678–686.

Wang, J.R., Zhong, A.L., Kimmins, J.P., 2002a. Biomass estimation errors associated with the use of published regression equations of paper birch and trembling aspen. Northern Journal of Applied Forestry 19, 128–136.

Wang, M., Borders, B.E., Zhao, D., 2007a. Parameter estimation of base-age invariant site index models: Which data structure to use? Forest Science 53, 541–551.

Wang, M., Borders, B.E., Zhao, D., 2008a. An empirical comparison of two subject-specific approaches to dominant heights modeling: The dummy variable method and the mixed model method. Forest Ecology and Management 255, 2659–2669.

Wang, M., Borders, B.E., Zhao, D., 2008b. Parameter estimation of base-age invariant site index models: Which data structure to use? A reply. Forest Science 54, 129–133.

Wang, S., Chen, W., Cihlar, J., 2002b. New calculation methods of diurnal distribution of solar radiation and its interception by canopy over complex terrain. Ecological Modelling 155, 191–204.

Wang, W., Peng, C., Zhang, S.Y., Zhou, X., Larocque, G.R., Kneeshaw, D.D., Lei, X., 2011. Development of TRIPLEX-Management model for simulating the response of forest growth to pre-commercial thinning. Ecological Modelling, doi: 10.1016/j.ecolmodel.2010.09.019.

Wang, X., Song, B., Chen, J., Crow, T.R., LaCroix, J.J., 2006. Challenges in visualizing forests and landscapes. Journal of Forestry 104, 316–319.

Wang, Y., LeMay, V.M., Baker, T.G., 2007b. Modelling and prediction of dominant height and site index of *Eucalyptus globulus* plantations using a nonlinear mixed-effects model approach. Canadian Journal of Forest Research 37, 1390–1403.

Wang, Y.P., 2003. A comparison of three different canopy radiation models commonly used in plant modelling. Functional Plant Biology 30, 143–152.

Wang, Y.-P., Jarvis, P.G., 1990. Description and validation of an array model – MAESTRO. Agricultural and Forest Meteorology 51, 257–280.

Wang, Y.-P., Leuning, R., 1998. A two-leaf model for canopy conductance, photosynthesis and partitioning of available energy I: Model description and comparison with a multi-layered model. Agricultural and Forest Meteorology 91, 89–111.

Wang, Y.P., Jarvis, P.G., Taylor, C.M.A., 1991. PAR absorption and its relation to above-ground dry matter production of Sitka spruce. Journal of Applied Ecology 28, 547–560.

Waring, R.H., 1983. Estimating forest growth and efficiency in relation to canopy leaf area. Advances in Ecological Research 13, 327–354.

Waring, R.H., 2000. A process model analysis of environmental limitations on the growth of Sitka spruce plantations in Great Britain. Forestry 73, 65–79.

Waring, R.H., McDowell, N., 2002. Use of a physiological process model with forestry yield tables to set limits on annual carbon balances. Tree Physiology 22, 179–188.

Waring, R.H., Landsberg, J.J., Williams, M., 1998. Net primary production of forests: A constant fraction of gross primary production? Tree Physiology 18, 129–134.

Waring, R.H., Coops, N.C., Landsberg, J.J., 2010. Improving predictions of forest growth using the 3-PGS model with observations made by remote sensing. Forest Ecology and Management 259, 1722–1729.

Warren, C.R., Adams, M.A., 2006. Internal conductance does not scale with photosynthetic capacity: Implications for carbon isotope discrimination and the economics of water and nitrogen use in photosynthesis. Plant, Cell and Environment 29, 192–201.

Watt, M.S., Whitehead, D., Richardson, B., Mason, E.G., Leckie, A.C., 2003. Modelling the influence of weed competition on the growth of young *Pinus radiata* at a dryland site. Forest Ecology and Management 178, 271–286.

Weber, L.A., Ek, A.R., Droessler, T.D., 1986. Comparison of stochastic and deterministic mortality estimation in an individual tree based stand growth model. Canadian Journal of Forest Research 16, 1139–1141.

Weiner, J., 1984. Neighbourhood interference amongst *Pinus rigida* individuals. Journal of Ecology 72, 183–195.

Weiner, J., 1990. Asymmetric competition in plant populations. Trends in Ecology and Evolution 5, 360–364.

Weinstein, D.A., Beloin, R.M., Yanai, R.D., 1991. Modeling changes in red spruce carbon balance and allocation in response to interacting ozone and nutrient stresses. Tree Physiology 9, 127–146.

Weiskittel, A., Maguire, D., 2004. Growth and mortality models that incorporate the effects of Swiss needle cast; an examination of SMC ORGANON biases and development of new equations using data from the Growth Impact, Pre-Commercial Thinning, and Commercial Thinning studies. In Mainwaring, D. (Ed.), Swiss Needle Cast Cooperative 2004 Annual Report. Oregon State University, College of Forestry, Corvallis, OR, pp. 64–70.

Weiskittel, A.R., Garber, S.M., Johnson, G.P., Maguire, D.A., Monserud, R.A., 2007a. Annualized diameter and height growth equations for Pacific Northwest plantation-grown Douglas-fir, western hemlock, and red alder. Forest Ecology and Management 250, 266–278.

Weiskittel, A.R., Maguire, D.A., Monserud, R.A., 2007b. Response of branch growth and mortality to silvicultural treatments in coastal Douglas-fir plantations: Implications for predicting tree growth. Forest Ecology and Management 251, 182–194.

Weiskittel, A.R., Maguire, D.A., Monserud, R.A., 2007c. Modeling crown structural responses to competing vegetation control, thinning, fertilization, and Swiss needle cast in coastal Douglas-fir of the Pacific Northwest, USA. Forest Ecology and Management 245, 96–109.

Weiskittel, A.R., Temesgen, H., Wilson, D.S., Maguire, D.A., 2008. Sources of within- and between-stand variability in specific leaf area of three ecologically distinct conifer species. Annals of Forest Science 65, 103.

Weiskittel, A.R., Hann, D.W., Hibbs, D.E., Lam, T.Y., Bluhm, A.A., 2009a. Modeling top height growth of red alder plantations. Forest Ecology and Management 258, 323–331.

Weiskittel, A., Gould, P., Temesgen, H., 2009b. Sources of variation in the self-thinning boundary line for three species with varying levels of shade tolerance. Forest Science 55, 84–93.

Weiskittel, A.R., Maguire, D.A., Monserud, R.A., Johnson, G.P., 2010. A hybrid model for intensively managed Douglas-fir plantations in the Pacific Northwest, USA. European Journal of Forest Research 129, 325–338.

Weiskittel, A.R., Crookston, N.L., Radtke, P.J., in press. Linking climate, gross primary productivity, and site index across forests of the western United States. Canadian Journal of Forest Research, in press.

Weller, D.E., 1987. A reevaluation of the -3/2 power rule of self-thinning. Ecological Monographs 57, 23–43.

Wells, C.G., Crutchfield, D.M., Trew, I.F., 1976. Five-year volume increment from nitrogen fertilization in thinned plantations of pole-size loblolly pine. Forest Science 22, 85–90.

Welty, J.J., Beechie, T., Sullivan, K., Hyink, D.M., Bilby, R.E., Andrus, C., Pess, G., 2002. Riparian aquatic interaction simulator (RAIS): A model of riparian forest dynamics for the generation of large woody debris and shade. Forest Ecology and Management 162, 299–318.

Wensel, L.C., Meerschaert, W.J., Biging, G.S., 1987. Tree height and diameter growth models for northern California conifers. Hilgardia 55, 1–20.

West, P.W., 1980. Use of diameter increment and basal area increment in tree growth studies. Canadian Journal of Forest Research 10, 71–77.

Westfall, J.A., Burkhart, H.E., 2001. Incorporating thinning response into a loblolly pine stand simulator. Southern Journal of Applied Forestry 25, 159–164.

Westfall, J.A., Burkhart, H.E., Allen, H.L., 2004. Young stand growth modeling for intensively-managed loblolly pine plantations in southeastern U.S. Forest Science 50, 823–835.

White, J., Harper, J.L., 1970. Correlated changes in plant size and number in plant populations. Journal of Ecology 58, 467–485.

White, M.A., Thorton, P.E., Running, S.W., Nemani, R.R., 2000. Parameterization and sensitivity analysis of the BIOME-BGC terrestrial ecosystem model: Net primary production controls. Earth Interactions 4, 1–85.

White, T.L., Adams, W.T., Neale, D.B., 2007. Forest Genetics. CABI Publishing, Cambridge, MA.

Williams, K., Percival, F., Merino, J., Mooney, H.A., 1987. Estimation of tissue construction cost from heat of combustion and organic nitrogen content. Plant, Cell and Environment 10, 725–734.

Williams, M.S., 1994. Evaluating the performance of weight functions with no constraints on function types. Forest Science 40, 787–793.

Williams, M.S., Reich, R.M., 1997. Exploring the error structure of taper equations. Forest Science 43, 378–386.

Williams, M.S., Williams, M.T., Mowrer, H.T., 2001. A boundary reconstruction method for circular fixed-area plots in environmental survey. Journal of Agricultural, Biological, and Environmental Statistics 6, 479–494.

Williams, R.A., Hoffman, B.F., Seymour, R.S., 1991. Comparison of site index and biomass production of spruce-fir stands by soil drainage class in Maine. Forest Ecology and Management 41, 279–290.

Wilson, D.S., Seymour, R.S., Maguire, D.A., 1999. Density management diagram for northeastern red spruce and balsam fir forests. Northern Journal of Applied Forestry 16, 48–56.

Wilson, F.G., 1946. Numerical expression of stocking in terms of height. Journal of Forestry 44, 758–761.

Wilson, F.G., 1979. Thinning as an orderly discipline: A graphic spacing schedule for red pine. Journal of Forestry 77, 483–486.

Wilson, G.F., Maguire, D.A., 1996. Simulation of early regeneration processes in mixed-species forests of Maine, USA: Germination, survival, and height growth. In Skovsgaard, J.P., Johannsen, V.K. (Eds.), Modelling regeneration success and early growth of forest stands: Proceedings from the IUFRO Conference. Danish Forest and Landscape Research Institute, Copenhagen, Denmark, pp. 530–539.

Wilson, J.S., McGaughey, R.J., 2000. Presenting landscape-scale forest information: What is sufficient and what is appropriate? Journal of Forestry 98, 21–27.

Wilson, J.S., Oliver, C.D., 2000. Stability and density management in Douglas-fir plantations. Canadian Journal of Forest Research 30, 910–920.

Wimberly, M.C., Bare, B.B., 1996. Distance-dependent and distance-independent models of Douglas-fir and western hemlock basal area growth following silvicultural treatment. Forest Ecology and Management 89, 1–11.

Woodall, C.W., Miles, P.D., Vissage, J.S., 2005. Determining maximum stand density index in mixed species stands for strategic-scale stocking assessments. Forest Ecology and Management 216, 367–377.

Woods, M.E., Penner, M., 2007. Revised FVS^{ONTARIO} model forms based on expanded Ontario data set. FVS Ontario Internal Report. Ontario Ministry of Natural Resources, North Bay, Ontario.

Woods, S., 2006. Generalized Additive Models: An Introduction with R. Chapman & Hall, Boca Raton, FL.

Woollons, R.C., 1998. Even-aged stand mortality estimation through a two-step regression process. Forest Ecology and Management 105, 189–195.

Wullschleger, S.D., 1993. Biochemical limitations to carbon assimilation in C_3 plants – a retrospective analysis of the A/C_i curves from 109 species. Journal of Experimental Botany 44, 907–920.

Wyckoff, P.H., Clark, J.S., 2000. Predicting tree mortality from diameter growth: A comparison of maximum likelihood and Bayesian approaches. Canadian Journal of Forest Research 30, 156–167.

Wyckoff, P.H., Clark, J.S., 2005. Tree growth prediction using size and exposed crown area. Canadian Journal of Forest Research 35, 13–20.

Wykoff, W.R., 1983. Predicting basal area increment for individual northern Rocky Mountain conifers. In Proceedings of the meeting on Growth Modelling and Simulation, IUFRO Subject Group S4.01, Vienna, Austria, pp. 127–143.

Wykoff, W.R., 1986. Supplement to the user's guide for the stand prognosis model – version 5.0. General Technical Report INT-208. USDA Forest Service, Intermountain Research Station, Ogden, UT.

Wykoff, W.R., 1990. A basal area increment model for individual conifers in the northern Rocky Mountains. Forest Science 36, 1077–1104.

Wykoff, W.R., 1997. Predicting basal area increment in irregular stands: A progress report. In Teck, R., Moeur, M., Adams, J. (Eds.), Proceedings of the Forest Vegetation Simulator conference. General Technical Report INT-GTR-373. USDA Forest Service, Intermountain Research Station, Ogden, UT, pp. 199–205.

Wykoff, W.R., Monserud, R.A., 1988. Representing site quality in increment models: A comparison of methods. In Ek, A.R., Shifley, S.R., Burk, T.E. (Eds.), Forest growth modeling and prediction. General Technical Report NC-120. USDA Forest Service, North Central Forest Experimental Station, St. Paul, MN, pp. 184–191.

Wykoff, W.R., Crookston, N.L., Stage, A.R., 1982. User's guide to the Stand Prognosis Model. General Technical Report INT-133. USDA Forest Service, Intermountain Forest and Range Experiment Station, Ogden, UT.

Wythers, K.R., Reich, P.B., Tjoelker, M.G., Bolstad, P.B., 2005. Foliar respiration acclimation to temperature and temperature variable Q_{10} alter ecosystem carbon balance. Global Change Biology 11, 435–449.

Xenakis, G., Ray, D., Mencuccini, M., 2008. Sensitivity and uncertainty analysis from a coupled 3-PG and soil organic matter decomposition model. Ecological Modelling 219, 1–16.

Xie, C.Y., Yanchuk, A.D., 2003. Breeding values of parental trees, genetic worth of seed orchard seedlots, and yields of improved stocks in British Columbia. Western Journal of Applied Forestry 18, 88–100.

Yang, F., Kant, S., 2008. Forest-level analyses of uneven-aged hardwood forests. Canadian Journal of Forest Research 38, 376–393.

Yang, Y., Monserud, R.A., Huang, S., 2004. An evaluation of diagnostic tests and their roles in validating forest biometric models. Canadian Journal of Forest Research 34, 619–629.

Yang, Y., Titus, S.J., 2002. Maximum size-density relationship for constraining individual tree mortality functions. Forest Ecology and Management 168, 259–273.

Yang, Y., Titus, S.J., Huang, S., 2003. Modeling individual tree mortality for white spruce in Alberta. Ecological Modelling 163, 209–222.

Yang, Y., Huang, S., Meng, S.X., Trincado, G., VanderSchaaf, C.L., 2009a. A multilevel individual tree basal area increment model for aspen in boreal mixedwood stands. Canadian Journal of Forest Research 39, 2203–2214.

Yang, Y., Huang, S., Meng, S.X., 2009b. Development of a tree-specific stem profile model for white spruce: A nonlinear mixed model approach with a generalized covariance structure. Forestry 82, 541–555.

Yao, X., Titus, S.J., MacDonald, S.E., 2001. A generalized logistic model of individual tree mortality for aspen, white spruce, and lodgepole pine in Alberta mixedwood forests. Canadian Journal of Forest Research 31, 283–291.

Yaussy, D.A., 2000. Comparison of an empirical forest growth and yield simulator and a forest gap simulator using actual 30-year growth from two even-aged forests in Kentucky. Forest Ecology and Management 126, 385–398.

Yoda, K., Kira, T., Ogawa, H., Hozumi, K., 1963. Self-thinning in overcrowded pure stands under cultivated and natural conditions. Journal of Biology, Osaka City University 14, 107–129.

Yue, C., Kohnle, U., Hein, S., 2008. Combining tree- and stand-level models: A new approach to growth prediction. Forest Science 54, 553–566.

Zeide, B., 1983. The mean diameter for stand density index. Canadian Journal of Forest Research 13, 1023–1024.

Zeide, B., 1987. Analysis of the 3/2 power law of self-thinning. Forest Science 33, 517–537.

Zeide, B., 1993. Analysis of growth equations. Forest Science 39, 594–616.

Zeide, B., 2004. Intrinsic units in growth modeling. Ecological Modelling 175, 249–259.

Zeide, B., 2005. How to measure stand density. Trees 19, 1–14.

Zeng, H., Talkkari, A., Peltola, H., Kellomäki, S., 2007. A GIS-based decision support system for risk assessment of wind damage in forest management. Environmental Modelling & Software 22, 1240–1249.

Zhang, L., Moore, J.A., Newberry, J.D., 1993. Disaggregating stand volume growth to individual trees. Forest Science 39, 295–308.

Zhang, L., Bi, H., Gove, J.H., Heath, L.S., 2005. A comparison of alternative methods for estimating the self-thinning boundary line. Canadian Journal of Forest Research 35, 1507–1514.

Zhang, L., Ma, Z., Guo, L., 2008. Spatially assessing model errors of four regression techniques for three types of forest stands. Forestry 81, 209–225.

Zhang, S., Burkhart, H.E., Amateis, R.L., 1996. Modeling individual tree growth for juvenile loblolly pine plantations. Forest Ecology and Management 89, 157–172.

Zhang, S., Burkhart, H.E., Amateis, R.L., 1997a. The influence of thinning on tree height and diameter relationships in loblolly pine plantations. Southern Journal of Applied Forestry 21, 199–205.

Zhang, S., Amateis, R.L., Burkhart, H.E., 1997b. Constraining individual tree diameter increment and survival models for loblolly pine plantations. Forest Science 43, 414–423.

Zhang, Y., Borders, B.E., 2001. An iterative state-space growth and yield modeling approach for unthinned loblolly pine plantations. Forest Ecology and Management 146, 89–98.

Zhao, D., Borders, B., Wilson, M., 2004. Individual-tree diameter growth and mortality models for bottomland mixed-species hardwood stands in the lower Mississippi alluvial valley. Forest Ecology and Management 199, 307–322.

Zhao, D., Borders, B., Wang, M., Kane, M., 2007. Modeling mortality of second-rotation loblolly pine plantations in the Piedmont/Upper Coastal Plain and Lower Coastal Plain of the southern United States. Forest Ecology and Management 252, 132–143.

Zianis, D., 2008. Predicting mean aboveground forest biomass and its associated variance. Forest Ecology and Management 256, 1400–1407.

Zianis, D., Mencuccini, M., 2004. On simplifying allometric analysis of forest biomass. Forest Ecology and Management 187, 311–332.

Zianis, D., Muukkonen, P., Mäkipää, R., Mencuccini, M., 2005. Biomass and stem volume equations for tree species in Europe. Silva Fennica Monographs 4.

Zumrawi, A.A., Hann, D.W., 1989. Equations for predicting the height to crown base of six tree species in the central western Willamette Valley of Oregon. Research Paper 52. Oregon State University, Forest Research Laboratory, Corvallis, OR.

Appendix 1: List of species used in the text

Scientific name	Common name	Abbreviation
Abies alba	White fir	WF
Abies balsamea	Balsam fir	BF
Abies grandis	Grand fir	GF
Abies lasiocarpa	Subalpine fir	SAF
Acer macrophyllum	Bigleaf maple	BM
Acer rubrum	Red maple	RM
Acer saccharum	Sugar maple	SM
Alnus rubra	Red alder	RA
Arbutus menziesii	Pacific madrone	PM
Betula alleghaniensis	Yellow birch	YB
Betula papyrifera	Paper birch	PB
Castanea dentata	Chestnut	CN
Castanopsis chrysophylla	Golden chinkapin	GC
Eucalyptus delegatensis	Eucalyptus	ED
Eucalyptus grandis	Eucalyptus	EG
Eucalyptus nitens	Eucalyptus	EN
Fagus grandifolia	American beech	AB
Fagus sylvatica	European beech	EB
Fraxinus americana	White ash	WA
Fraxinus nigra	Black ash	BA
Juglans cinerea	Butternut	BN
Juniperus virginiana	Eastern red cedar	RC
Larix decidua	European larch	EL
Larix laricina	Tamarack	TA
Libocedrus decurrens	Incense-cedar	IC

(*continued*)

Forest Growth and Yield Modeling, First Edition. Aaron R. Weiskittel, David W. Hann, John A. Kershaw, Jr. and Jerome K. Vanclay.

Scientific name	Common name	Abbreviation
Liriodendron tulipifera	Yellow-poplar	YP
Lithocarpus densiflorus	Tanoak	TO
Nyssa sylvatica	Blackgum	BG
Picea abies	Norway spruce	NS
Picea engelmannii	Engelmann spruce	ES
Picea glauca	White spruce	WS
Picea mariana	Black spruce	BS
Picea rubens	Red spruce	RS
Picea sitchensis	Sitka spruce	SS
Pinus banksiana	Jack pine	JP
Pinus contorta	Lodgepole pine	LP
Pinus lambertiana	Sugar pine	PL
Pinus pinea	Stone pine	TP
Pinus ponderosa	Ponderosa pine	PP
Pinus radiata	Radiata pine	RP
Pinus strobus	White pine	WP
Pinus sylvestris	Scots pine	SP
Populus balsamifera	Balsam poplar	BP
Populus grandidentata	Bigtooth aspen	BT
Populus tremuloides	Quaking aspen	QA
Prunus serotina	Black cherry	BC
Prunus virginiana	Choke cherry	CC
Pseudotsuga menziesii	Douglas-fir	DF
Quercus alba	White oak	WO
Quercus chrysolepis	Canyon live oak	LO
Quercus garryana	Oregon white oak	OO
Quercus kelloggii	California black oak	BO
Quercus petraea	Sessile oak	SO
Quercus rubra	Northern red oak	RO
Salix spp.	Willow	WL
Taxus brevifolia	Pacific yew	PY
Thuja occidentalis	Northern white-cedar	NC
Thuja plicata	Western red cedar	WRC
Tilia americana	Basswood	BW
Tsuga canadensis	Eastern hemlock	EH
Tsuga heterophylla	Western hemlock	WH
Ulmus americana	American elm	AE

Appendix 2: Expanded outline for ORGANON growth and yield model

Operational steps of ORGANON EDIT

I. Input raw tree list
   A. Import external tree data file
      1. Plot/point number
      2. Species code number
      3. Diameter at breast height
      4. Total tree height
      5. Crown ratio
      6. Expansion factor
      7. Radial growth
      8. User code
   B. Enter tree data by keyboard
      1. Plot/point number
      2. Species code number
      3. Diameter at breast height
      4. Total tree height
      5. Crown ratio
      6. Expansion factor
      7. Radial growth
      8. User code

II. Raw tree list data are edited
   A. If problems, user edits
   B. If no problems, go to III

Forest Growth and Yield Modeling, First Edition. Aaron R. Weiskittel, David W. Hann, John A. Kershaw, Jr. and Jerome K. Vanclay.

III. Input
  A. Site index of stand by species
  B. Stand structure
    1. Even aged
      a. Enter breast height age
        i. User enters number of years to breast height
        ii. ORGANON determines number of years to breast height
  C. Maximum SDI by species

IV. Calibration of select regional equations to input data
  A. Height/diameter equations
  B. Crown ratio (height to crown base) equations
  C. Diameter increment equations

V. ORGANON fills in missing values
  A. Total tree heights
  B. Crown ratios (heights to crown base)
  C. Expansion factors
    1. User describes sampling unit design
    2. Calculation of expansion factors

VI. Final check of uncalibrated and calibrated data by graphical analysis

VII. Output of prepared data file for the stand (.INP file)

Operational steps of ORGANON RUN

I. Input edited and prepared file for the stand (.INP file)

II. Input run specifications
  A. Use record tripling or not
  B. Use calibration value(s) or not
    1. On height–diameter equations
    2. On crown ratio (height to crown base) equations
    3. On diameter increment equations
  C. Use limit on maximum SDI or not
    1. If yes, then mortality equations will be adjusted
  D. Output wood quality data or not
    1. If yes, then wood quality data are calculated

- E. Choose definition of juvenile wood core for wood quality calculations
  - 1. Definition based on age of tree
  - 2. Definition based on crown grown wood
- F. Use printer form feed or not on tabular output
- G. Output detailed tree-list data or not
- H. Set source of volume/taper equations to use
  - 1. Oregon State University equations
    - a. Cubic foot volume defaults
      - i. Top diameter
      - ii. Stump height
    - b. Board foot volume defaults
      - i. Top diameter
      - ii. Log length
      - iii. Minimum log length
      - iv. Trim allowance
      - v. Stump height
  - 2. Bureau of Land Management equations
    - a. Choose resource area
    - b. Cubic foot volume defaults
      - i. Top diameter
      - ii. Stump height
    - c. Board foot volume defaults
      - i. Top diameter
      - ii. Log length
      - iii. Minimum log length
      - iv. Trim allowance
      - v. Stump height

III. Input genetic worth values for Douglas-fir if desired
- A. Diameter increment
- B. Height increment

IV. Input level of Swiss needle cast impact on Douglas-fir if desired

V. Calculate initial tree volumes
- A. Cubic foot volume of each tree
- B. Board foot volume of each tree

VI. Calculate initial wood quality variables if requested

VII. Output detailed initial tree-list data to file if requested

VIII. Process one or more growth cycles
- A. Calculate predictor variables at start of growth period
  1. Basal area in larger trees for each tree
  2. Crown closure at top of tree for each tree
     - a. Calculate maximum crown width for each tree
     - b. Calculate largest crown width for each tree
     - c. Calculate crown width for all trees at subject tree's height
  3. Crown competition factor in larger trees for each tree
     - a. Calculate maximum crown width for each tree
  4. Basal area of the stand
- B. Examine or output selected tables and graphs before growing
- C. Treat stand before growing
  1. Thin stand
  2. Fertilize an even-aged Douglas-fir stand
  3. Prune stand
  4. Add ingrowth to stand
  5. Harvest stand
     - a. Even-aged clearcut (no growth possible after execution)
     - b. Uneven-aged
     - c. Overstory removal
- D. Grow stand one or more growth cycles
  1. Calculate diameter increments
     - a. Calculate regional estimates of diameter increment
     - b. Apply calibration values if requested
     - c. Calculate and apply thinning modifier if appropriate
     - d. Calculate and apply fertilization modifier if appropriate
     - e. Calculate and apply genetic worth modifier if requested
     - f. Calculate and apply Swiss needle cast modifier if requested
     - g. Triple tree list if requested and possible
  2. Calculate height increments
     - a. Calculate regional estimates of height increment
       - i. Calculate potential height increment
       - ii. Calculate and apply height increment modifier
     - b. Calculate and apply thinning modifier if appropriate
     - c. Calculate and apply fertilization modifier if appropriate
     - d. Calculate and apply genetic worth modifier if requested
     - e. Calculate and apply Swiss needle cast modifier if requested
     - f. Triple tree list if requested and possible
  3. Calculate mortality rates
     - a. Calculate regional estimates of mortality rate
     - b. Calculate and apply additional mortality to keep stand on the maximum size–density trajectory if requested and appropriate

4. Update DBH, total height, and expansion factors to end of growth period values
5. Calculate crown recession rate
   a. Calculate height to crown base at start of growth period
   b. Calculate height to crown base at end of growth period
   c. Use differences to calculate crown recession
   d. Constrain crown recession to be non-negative
6. Update crown ratio (height to crown base) to end of growth period values
7. Calculate wood quality values if requested and write to wood quality File
8. Calculate tree volumes at end of growth period
9. Output detailed tree-list data to file if requested
10. Calculate and write out yield table values to output file
11. If more than one growth cycle requested, then:
    a. Reset DBH, total tree height, crown ratio (height to crown base, expansion factor, and volume values for start of growth period to the end of growth period values
    b. Go to D.

IX. If processing a set of growth cycles is complete, then either:
   A. Process another set of growth cycles
      1. Reset DBH, total tree height, crown ratio (height to crown base, expansion factor, and volume values) for start of growth period to the end of growth period values
      2. Go to VIII
   B. Restart program with a new tree list
   C. Exit program

Index

Forest Growth and Yield Modeling, First Edition. Aaron R. Weiskittel, David W. Hann, John A. Kershaw, Jr. and Jerome K. Vanclay.

Printed and bound by CPI Group (UK) Ltd, Croydon, CR0 4YY
18/09/2026
14965015-0005